D1599113

Satellite Communications Systems

Other McGraw-Hill Communications Books of Interest

*To order or receive additional information on these or any other
McGraw-Hill titles, in the United States please call 1-800-822-8158.
In other countries, contact your local McGraw-Hill representative.*   **BC14BCZ**

Satellite Communications Systems

Design Principles

M. Richharia
BSc(Eng), MSc(Eng), PhD, CEng, MIEE

McGraw-Hill, Inc.
New York   San Francisco   Washington, D.C.   Bogotá
Caracas   Mexico City   San Juan   Toronto

5 6 7 8 9 0   QBP/QBP   9 0 9 8

ISBN 0-07-052374-6

First published 1995 by THE MACMILLAN PRESS LTD
Houndmills, Basingstoke, Hampshire RG21 2XS
and London.

Printed and bound by Quebecor-Book Press

Acknowledgements

The author and publishers wish to thank the following for permission to use copyright material:

AT&T Bell Laboratories for Fig. 3.4, from T. S. Chu, 'Rain-induced cross-polarization at
centimeter and millimeter wavelengths', *Bell System Technical Journal*, Vol. 53, No. 8, October
1974, pp 1557–79. Copyright © 1974 AT&T.
The Institute of Electrical and Electronics Engineers, Inc. for Fig. 3.5, from H. W. Arnold,
D. C. Cox, H. H. Hoffman and R. P. Leck, 'Characteristics of rain and ice depolarisation for
19 & GH propagation paths from a COMSTAR satellite', *IEEE Transactions Antenna and
Propagation*, Vol. AP-28, pp 22–8, January 1980. Copyright © 1980 IEEE; Fig. 8.14, from
S. S. Lam, 'Satellite multiaccess schemes for data traffic', *International Conference on Communications*, Vol. 111, June 12–15, 1977, Fig. 1. Copyright © 1977 IEEE; Fig. 8.16, from
V. O. K. Li, 'Multiple access communications networks', *IEEE Communications Magazine*,
Vol. 25, No. 6, 1987, Fig. 2. Copyright © 1987 IEEE; Fig. 3.2, from L. J. Ippolito, 'Radio
propagation of space communication systems', *Proc. IEEE*, Vol. 69, No. 6, June 1981, Fig.
8. Copyright © 1981 IEEE: and Fig. 6.5, from M. P. Ristenbatt, 'Alternatives in digital
communications', *Proc. IEEE*, Vol. 61, June 1973, Fig. 5.
International Telecommunications Satellite Organization (INTELSAT) for Fig. 9.6, from
T. Abdel-Nabi, E. Koh and D. Kennedy, 'INTELSAT VII communications capabillities and
performance', *AIAA 13th International Communication Satellite Conference*, March, Paper
AIAA-90-0787-CP, pp 84–94.
International Telecommunications Union for Figs 3.1 and 5.2 taken from their CCIR Report 390-4, p 221 and p 103. Copyright © International Telecommunication Union. Complete volumes of the ITU material can be obtained from International Telecommunication
Union, General Secretariat – Sales and Marketing Service, Place des Nations, CH-1211 Geneva 20.
John Wiley and Sons Ltd for Fig. 9.19, from W. L. Pritchard, 'Estimating the mass and
power of communication satellites', *International Journal of Satellite Communications*, Vol.
2, 1984, Fig. 2.

Every effort has been made to trace all the copyright holders, but if any have been inadvert-
ently overlooked the publishers will be pleased to make the necessary arrangement at the
first opportunity.

To my parents

Contents

About the Author

M. Richharia has been involved in the field of satellite communications since 1974. He is presently a senior systems engineer with the International Maritime Satellite Organization (INMARSAT) in London, UK, a consortium of 64 countries providing world-wide communications via satellite to ships, aircraft, and land vehicles. The author of numerous publications on satellite communications technology, Dr. Richharia was previously a research fellow at Southampton University, and prior to that, was employed by the Indian Space Research Organization (ISRO).

Preface

Satellite communication systems are now an integral part of most major wide-area telecommunication networks throughout the world. The purpose of this book is to introduce the various elements of a satellite communication system and to develop the principles of system design. The understanding of a complete satellite system requires coverage of a broad range of topics. An attempt has been made to treat each topic at a system level in sufficient depth so as to develop a sound understanding of each element and its relationship to the overall system. Problems and examples have been included to illustrate the applicability of the concepts introduced.

The book consists of 11 chapters. The introductory chapter provides an overview of satellite communications, including a brief history. Chapter 2 discusses the fundamentals of satellite orbits, focusing mainly on the principal features of the geostationary orbit, currently the most commonly used orbit for satellite communications. An appendix summarizes various useful orbit-related formulas. Chapter 3 develops a basic understanding of issues involved in selecting frequency for a given satellite communication system. Chapter 4 introduces essential topics such as antenna characteristics, the transmission equation and the effects of noise. The final section of the chapter ties together these concepts, forming the basis of the overall system design. Chapters 5 and 6 discuss, respectively, the topics of modulation and coding applied to satellite communications. The treatment is at a system level so as to assist a system designer in the selection of appropriate modulation and coding schemes for a given application. Chapter 7 characterizes various types of baseband signals commonly used in satellite communications. A satellite is a common resource which must be shared efficiently by a large number of users. Chapter 8 discusses various techniques used for accessing a satellite. Chapters 9 and 10 describe the main sub-systems and outline issues related to the design of communication satellites and earth stations respectively. The concluding chapter of the book examines the likely future evolution in various areas of satellite communication system on the basis of current trends and the impact of other developments in telecommunications.

The author expresses his gratitude to Professor B.G. Evans of Surrey University for his encouragement, especially in the initial phases of the project. Thanks are also due to Mr Malcolm Stewart of The Macmillan Press and to Dr Paul Lynn, the series editor, for their continued support throughout.

The author expresses appreciation to his wife Kalpana for her patience and support. Finally, it is a pleasure to acknowledge the enthusiastic support of my son Anshuman and daughter Meha. Anshuman's excellent assistance in preparing the typescript and illustrations are commendable.

Guildford, UK   Madhavendra (Manu) Richharia

1 Introduction

1.1 General

Satellite communication networks are now an indispensable part of most major telecommunication systems. Satellites have a unique capability for providing coverage over large geographical areas. The resulting inter-connectivity between communication sources provides major advantages in applications such as interconnecting large traffic nodes (e.g. telephone exchanges), provision of end-to-end connections directly to users, mobile communications, television and sound broadcasts directly to the public. The advantages offered have enabled this technology to mature within just three decades. To date, most benefit within the telecommunications area has been achieved for point-to-point communication within the international and domestic systems. In recent years, satellite communication systems have begun to face competition from optical fibre systems for point-to-point communication between large concentrated traffic sources. To retain a competitive edge, it has been necessary to develop various new techniques. Thus the phenomenal growth of satellite technology continues, the major growth now being in those areas where satellites can provide unique advantages. Such applications include service provision directly to customers using small, low-cost earth stations; mobile communication to ships, aircrafts and land vehicles; direct-to-public television/sound broadcasts and data distribution/gathering from widely distributed terminals. In many applications, such as video distribution, service-providers are combining the benefits of satellite communications with optical fibre systems to produce the best solution to users' needs. Similarly, there is a growing interest in combining the advantages of satellite mobile communication systems with those of terrestrial mobile systems to provide a seamless coverage – terrestrial systems providing service in populated areas and satellite systems in those areas unserved by the terrestrial system.

The purpose of the book is to give the reader a sound understanding of the various issues involved in designing satellite communication systems, be it for a new system design, a trade-off study between various transmission media, an understanding of an existing satellite system or academic pursuit. This chapter begins with a brief history of satellite communications, and an overview of the main components of a satellite communication system follows. Major design considerations are then discussed and the present status and future trends of the technology are briefly reviewed.

1

1.2 Background

The first known use of a device resembling a rocket is said to have been in China in the year 1232. A number of instances of the use of such devices were subsequently recorded over the next few centuries. However, significant progress in the field was not made until the Russian school teacher, Konstantin E. Tsiolkovsky (1857–1935), propounded the basis of liquid propelled rockets. He went further and put forward ideas for multi-stage launchers, manned space vehicles, space walks by astronauts and large platforms that could be assembled in space complete with their own biological life support systems. His theoretical work on liquid propelled rocket engines was verified when in 1926 Robert H. Goddard launched the first liquid propelled rocket in the United States. Other rocket pioneers also corroborated and extended Tsiolkovsky's ideas. However, the work of a small German amateur group provided the breakthrough which laid the foundations of the present rocket technology. Their work was later supported by the German military, leading to the succesful launch of V-2 rockets in 1942. (Gatland, 1975).

Recognizing the potential of V-2 rockets, in 1945 Arthur Clark suggested the use of geostationary satellites for world-wide voice broadcasts (Clark, 1945). The work on V-2 rockets was extended in the United States and the former Soviet Union after the second world war. This led to the development of the first satellite launchers. The satellite era began in October 1957 with the launch of *Sputnik-1*, a Russian satellite. This was soon followed by the launch of a US satellite, *Explorer-1* in 1958.

Communication by moon reflections was demonstrated and used in the United States in the 1940s and 1950s. The early years of satellite communications exploited similar techniques. Satellites were used as passive reflectors of radio waves for establishing communication. An immediate problem with the use of satellites as passive reflectors was the extremely low level of signal strength, resulting in a need for very sensitive and hence costly receivers. The main reason favouring the use of passive satellites at the time was the lack of space-qualified electronics. It was recognized that the use of satellites capable of amplifying the received signal on-board before retransmission, could greatly enhance the capability of satellites for communications. Therefore considerable research and development effort was spent during the next few years in the development of space-qualified electronics, eventually leading to the introduction of active repeaters. In 1963 the first geostationary satellite, *Syncom III*, was successfully launched. The first commercial satellite for international communication, *Early Bird*, was launched in 1965.

The breakthrough provided by satellites in telecommunications resulted in a major research and development effort in all the related technologies. Most of the early work concentrated on international point-to-point telecommunications applications. Later, the application of satellite communication extended to direct satellite broadcasts (1970s) and mobile communications (1980s).

There was a phenomenal increase in international telecommunications traffic in the period 1965–90, and growth is predicted to continue (Bennet and Braverman, 1984). The increase in the number of global circuits in the INTELSAT network was about a thousand fold in the same period. The revenue generated has funded large projects for research and development into both space and ground segment hardware. As a result, there has been a steady reduction in cost per circuit. Similarly, in the area of satellite mobile communications, the total space segment capacity of the International Maritime Satellite Organization (INMARSAT) has increased severalfold during the decade from 1981 to 1991. INTELSAT and INMARSAT are international organizations which operate global fixed and mobile communications networks respectively. Most countries of the world are members of these organizations.

In addition to providing international telecommunications, satellite technology is used in regional and domestic telecommunication networks. Regional networks refer to those networks formed to serve a group of countries in a region (such as Europe), and domestic networks refer to those designed for intra-country use.

At present, international point-to-point communication is mainly provided by INTELSAT which was founded in 1964 to develop the commercial uses of satellite communications. Examples of other international operators are INTER-SPUTNIK which provides communication to ex-Soviet block countries, and the International Maritime Satellite Organization (INMARSAT) which provides worldwide mobile communications. Examples of regional system operators are the European Telecommunication Satellite Organisation (EUTELSAT) and the Arabian Satellite Communication Organisation (ARABSAT) which provide services in the European and the Middle East regions, respectively. Several countries now use domestic satellite systems – for example, North America, Canada, India and Indonesia.

A summary of the major milestones in the development of satellite communications is given at the end of the chapter.

The fact that satellites have a wide view of the Earth makes them useful for a variety of applications besides telecommunication. These applications are in the fields of meteorology, navigation, astronomy, management of earth resources such as forestry and agriculture, military reconnaissance, amateur radio and others. To coordinate the working of various radio systems, an international body, the International Telecommunications Union (ITU), has categorized these services and set guidelines for the design and operation of each satellite service. Telecommunication is provided by the Fixed Satellite Service (FSS) used for communication between fixed points on Earth; Mobile Satellite Service (MSS) used for communication with moving terminals; and the Broadcast Satellite Service (BSS) which deals with television and sound broadcasts directly to customers. In this book we are concerned with the design principles of telecommunications services. However, the system design concepts developed are generally also applicable to other services.

1.3   Basic satellite system

A basic satellite system consists of a space segment serving a specific ground segment. The characteristics of each segment depend on whether the system is for fixed, mobile or direct broadcast applications. The main features of these services and the main system related issues are briefly addressed in this section.

The main elements of a satellite communication system are shown in figure 1.1. Ground stations (or earth stations) in a network transmit radio frequency (RF) signals to the operational satellite. The received signals are amplified, translated into another frequency and, after further amplification, retransmitted towards the desired regions of the Earth. Communication can be established between all the earth stations located within the coverage region.

Space segment

The space segment consists of a satellite in a suitable orbit. Satellite and orbital characteristics depend on the application needs. The satellite is controlled and its performance monitored by the Telemetry Tracking and Command (TT&C) stations. In an operational system the satellite in use is usually backed up by an in-orbit spare satellite. Most of the present communication satellites are in the geostationary orbit. A circle at an altitude of approximately 35 786 km above the equator is known as the geostationary arc. Satellites orbiting the Earth in such an orbit rotate in unison with the Earth, appearing almost stationary to an observer on the ground. A satellite which appears stationary minimizes the operational requirements of earth stations. The design of ground terminals is simplified because simple tracking systems can be used and RF signals do not suffer significant Doppler frequency shift (chapter 2). Further, a single geostationary satellite can provide communication to large areas (over about one-third of the Earth), permitting easy interconnection between distant ground terminals. Thus three geostationary satellites placed 120° apart can provide coverage to almost all the populated areas of the world, as shown in figure 1.2. It is therefore not surprising that most communication satellites use the geostationary orbit.

One disadvantage of geostationary satellites is that they appear almost at

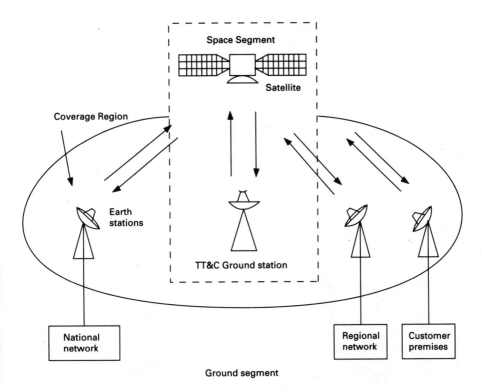

Figure 1.1 The main elements of a satellite communications network.
Ground segment of a fixed satellite service is shown.

the horizon at latitudes above approximately ±75°. Above about ±84° latitude the satellites are no longer visible. They become unusable (or unreliable) below about 5° elevation because the received signal quality is inadequate
for good-quality communication. This is caused by a combination of increased degradation in the troposphere and ground reflections which cause
rapid signal fluctuations.

Thus, for providing service to high-latitude locations other types of orbits are necessary. An elliptical orbit inclined at 63.4° to the equatorial
plane has been in use for this purpose. Satellites in this type of orbit appear
almost stationary to higher-latitude earth stations for periods of 8–12 hours.
Therefore several satellites are necessary to provide a 24-hour service. The
use of this type of orbit has also been proposed for providing land mobile
communication in high-latitude areas. Figure 1.2 shows the coverage areas
provided by a geostationary satellite with a 63.4° inclined elliptical orbit.
Chapter 2 covers the topic of orbit in detail.

Renewed interest in the use of low earth orbit for mobile communications has been generated recently (e.g. Richharia *et al.*, 1989; Maral *et al.*,

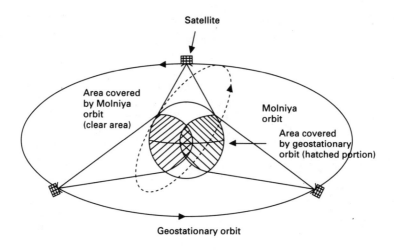

Figure 1.2   Three geostationary satellites can provide coverage within about ±75° latitude. Locations at higher latitude may be served by Molniya orbit.

1992). The use of low and medium earth orbits was considered during the early stages of development of satellite communications, but abandoned because of the resulting complexity in the earth station and network architecture. The main problems are the requirement for complex earth stations; the need for a relatively large constellation of satellites to cover the Earth; complex handover procedures as satellites continuously appear and set at a particular location on the ground; and complex techniques to interconnect terminals which cannot view the same satellite simultaneously. However as technology has matured, solutions to many of these problems appear feasible. Some of the main advantages claimed for the use of low earth orbit include: (a) the possibility of using hand-held receiver terminals because satellites are closer to the Earth and can therefore provide stronger signals at the ground (however this advantage can best be exploited by using complex 'spot' beam antennas and network architecture); (b) the possibility of reusing the frequencies more often than is possible with geostationary orbit because the geographical area covered by low earth orbit satellites is much smaller; (c) the possibility of reduction in transmission delay.

Various types of communication satellites are in use, from small satellites for domestic communications to large and complex satellites, serving international traffic. Chapter 9 discusses satellite technology in detail. In general, satellites serving the mobile and broadcast sectors need to transmit at higher power than do satellites serving the fixed network.

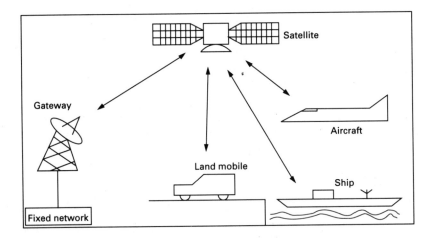

Figure 1.3   The main elements of a mobile satellite service. Satellite communications provide a unique advantage for mobile applications because of the wide geographical coverage made possible by satellites.

Ground segment

The ground segment of each service has distinct characteristics. The ground segment of a fixed satellite service (FSS) consists of several types of fixed earth stations. The size and characteristics of earth stations depend on the application. For example, earth stations for handling international traffic use large antennas (11–30 m) together with complex RF and baseband subsystems, whereas terminals for providing communication directly to customers' premises employ non-tracking antenna of 1–3 m diameter with simple RF and baseband hardware. Each earth station is interfaced to the user either directly or through a regional or national public switched network. The interface between an earth station and the user is an important consideration in the design of a FSS network.

The main elements of a mobile satellite service (MSS) are shown in figure 1.3. The ground segment consists of several types of mobile terminals connected to the fixed telecommunication networks via the satellite. Mobile satellite communication is categorized into three classes according to the environment served – maritime, aeronautical and land based. Ground terminals serving each category of mobile differ to varying extents. Differences occur because: the physical space available for mounting the antenna and receiver depends on the type of mobile (e.g. a ship has more available space than does a truck); the communication needs and affordable terminal cost differ (e.g. a personal communicator must be low cost with a moderate communication capability relative to, say, an aeronautical terminal); the behaviour of the transmit/receive signals is significantly affected by the environment in the proximity of the mobile. However, this does not imply that a terminal designed for one environment cannot operate in the other. If the application so demands, a terminal designed for use in one environment can indeed be used in the other. For example, ship terminals can be readily altered to operate as a land terminal, or land terminals can be used in aircraft. In general, ship terminals use the largest antenna and land mobile terminals the smallest. Consequently, maritime terminals can provide the largest communication capability, followed by aeronautical terminals. At present, land mobile terminals provide a rather limited capability because of severe limitations on the antenna size, and the harsh propagation environment coupled with the technological limitations of the space segment. A radical change in space segment design is essential if voice communication directly to hand-held communicators (resembling a cellular telephone) is to become feasible.

A rescue coordination centre can also be incorporated in the MSS network, to respond to distress calls from mobiles in the network.

Figure 1.4 shows the main elements of a direct broadcast system (DBS). Programmes (originating in a studio or 'live') are transmitted through a large 'gateway' earth station and a high-power satellite to small terminals dispersed throughout the service area. Terminals typically use 50–100 cm antennas with a facility for interfacing to television sets used to receive local terrestrial transmissions. In recent years the cost of direct broadcast receivers has become comparable to other home entertainment appliances. Cable television systems have also exploited the benefits of direct broadcasts. Cable television companies can afford to use larger and more sensitive receivers, and therefore provide a better signal quality and often a larger choice of programme to their subscribers.

1.4 System design considerations

The design of a satellite network is based on the service to be provided, e.g. voice, data, video, etc. The objective of the design is to meet the desired signal quality within all system constraints such as cost and the state of technology. Moreover, a satellite link may only be a sub-set of a larger network consisting of many other types of link (microwave radio relays, optical fibres, etc.). In such cases the satellite portion cannot be considered in isolation from the rest of the network. Further, optimization criteria are different for fixed, mobile and direct broadcast satellite services. Hence the design of a satellite system is complex and involves many variables which are traded-off to achieve an optimum configuration. Chapter 4 discusses issues related to the link design.

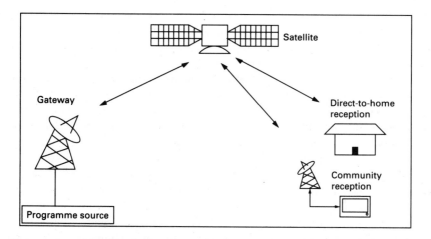

Figure 1.4   The main elements of a direct broadcast system. Programmes transmitted by an earth station are received directly by the audience.

An important parameter in the design of a satellite system is the selection of RF for a given application. The choice is mainly governed by the special needs of the application, the expected propagation characteristics, the state of the technology, the prevalent radio regulations and the availability of spectrum.

To enable radio systems to coexist without interfering with each other, the use of RF spectrum is broadly managed by the International Telecommunications Union (ITU). This is achieved by allocating frequencies for each service and laying down procedures for the use of RF spectrum according to the requirements of various services and regions. Radio regulations issued by the ITU specify detailed frequencies for various services, and methods to avoid excessive interference between users. Although individual governments regulate the use of radio systems within their territory, international agreements are necessary to harmonize the use of radio on an international basis. Chapter 3 discusses the issues involved in the selection of frequency, and reviews propagation effects.

A further system design consideration is the selection of an optimal modulation and coding scheme. The choice is governed by the type of message (e.g. video, telephony, data) together with the radio link characteristics and permitted complexity in earth stations. Chapters 5 and 6 respectively discuss various aspects of modulation and coding schemes used in satellite communications. It is worth mentioning here that the characteristics of baseband signals also influence the selection of modulation/coding schemes. This aspect is covered in the first part of chapter 7. A satellite network may consist of a few to several thousands of earth stations and provision has to be made so that all earth station in the network may share the satellite resource equitably. Thus, it is necessary to use appropriate methods for accessing the satellite. Several types of accessing scheme have been developed. These techniques attempt to make an equitable use of the available satellite bandwidth and power for the given application. Chapter 8 discusses the details of various accessing schemes used in satellite communications.

Other dominating factors are: the type of service; permitted earth station size and complexity; the size and shape of the service area; and the state of the prevailing technology related both to spacecraft and ground stations. Chapters 9 and 10 respectively discuss the topics of earth stations and satellites.

All the aspects mentioned above must be tied together in the best possible way in order to provide an acceptable system design solution. The trade-offs are complex and depend on the existing constraints. An adequate solution may not always be possible. The optimal solution for one administration may not be so for another. The necessary trade-offs will become clearer as we develop further understanding of the various components of satellite systems in subsequent chapters. Examples included throughout the book should be useful in relating the concepts to reality.

1.5 Applications

Some of the main applications of satellite communication systems have been mentioned in the introductory paragraph of the chapter. Here the application areas are expanded to provide a wider perspective. In addition to the applications mentioned, numerous other applications are possible and interested readers are encouraged to consider how satellite communications can be utilized for their needs, based on the fundamental advantages of satellite communications mentioned here. The main limitations of satellite communications are also included to develop a balanced perspective.

Several types of transmission system are used in a fixed telecommunications network. These include coaxial or fibre optic cables, line-of-sight radio relays and satellite links. At present, coaxial and optical fibre cables are used in subscriber loop and junction lines, which typically run for tens of kilometres. Trunk routes can however be much longer and, hence, several types of transmission system are used in such cases. These include radio relays and, increasingly, optical fibres. On such routes the use of satellite communications becomes economic when the distances involved are large (several hundreds of kilometres), and under special situations such as the presence of difficult terrain or sea between switching centres. The main switching centres are connected to the international switching centres (ISC) and usually these are connected to the ISC of other countries via satellites or cables. Satellite communication is used extensively on such international routes, where the distances involved are usually very great.

Satellites also prove attractive for the rapid deployment of telecommunication services between or with isolated communities. Hence several countries, where populated areas are separated by vast distances or difficult terrain or sea (e.g. an archipelago), use satellites on trunk routes. In such cases, satellites also provide a useful back-up for existing terrestrial services.

With advancing technology, ground terminal antenna and other hardware sizes have reduced to an extent where terminals can now be installed in customers' premises. These terminals are commonly known as Very Small Aperture Terminals (VSAT). By directly interconnecting VSATs, entire public switched networks can be bypassed. In several countries such bypass networks are gaining in popularity because: (a) for many applications the use of a VSAT network is more cost-effective than the use of the public switched network (PSN); (b) VSATs eliminate delays/faults associated with the PSN; and (c) a user can obtain access to the network quickly. Hence it is not surprising that VSAT networks are growing very rapidly in several developed and developing countries and numerous applications are emerging (e.g. in the auto, retail and business sectors).

One of the fastest growth areas is that of mobile satellite communications. Telecommunication services to mobiles in the form of terrestrial cellular radio is now well established. Whereas coverage by cellular radio is limited to populated areas of the developed world, satellites are capable of providing coverage over a much wider area such as oceans, inter-continental flight corridors and large expanses of land mass. Additionally, a satellite mobile system is capable of providing full international 'roaming' to mobiles. At present, satellites are used to provide voice and data communications to aircraft, ships and land vehicles. Some of the novel applications are provision of live television to ships using advanced picture compression techniques; messaging and voice communications to briefcase size terminals; tracking movements of truck fleets or yachts; supporting relief operations in remote areas; instant contact with news reporters from inaccessible areas; responding to distress calls from ships; and worldwide paging service.

Satellites are also well suited for rapid provision of television and sound broadcasts to large otherwise unserved areas. A satellite communication system can provide coverage over wide areas in a relatively short period. From the planning stage to actual implementation of the system typically takes 4–5 years and, once the satellite is deployed, provision of service is available throughout simultaneously. To provide coverage by terrestial means in such areas could take a very long period and even then some areas could remain unserved. Hence it is not surprising that some developing countries, such as India, have chosen to use satellite communication to provide rapid television coverage. Direct broadcast by satellite has also proven popular in developed regions such as Europe, despite strong competition from other television delivery media.

Other applications for which satellites are well suited are: (a) provision of service to regions of sparse population spread over large areas (e.g. areas of the American continent); (b) augmenting the communication infrastructure of developing countries that possess limited terrestrial infrastructure (e.g. India).

Let us now summarize the main limitations of satellite communications. One of the main disadvantages of a satellite communication network is the high cost associated with its introduction. A careful techno-economic study is therefore essential before taking the decision to introduce such a system.

Another consideration is the loss of service to large areas if a satellite meant to serve the region was to fail or malfunction during launch or after deployment. Fortunately, the reliability of launchers has improved greatly and present launchers typically have reliabilities better than 99%. Similarly the reliability of satellites has increased considerably. Many satellites continue to provide good service far beyond their design lifetime. The problem of in-orbit catastrophic failure is solved by deploying an in-orbit spare, often leased from another organization to minimize cost. To make better utilization of the spare satellite it is sometimes used to augment the communication capacity of the network.

Another limitation of geostationary satellite systems is transmission delay caused by the long propagation path. The problem is exacerbated when the delay is accompanied by echo caused by mismatch at the terrestrial/satellite system interface. It is under these circumstances that telephone users suffer the worst annoyance. A great deal of progress has been made in echo suppressors/cancellers technology, reducing the echo problem to a minimum. Users have now become used to delay in single hop satellite links, but it should be noted that the delay problem is an inherent limitation of satellite communication systems. In particular, the delay associated with two satellite hops can become excessive. Fortunately, transmission delay does not pose too much of a problem to the various types of data transmission applications that are insensitive to transmission delays – such as the transfer of large data files.

1.6 Future trends

By considering trends and other influencing factors it is possible to project future growth areas. Considering the FSS first, on the global scene the traffic demand is likely to continue to grow, although the advent of trans-oceanic fibre optic cables may have an impact on the rate of growth (Pelton and Wu, 1987). The use of very small aperture terminals for business and rural applications is expected to grow in many parts of the world.

Satellites are likely to play a greater role in mobile communications. According to forecasts, mobile communication by satellites is expected to be a major growth area throughout the 1990s. Mobile satellite systems using non-geostationary orbit may begin to emerge towards the second half of the 1990s. The development of small hand-held terminals which communicate via satellites has been initiated by large service-providers such as INMARSAT. Hand-held terminals are likely to appear before the year 2000. Several different regional mobile satellite systems are likely to emerge.

The use of direct-to-home broadcasting is also expected to rise in many parts of the world, although in certain areas satellite broadcast systems may have to compete with other television delivery systems. However it is envisaged that the direct broadcast system will complement other types of well-established delivery systems, at least in some cases. Continuing growth is expected where television distribution and community broadcasting are established.

Several developments in technology are likely to further enhance the applicability of satellite communications. For applications such as VSATs or personal mobile terminals, simple inexpensive ground receivers are essential. One possible technical solution is the use of satellites with regenerative repeaters. Such repeaters are more intelligent than the simple repeaters used at present and are equipped with functions such as demodulation and switching, etc. (Evans, 1984; Bartholome, 1987). Intelligent satellites together with multiple beam coverage are likely to play an increasing role in the future.

Other areas being investigated include reduction in the coding bit rate of speech signals, which will result in greater bandwidth utilization; the use of as yet un-utilized high radio frequency bands, such as 20/30 GHz, to alleviate frequency congestion problems of existing bands; the use of non-geostationary orbits for specific applications; inter-satellite links in space to increase space segment capacity and connectivity; advanced antenna concepts; and others (see the literature, for example Idda *et al.*, 1985).

1.7 Some important milestones in the development of satellite communications

1000 AD   Chinese invent rocket.
1903   Russian school teacher (Konstantin Tsiolkovsky publishes his ideas on space flight).
1926   First liquid propellant rocket launched by R.H. Goddard in the USA.
1942   First successful launch of a V-2 rocket in Germany.

1945	Arthur Clarke publishes his ideas on geostationary satellites for worldwide communications.
1957	First man-made earth satellite launched by the former Soviet Union.
1958	First US satellite launched. First voice communication established via satellite.
1960	First communication satellite (passive) launched into space.
1962	First active communication satellite launched.
1964	First satellite launched into the geostationary orbit. INTELSAT founded.
1965	First satellite launched into the geostationary satellite for commercial use (INTELSAT 1).
1972	First domestic satellite system operational (Canada). INTER-SPUTNIK founded.
1975	First successful direct broadcast experiment (one year duration; USA–India).
1977	A plan for direct-to-home satellite broadcasting assigned by the International Telecommunication Union (ITU) in regions 1 and 3 (most of the world except the Americas).
1979	International maritime satellite organization (INMARSAT) established.
1981	First reusable launch vehicle flight (American Space Shuttle).
1982	International maritime communications made operational.
1983	ITU direct broadcast plan extended to region 2.
1984	First direct-to-home broadcast system operational (Japan).
1987	Successful trials of land mobile communications (INMARSAT).
1989–90	Global mobile communication service extended to land mobile and aeronautical use (INMARSAT).
1990–92	• Several organizations/companies propose the use of non-geostationary satellite systems for mobile communications.
	• Plans for provision of service to hand-held telephones by the year 200 announced by INMARSAT and other organizations/companies.
	• Continuing growth of VSATs in diverse regions of the world.
	• WARC allocates new frequencies for mobile satellite communication.
	• Continuing growth of direct broadcast system in Asia and Europe.

References

Bartholome, P. (1987). 'Future trends in satellite communications', chapter 22, in Evans, B.G. (ed.), *Satellite Communications*, Peter Peregrinus, London.

Bennet, S.B. and Braverman, D.J. (1984). 'INTELSAT VI-A continuing evolution', *Proc. IEEE*, Vol. 72, No. 11, November.

Clarke, A.C. (1945). 'Extra terrestrial relays', *Wireless World*, October.

Evans, B.G. (1984). 'Satellite on-board processing', *Electronics and Power*, July, pp 533–536.

Gatland, K. (1975). *Missiles and Rockets*, Blandford Press, London.

Idda, T., Shimoseko, S., Iwasaki K. and Shimod, M. (1985). 'Satellite communications in the next decade', *Space Communication and Broadcasting*, No. 3, pp 27–38.

Maral, G., Ridder, J.D., Evans, B.G. and Richharia, M. (1991). 'Low earth orbit satellite systems for communications', *International Journal of Satellite Communications*, Vol. 9, pp 209–225.

Pelton, J.N. and Wu, W.W. (1987). 'The challenge of 21st century satellite communications: INTELSAT enters the second millenium', *IEEE Journal on Selected Areas in Communication*, Vol. SAC-5, No. 4, p 571.

Richharia, M., Hansel, P., Bousquet, P.W. and O'Donnel, M. (1989). 'A feasibility study of a mobile communication network using a constellation of low earth orbit satellites', *IEEE Global Telecommunications Conference, GLOBECOM '89*, Dallas, 27–30 November.

2 Satellite Orbits

2.1 Introduction

Communication via a satellite begins when the satellite is positioned in the desired orbital position. Several types of orbits are possible, each suitable for a specific application. However, only a few types of orbits are well suited for communication. In this chapter the basics of orbital mechanics are reviewed. After a general treatment, emphasis is put on the geostationary orbit because of its importance for satellite communication. A section on launching of satellites into geostationary orbit is also included. In addition, several useful formulas, including some interesting results on low earth orbit constellation design, are summarized in appendix B.

2.2 Laws governing satellite motion

Satellite orbits follow the same laws that govern the motion of the planets around the Sun. These laws are due to Johann Kepler who postulated the first two laws in 1609 followed by the third law in 1619, based on the experimental observations of Tycho Brahe (taken around the year 1600). These laws may be stated as follows.

First law

The orbit of each planet follows an elliptical path in space, the Sun being the focus.

Second law

The line joining a planet to the Sun sweeps out equal areas in equal time.

Third law

The square of the period of a planet is proportional to the cube of its mean distance from the Sun.

Kepler's laws only describe the planetary motion without attempting to suggest any explanation as to why the motion takes place in that manner.

Newton characterized the forces that give rise to Kepler's laws. Newton's laws of motion may be stated as follows.

16

First law

Every body continues in its state of rest or of uniform motion in a straight line unless it is compelled to change that state by forces impressed on it.

Second law

The rate of change of momentum of a body is proportional to the force impressed on it and is in the same direction as that force.

The second law may be expressed mathematically as

$$\Sigma \mathbf{F} = m\ddot{\mathbf{r}} \tag{2.1}$$

where $\Sigma \mathbf{F}$ = vector sum of all forces acting on mass m
$\ddot{\mathbf{r}}$ = vector acceleration of the mass.

Third law

To every action there is an equal and opposite reaction.

Besides these three laws, Newton also postulated the law of gravitation which states that any two bodies attract one another with a force proportional to the product of their masses and inversely proportional to the square of the distance between them. The law may be expressed mathematically as

$$\mathbf{F} = -\frac{GMm}{r^2}\frac{\mathbf{r}}{r} \tag{2.2}$$

where $\mathbf{F}$ = force on mass m due to mass M
r = distance between masses
$\mathbf{r}$ = vector from M to m.

The universal gravitational constant, G, has the value 6.67×10^{-8} dyn cm^2 g^{-2}.

Newton's laws, in conjunction with Kepler's laws, can completely explain the motions of planets around the Sun. Kepler's second law may be satisfied if planets were being acted on by a central attractive force, the Sun. His first law is satisfied if this force varied as $1/r^2$ (r being the distance between the Sun and a planet). For Kepler's third law to be valid the force must be proportional to the mass of the planet.

The motion of a satellite around the Earth follows the same basic laws, since the forces acting on an Earth-orbiting satellite are similar.

2.3 Useful concepts

Before proceeding further, let us first discuss some basic concepts which are useful in any study of satellite orbits.

Coordinate systems (Bate et al., 1971)

Several coordinate systems are used by astrodynamics, but only some are useful for our requirements. It should, however, be noted that a point in any coordinate system can be transformed into other coordinate systems.

(a) *Heliocentric–ecliptic coordinate system*
The heliocentric–ecliptic coordinate system, shown in figure 2.1, is used for describing orbits around the Sun – such as the motions of planets or deep space probes around the Sun. The origin of the system is considered at the centre of the Sun. The fundamental *XY* plane coincides with the ecliptic plane which is the plane of the Earth's revolution around the Sun. The *X*-axis is defined as the line joining the origin and the intersection of the ecliptic and the Earth's equatorial planes. The positive direction of this axis is in the direction of the Vernal Equinox, the direction of a line joining the centre of the Earth (geocentre) and the Sun on the first day of spring. The positive *Y*-axis is to the east of the positive *X*-axis, and the positive *Z*-axis to the north of the origin.

(b) *Geocentric–equatorial coordinate system*
The geocentric–equatorial coordinate system is shown in figure 2.2. Here the equator is the fundamental plane; the geocentre constitutes the origin; the positive *X*-axis is in the direction of the Vernal Equinox, the positive *Y*-axis to the east of the Vernal Equinox, and the *Z*-axis points in the direction of the North Pole.

(c) *Right ascension–declination coordinate system*
In such a coordinate system (see figure 2.3) the position of an object in space is projected onto a sphere of infinite radius, called a *celestial sphere*. The celestial equator is considered as the fundamental plane. An object is described by the *right ascension angle*, α, and a *declination*, δ. The right ascension is measured in the eastward direction from the direction of the Vernal Equinox and the declination in the northward direction from the celestial equator.

(d) *Perifocal coordinate system*
The fundamental plane (see figure 2.4) in this system is the plane of satellite motion and the geocentre is defined as the origin. The *X*-axis is in the

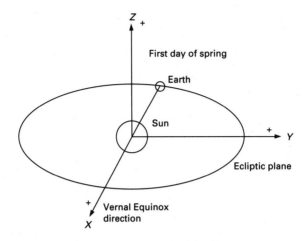

Figure 2.1   Heliocentric–ecliptic coordinate system.

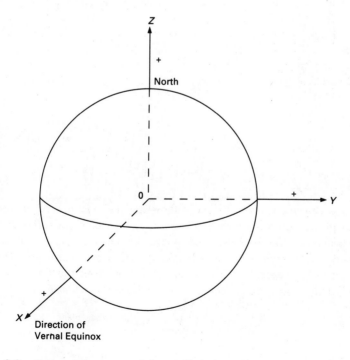

Figure 2.2   Geocentric–equatorial coordinate system.

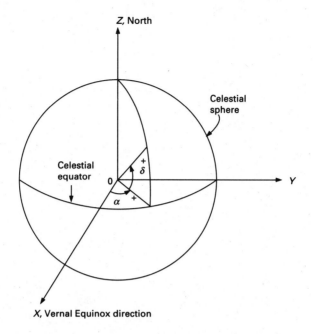

Figure 2.3   Right ascension–declination coordinate system.

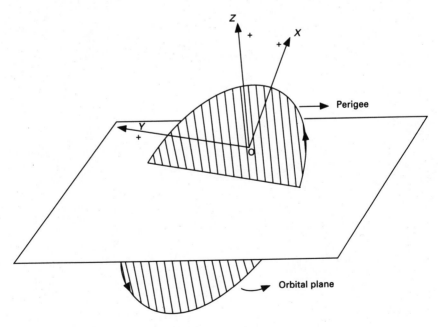

Figure 2.4   Perifocal coordinate system.

direction of the perigee, the Y-axis is rotated by 90° in the direction of satellite motion and the Z-axis is in a direction so as to complete a right-handed coordinate system.

(e) *Celestial horizon coordinate system*

The position of a geostationary satellite, with respect to a point on Earth, is usually specified by such a coordinate system. The observer's horizon becomes the reference plane and his position, the origin. One coordinate is the angular distance, η, along the vertical circle, from the horizon to the satellite location (vertical circles are all great circles which pass through the zenith and the nadir). This angle is termed the *elevation* of a satellite. The other coordinate, ξ, measured in the horizontal plane, is the angle between the direction of *true* north and the direction of the satellite measured in the clockwise direction. This angle is referred to as the *azimuth* of the satellite from the point. Figures 2.5(a) and (b) show the azimuth and the elevation respectively of a satellite from a point T on Earth.

Coordinates of a point on Earth

Any point on Earth is specified by its angular coordinates – latitude and longitude. The latitude of a point is the angle between the equatorial plane and the line joining the specified point to the geocentre. A suffix N (North) or S (South) is added to differentiate between the northern and southern hemispheres respectively. The longitude of a point is related to the great circle (or meridian) containing the north–south axis of the Earth on which the point lies. The original site of the Royal Greenwich Observatory in England was chosen as the reference meridian or 0° longitude. The longitude of a given point is the angle included between the lines joining the geocentre to the intersection of the equator and the reference meridian and the intersection of the equator and the meridian passing through the specified point. To avoid ambiguity, a suffix of East or West is added to locations to the east and west of the reference meridian respectively. As an example, figure 2.6 shows the latitude, $\theta°N$ and longitude, $\phi°E$, of a location A in the northern hemisphere.

Solar and sidereal day

In daily life we are used to time measured with reference to the Sun. A solar day is defined as the time between the successive passage of the Sun over a local meridian. On closer examination it becomes apparent that the Earth revolves by more than 360° for successive passage of the Sun over a point because the Earth itself travels, on average, a further 0.986°/day around

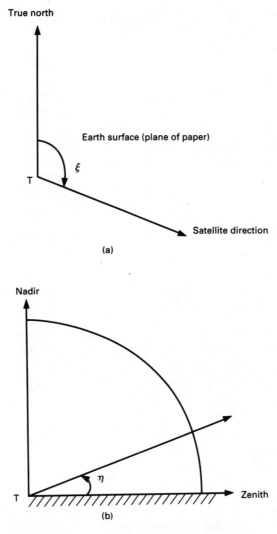

Figure 2.5 Celestial horizon coordinate system showing (a) azimuth, ξ, and (b) elevation, η, of a satellite from a point T.

the Sun. This concept is shown in figure 2.7. A sidereal day is defined as the time required for the Earth to travel 360° around its axis. As shown in the figure, this time is somewhat less than the solar day – owing to the Earth's movement around the Sun. The sidereal day, in fact, consists of 23 h 56 min and 4.1 s. A geostationary satellite (see section 2.6), therefore, must have an orbital period of one sidereal day in order to appear stationary to an observer on Earth.

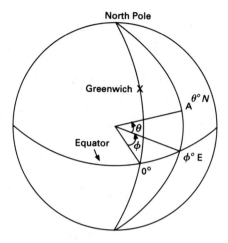

Figure 2.6   Latitude ($\theta°$N) and longitude ($\phi°$E) of a point A.

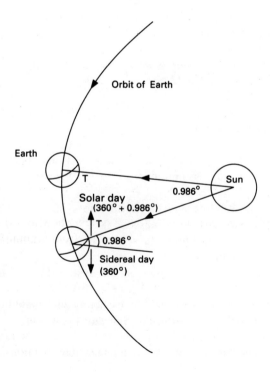

Figure 2.7   Solar and sidereal days – basic geometry.

Orbital parameters

In order to completely describe the position of a satellite in space at any given time, a classical set of six parameters, called the orbital parameters, is used. These parameters, shown in figures 2.8 and 2.9, are briefly described below.

1. The *semi-major* axis, *a*, describes the size of a conic section (ellipse or circle) for satellite communications (figure 2.8).
2. The *eccentricity*, *e*, shows the ellipticity of the orbit (figure 2.8).
3. The *inclination*, *i*, is the angle between the plane of the orbit and the equatorial plane measured at the ascending node in a northward direction. An ascending node is the point at which the satellite crosses the equatorial plane moving from south to north. Similarly a descending node is the point where the satellite crosses the equatorial plane moving in the direction from north to south. The line joining these two nodes is called the line of nodes (figure 2.9).
4. The *right ascension* of an ascending node, Ω, is the angle between the X-axis (the direction of Vernal Equinox) and the ascending node (figure 2.9).
5. The *argument of perigee*, ω, is the angle in the orbital plane between the line of nodes and the perigee of the orbit.
6. Time, t_p, is the time elapsed since the satellite passed the perigee. Sometimes this parameter is given as *mean anomaly M*. The parameter t_p can be readily calculated from *M* (see next section).

2.4 Satellite path in space

A partial solution to the path of a satellite in space may be obtained under the following simplifying assumptions:

(1) The bodies (the satellite and the Earth) are symmetric spherically and may therefore be treated as point masses.
In reality the Earth is not strictly a sphere but has a slightly ellipsoidal shape with a bulge at the equator.
(2) There are no other forces acting on the system besides their gravitational forces.
In reality there are several external forces acting on a satellite. These include the gravitational forces from the Sun and the Moon, drag due to the atmosphere, solar radiation pressure, etc.
(3) The mass of the Earth is much greater than that of the satellite.

These assumptions lead to the 'two-body' problem. Applying Newton's laws

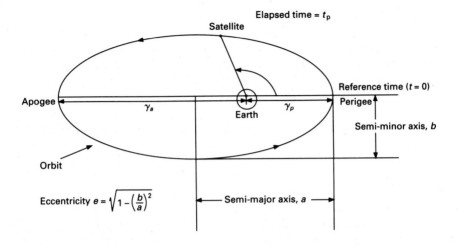

Figure 2.8   Major parameters of an elliptical orbit.

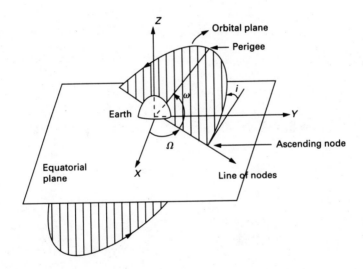

Figure 2.9   Orbital parameters.

(equations 2.1 and 2.2) to such a system, we obtain the following equation of motion

$$\ddot{\mathbf{r}} + \frac{\mu}{r^3} \, \mathbf{r} = 0 \tag{2.3}$$

where $\ddot{\mathbf{r}}$ = vector acceleration in the given coordinate system
             $\mathbf{r}$ = vector from M (mass of the Earth) to m (mass of satellite)

r = distance between M and m

μ = GM (termed gravitational parameter)

  = 398 613.52 km^3 s^{-2}.

The complete solution of this equation is not simple. However, a partial solution is easy to obtain and is adequate for illustrating the size and the shape of an orbit. The resulting trajectory equation has the general form of a conic section

$$r = \frac{p}{1 + e \cos(\theta)} \qquad (2.4)$$

where r = the distance of any point on the trajectory from the geocentre

p = a geometrical constant − termed parameter of the conic

e = the eccentricity which determines the type of conic section

θ = the polar angle between r and the point on the conic nearest the focus.

The orbit followed by a satellite is governed by the initial conditions which depend on factors such as the velocity and the position of the satellite at the final stage burnout of the launcher. For satellite communications we are concerned with elliptical orbits ($0 \leqslant e < 1$) and in particular, circular orbit ($e = 0$). For such orbits

$$p = a (1 - e^2) \qquad (2.5)$$

Figure 2.8 shows the section of an elliptical orbit together with some major parameters of the orbit.

Satellite period

Applying Kepler's third law, the period T of a satellite and the semi-major axis, a, are related as follows

$$T^2 = 4 \frac{\pi^2}{\mu} a^3 \qquad (2.6)$$

For a satellite in a circular orbit around the Earth

$$T^2 = 4\pi^2 \frac{(R + h)^3}{\mu} \qquad (2.7)$$

where R = the radius of the Earth

h = the satellite altitude.

Satellite velocity

The total mechanical energy of a satellite is constant but there is an inter-change between the potential and the kinetic energies. As a result, a satel-lite slows down when it moves up and gains speed as it loses height. The velocity of a satellite in an elliptic orbit may be obtained as follows

$$V^2 = \mu \left(\frac{2}{r} - \frac{1}{a} \right) \tag{2.8}$$

For a circular orbit, equation (2.8) reduces to

$$V^2 = \frac{\mu}{r} \tag{2.9}$$

Satellite position

It is convenient to use the perifocal coordinate system to describe the posi-tion of a satellite as a function of time. Figure 2.10 shows the motion of a satellite in the orbital plane. The origin, O, of the coordinate system is the geocentre. The satellite, at any instant, t_p, is assumed to be at S. In practice, it is necessary to determine the position of a satellite at a given time with respect to a fixed point in the orbit, usually the perigee P. Assume that a circle (shown by a broken line) is drawn from the centre C of the ellipse with a radius equal to the semi-major axis and a perpendicular BM is drawn passing through the point S. The angle E (BCM) is then called the *eccentric anomaly* and the angle θ (SOM) is the *true anomaly*. *Mean anomaly* is de-fined as the angle from the perigee that the satellite would traverse in the same time, t_p, moving at the average orbital angular velocity. For an ellipti-cal orbit, the time t_p elapsed from a perigee-pass may be derived by the use of Kepler's second law (Bate *et al.*, 1971; Escobal, 1976) and is given as

$$t_p = \frac{T}{2\pi} \left| E - e \sin(E) \right| \tag{2.10a}$$

$$= \frac{T}{2\pi} M \tag{2.10b}$$

where M is the mean anomaly given as

$$M = E - e \sin(E) \tag{2.11a}$$

Eccentric anomaly (E) may be defined in several ways. One possible

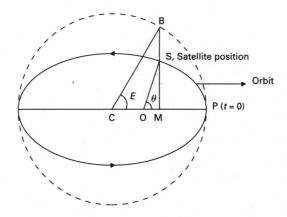

Figure 2.10   Satellite position with time – the associated geometry
(θ = true anomaly; E = eccentric anomaly).

definition of E (in radians) is

$$E = \arccos\left[\frac{e + \cos(\theta)}{1 + e\,\cos(\theta)}\right] \tag{2.11b}$$

where θ = true anomaly.
The true anomaly is given as

$$\theta = 2\,\tan^{-1}\left[\left(\frac{1 + e}{1 - e}\right)^{\frac{1}{2}}\tan\left(\frac{E}{2}\right)\right] \tag{2.11c}$$

When the eccentricity = 0, the eccentric, mean and true anomalies are all equal.

Finally, the distance between a satellite and the geocentre is given as

$$r = \frac{a\,(1 - e^2)}{1 - e\,\cos\,\theta} \tag{2.12}$$

The satellite velocity and position given by equations (2.8) to (2.12) are applicable to the two-dimensional perifocal coordinate system (see section 2.3(d)). However, an earth station operator is interested in determining the look angle of a satellite from the ground. To estimate the satellite position in three-dimensional space, it is necessary to transform the satellite position vector from the perifocal coordinate system to the right ascension–declination coordinate system. The transformation relates the two coordinate system via the orbital parameters describing the orientation of the orbit in three-dimensional space (i.e. inclination, right ascension of the ascending node and the argument of perigee). The position of the satellite relative to an earth station is usually specified in a celestial horizon coordinate system, as satellite azimuth and elevation. Again, a coordinate transformation becomes essential – in this instance from the right ascension–declination coordinate system to the celestial horizon coordinate system. Appendix B summarizes the equation set used in obtaining satellite look angles and satellite range from any specified ground location and orbital parameters. Orbital parameters are readily available from the agency operating the satellite.

It may be of interest to note that certain earth station tracking systems have a capability of developing a model of satellite motion as viewed from the earth station, from real-time tracking data stored in their memory. The system utilizes the model to improve its tracking performance (see chapter 10).

2.5 Corrections to simplified model

In a real environment, a satellite is acted upon by several additional forces, leading to some important effects not predicted by the simplified two-body model discussed above. Thus the two-body problem is modified to an n-body problem. The effects of such perturbations depend on the characteristics of the orbit.

Effect of Earth

The gravitational force around the Earth is non-uniform because of the non-uniform distribution of the Earth's mass. Further, the Earth may be approximated as an ellipsoid with a slight bulge at the equator. The difference between the polar and equatorial radius (about 6378 km) is about 21 km. This deviation from the assumption of a spherical Earth causes additional forces on the satellite because the gravitational pull is offset from the centre of the Earth. A more precise description of the effect of Earth's gravitational field is obtained by expressing the field as a harmonic series. The first term in the series represents the principal gravitational law and the higher-order terms the perturbations. The main effects of perturbations on a satellite are discussed below.

(i) The component of perturbation in the orbital plane causes the perigee of an elliptical orbit to rotate in the orbital plane. The rate of change of the argument of perigee, $\dot{\omega}$, is given as (Martin and Davidoff, 1985)

$$\dot{\omega} = 4.97 \left[\frac{R_m}{a}\right]^{3.5} \frac{5 \cos^2(i) - 1}{(1 - e^2)^2} \tag{2.13}$$

where R_m = mean equatorial radius ($\sim$6378 km)
$\quad\quad a$ = semi-major axis
$\quad\quad i$ = inclination
$\quad\quad e$ = eccentricity.

Let us focus our attention on the $5\cos^2(i) - 1$ term of equation (2.13). It is observed that when the inclination is 63.4°, $\dot{\omega}$ reduces to zero implying that the perigee remains fixed in space. This is a useful property. Several variations of such orbits are possible. For example, when the apogee is $\sim$40 000 and the perigee is $\sim$1000 km the orbit is known as a *Molniya* orbit – after the Molniya series of satellites first used by the former Soviet Union. A *Tundra* orbit has an apogee of $\sim$46 300 km and a perigee of $\sim$25 250 km. These orbits are useful in providing coverage to higher latitude locations which cannot be served well by geostationary satellites. This type of orbit has also been suggested for provision of land mobile communication service to high latitude locations. A high elevation angle visibility is a desirable feature for this application since land mobile–satellite links tend to remain heavily shadowed at low elevation angles and this type of orbit can provide high elevation angles at high latitudes.

(ii) Another effect of the perturbations is that the orbital plane rotates around the Earth's north–south axis. The rate of precession of the ascending node of a circular orbit/day, $\dot{\Omega}$, is given as

$$\dot{\Omega} = 9.95 \left[\frac{R_m}{r}\right]^{3.5} \cos(i) \tag{2.14}$$

where r = satellite–geocentre distance in same units as R_m.

The rotation is in a direction opposite to the motion of the satellite. For a geostationary orbit, the magnitude is $\sim$4.9°/year, implying that the ascending node rotates around the Earth in $\sim$73 years.

This precession is utilized in the Sun-synchronous orbit. An orbital plane fixed with respect to the Earth effectively makes a 360° rotation in space in a year (about 365.25 days), since the Earth itself rotates by 360° around the Sun. This is equivalent to a rotation of the orbital plane by about 0.986°/day. If r and i in equation (2.14) are adjusted such that the resulting precession compensates the orbital rotation due to the Earth's movement around the Sun, then a satellite in this orbit maintains a constant angular relationship with the Sun throughout. Consequently, such satellites attain the property of appearing at about the same time each day over any given point on the Earth. This is useful in applications such as earth observations, reconnaissance etc. because the satellite always views the Earth under a similar illumination.

(iii) In addition to causing rotation of perigee, the component of perturbating force along the orbital plane imparts a force vector on a satellite. For most orbits, such components tend to cancel out as the satellite position relative to the Earth changes continuously. However, as explained below, this is not applicable to a specific type of equatorial orbit of interest for satellite communications – the geostationary orbit. (The geostationary orbit is discussed in section 2.6.) As already mentioned, the equatorial cross-section of the Earth can be approximated as an ellipse. The ellipse has a semi-minor axis approximately along the line joining 75°E and 255°E (105°W) and semi-major axis approximately along the line 165°E and 345°E (15°W). As a consequence, the gravitational force on a satellite is directed towards the nearest equatorial bulge instead of the Earth's centre, producing a component of force in the orbital plane. But unlike satellites in other types of orbits, a satellite in a geostationary orbit maintains a constant relationship with the Earth. Therefore the resultant perturbation components do not cancel but cause a satellite to drift towards one of the two nearest stable points of the orbit. The stable positions are approximately (but not quite) on the minor axis, showing that the elliptical approximation of the earth is not precisely accurate. Data obtained from observations on drift behaviour of free-moving geostationary satellites show the stable positions to be at 79°E and 252.4°E (Morgan and Gordon, 1989). In addition there are two semi-stable points which lie on the semi-major axis.

The magnitude and direction of drift acceleration depends on the location of the satellite with respect to the stable points and may be approximated as a slightly decaying sinusoid, with zero crossings at each stable and semi-stable point. The maximum amplitude of drift acceleration is $\sim \pm 0.0018°/$day^2. The positive-to-negative crossing of the sinusoid occurs at 79°E.

To maintain a satellite at a specified orbital location in the geostationary orbit, the drift is cancelled by firing thrusters regularly. Without such 'station-keeping' manoeuvres, a satellite drifts towards the stable point but because of inertia it overshoots the stable point. The satellite then experiences a negative acceleration, causing it to drift back towards the stable point and the process continues. Such oscillations have a period of approximately 3 years, the exact period depending on the longitudinal difference between the stable point and the initial position of the satellite.

Gravitational effects from heavenly bodies

For low Earth orbit satellites the effects of gravitational forces from the Sun and the Moon are small when compared with the gravitational force of the Earth. However, these forces cause noticeable perturbations to a geostationary satellite orbit.

The main sources of perturbations on a geostationary satellite, in order of magnitude, are the gravitational forces of the Moon and the Sun. The satellite receives a stronger gravitational pull in the direction of the heavenly body when nearer to it, causing a gravity gradient. The main effect of such a gravity gradient is to change the inclination of the satellite orbit. The average inclination change causes the orbit normal (a line drawn perpendicular to the orbital plane) to move in the direction of the Vernal Equinox. (Variations in orbital inclination are often treated as variations in the orbit normal.)

The combined effect of the Sun and the Moon is to cause a change in the inclination of a geostationary satellite between 0.75° and 0.94°.

The inclination of the orbital plane caused by the Moon changes cyclically between ~0.48° (minimum) and ~0.67° (maximum) with a period of 18.6 years (6798 days). The maximum inclination change due to the Moon occurred in the year 1987 and the minimum is predicted to occur in February 1997. The cyclic variations occur because the inclination of the Moon orbit itself varies because of the influence of the Sun. Maximum perturbation to a satellite in the geostationary orbit occurs when the orbit normal of the Moon and the satellite are furthest apart.

The change in the inclination of the orbit caused by the Sun is ~0.27°/year. The change is steady for the accuracy considered here as the relative position between the satellite and the ecliptic (plane of the Earth around the Sun) remains almost fixed.

Note that there are mainly three forces affecting the inclination of a geostationary satellite – the gravitational pull of the Sun and the Moon acting in the same average direction and a force caused by the non-spherical nature of the satellite (i.e. the force causing precession of the ascending node). The latter force has a component in a direction opposite to the former two forces. The forces cancel out at an inclination of about 7.5°. As a consequence, the inclination of a satellite in a geostationary satellite, when left uncorrected, oscillates around this stable inclination with a period of about 53 years, reaching a maximum inclination of 15°.

Solar radiation pressure

When considering geostationary satellites, the effects of solar radiation pressure must also be considered. The effect increases as the surface area of the satellite projected in the direction of the Sun is increased. This is the case for large powerful satellites which use large solar arrays. The net effect of the solar radiation pressure on a geostationary satellite is to increase the orbital eccentricity and to introduce a disturbing torque affecting the north–south axis of the satellite. Such perturbations are corrected periodically.

Other effects

There are several second-order effects which need to be considered for a precise treatment of satellite motion in the geostationary orbit. These include the effects of the Earth's magnetic field, meteorites, self-generated torques and pressures caused by RF radiation from the antenna (Maral and Bousquet, 1987).

Atmospheric drag

A satellite suffers atmospheric drag due to the friction caused by collision with atoms and ions. Atmospheric drag affects low-orbit satellites the most. The effect of drag is to reduce the ellipticity of an elliptical orbit, making it more circular, and to cause a loss of altitude of a circular orbit. At very low orbital altitudes the friction causes excessive heat on a satellite which finally results in its loss by burning. The lower limit on the altitude of a satellite, owing to the effects of drag, is about 100 nautical miles. The orbital lifetime of a satellite (limited by drag) is a complex function of the initial orbit, geometry and mass of the satellite and the ionospheric conditions (King-Hele, 1978). However in predicting a satellite's life the orbital life must be distinguished from the operational life. The latter is the period during which a satellite performs the planned mission successfully. A satellite can however continue to orbit even after ceasing to function.

For example, the orbital life of a small satellite in a 400 km circular earth orbit is typically a few months, whereas the orbital life of a similar satellite in a 800 km circular orbit could be several decades. In the former case (i.e. 400 km orbit) the functional life of the satellite is mainly governed by the orbital lifetime (except for the less likely situation where the satellite's equipment fails earlier), whereas in the latter the functional life depends on the lifetime of the satellite equipment.

The functional lifetime of a geostationary satellite is governed by the equipment life and the fuel capacity of the satellite, typically of the order of 10–15 years at present.

Doppler effect

Johann Doppler (1803–53) explained and quantified the phenomenon of apparent change in frequency of sound waves at the receiver when the sound source moves with respect to the receiver. A classical example is the variation in the pitch of the whistle of a moving train to a stationary observer. The whistle appears at a higher pitch when the train is approaching the stationary observer compared with when the train is moving away, although the pitch would appear unchanged to an observer travelling in the train. This phenomenon, called the *Doppler effect*, is also observed at radio frequencies.

The frequency of satellite transmissions received on the ground increases as the satellite is approaching the ground observer and reduces as the satellite moves away. When dealing with satellite communication systems, it is necessary to estimate the frequency shift caused by the Doppler effect (or simply, the Doppler shift) for the following main reasons:

1. To isolate the Doppler component of frequency uncertainties at the receiver. When developing specifications for a receiver, a frequency 'budget' is prepared to estimate the receiver filter bandwidth and the phase lock loop capture range of the local oscillator. The budget includes frequency uncertainties/drifts in the earth stations and the satellite local oscillator, together with the Doppler shifts associated with the uplink and the downlink.
2. To eliminate the Doppler component in the measurement of transmission frequency of the earth stations or the spacecraft from measurements taken at monitoring stations. Essentially, the Doppler component is removed from the measured frequency. (The function of a monitoring station is to maintain an operational vigil by measuring the transmission parameters of each carrier in the network.)
3. To estimate the received downlink frequency, when the transmission frequency of an earth station and the satellite local oscillator frequency are known. This knowledge is used by radio amateurs to tune their receivers. Such knowledge is usually not essential for an earth station operating with a geostationary satellite, because the bandwidths of the signals are often much larger than the Doppler shift. However, if the Doppler component is likely to cause difficulties, the uplink transmission frequency is offset appropriately to correct the Doppler shift (and satellite translation frequency errors), giving the desired downlink frequency.

Doppler shift can also be used to estimate the position of an observer provided that the orbital parameters of the satellite are known precisely. Alternatively, Doppler shift can be used to estimate the orbital parameters of a satellite (Bate *et al*, 1971; Escobal, 1976).

The Doppler shift Δf_d at a frequency f_t is given by

$$\Delta f_d = \pm \frac{v_r}{c} f_t \tag{2.15}$$

where v_r = relative radial velocity between the observer and the transmitter
c = velocity of light
f_t = transmission frequency.

The sign of the Doppler shift is positive when the satellite is approaching the observer. Formulas for estimating the Doppler shift are summarized in appendix B.

2.6 Geostationary satellites

Using equation (2.9) it can be readily calculated that the orbital period exactly matches the period of a sidereal day at an altitude of $\sim$37 786 km. The satellite velocity in this orbit is 3075 m/s. Such an orbit is called the *geosynchronous* orbit. When the inclination and the eccentricity of this orbit are zero (i.e. circular, equatorial orbit) the satellite appears stationary to an observer on the ground and the orbit is termed as a *geostationary* orbit. In practice, however, the inclination and the eccentricity are rarely zero and, in consequence, satellites make small daily excursions relative to the Earth (typically less than 0.5°).

There are several advantages of the geostationary orbit and therefore this orbit is now well established for communications. A satellite which appears stationary to all the ground stations within its coverage region minimizes the operational demands on the terminals since the tracking requirements are minimal and the transmission parameters such as path loss are invariant. Moreover the coverage offered by satellites in such an orbit is adequate for most populated areas (±75° latitude). Other advantages include minimal Doppler shift and predictable interference to and from other radio systems due to its constant geometry.

However, there are also a few inherent disadvantages in this type of orbit. Propagation delays are quite significant ($\approx$250 ms in one direction) because of the large satellite range. This delay is adequate for a coherent telephonic conversation over a single satellite hop, but a two-hop telephonic conversation is extremely difficult. Further, there is a degradation in the communication quality for short durations (which can, however, be predicted) when the Sun appears within the beamwidth of an earth station antenna because the Sun is a strong source of noise. Another disadvantage of this orbit is its inability to provide adequate coverage to locations beyond $\approx$±75° latitude. Geostationary satellites appear at very low elevation angles beyond these latitudes, resulting in a very poor signal quality. These locations, therefore, must be served by satellites using a different type of orbit such as the Molniya orbit. A limitation of the geostationary orbit which has been mentioned recently is related to its applicability for mobile communication in high latitude regions such as Europe. The relatively low elevation angles of satellites from such locations result in a large propagation loss due to shadowing from obstructions such as building, trees, etc., limiting the capacity and reliability of such links (see chapter 3).

It should also be mentioned that the advantages of the orbit far outweigh the disadvantages in most applications, and hence the geostationary orbit is used in most existing satellite communication systems. Hence attention in the remaining part of the chapter will be focused on this orbit.

A knowledge of a number of orbit-related parameters is vital for reliable communication satellite system design. It is essential to estimate a satellite's range and look angles from the ground. Furthermore, a geostationary satellite is periodically eclipsed by the Earth and the Moon. The time of occurrence and the duration of eclipse have an impact on a spacecraft's design and operations. Several times during a year the Sun appears in the beamwidth of earth station antennas, causing a considerable increase in system noise temperature. These topics are addressed in the rest of this section. A summary of useful associated topics and formulas, applicable to both geostationary and low earth orbits, is presented in appendix B.

Geometric considerations

For most engineering studies of satellite communications, simple geometric solutions involving spherical and planar trigonometry provide the required accuracy. It is possible to obtain simple working relationships by assuming a spherical Earth with the satellite at the equatorial circle (Siocos, 1973).

Satellite elevation

The elevation of a satellite (shown on figure 2.11a as η) is the angle which a satellite makes with the tangent at the specified point on the Earth. The elevation is given as

$$\eta = \arctan\left(\frac{\cos(\beta) - \sigma}{\sin(\beta)}\right) \qquad (2.16)$$

where

$$\text{coverage angle, } \beta = \arccos\left(\cos(\theta_e)\cos(\phi_{es})\right) \qquad (2.17a)$$

$$\phi_{es} = |\phi_e - \phi_s|$$

$$\sigma = \frac{R}{(R + h)} \approx 0.151$$

In terms of elevation angle

$$\beta = 90° - \eta - \sin^{-1}(\cos(\eta)/6.6235) \qquad (2.17b)$$

In terms of tilt angle (see equation 2.18)

$$\beta = \sin^{-1}(6.6235 \sin(\gamma)) - \gamma) \qquad (2.17c)$$

where θ_e = latitude of earth station
   ϕ_e = longitude of earth station
   ϕ_s = longitude of sub-satellite point
   R = radius of the Earth (6378 km)
   h = satellite height above equator (35 787).

The tilt angle, γ (see figure 2.11a), is given as

$$\gamma = \arctan\left(\frac{\sin(\beta)}{6.6235 - \cos(\beta)}\right) \qquad (2.18)$$

Note that some authors refer to tilt angle as nadir angle.

Azimuth

The azimuth of a satellite from a given point (shown on figure 2.11b as ξ) is the angle which the satellite direction makes with the direction of true north measured in the clockwise direction. Note that there is a difference (called *magnetic variation* or *magnetic deviation*) between true north and magnetic north (i.e. north shown by a compass). Magnetic variation is a function of location and year. In the UK its value can be obtained from the British Geological Society. The azimuth, ξ, can be obtained as

$$A = \arctan\left(\frac{\tan(\phi_{es})}{\sin(\theta_e)}\right) \qquad (2.19)$$

In the northern hemisphere

   $\xi = 180° + A°$; when the satellite is to the west of the earth station
   $\xi = 180° - A°$; when the satellite is to the east of the earth station

In the southern hemisphere

   $\xi = 360° - A°$; when the satellite is to the west of the earth station
   $\xi = A°$; when the satellite is to the east of the earth station

Range

The range d to a geostationary satellite can be expressed by a number of formulas, some of which are given below.

$$d = 35786\left[1 + 0.4199\left|1 - \cos(\beta)\right|\right]^{1/2} \text{ km} \qquad (2.20a)$$

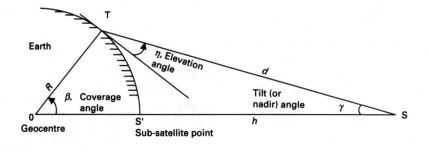

Figure 2.11   Geometry of a geostationary link showing (a) elevation, (b) azimuth.

In terms of radius of the Earth $\left(\text{i.e. } d_{er} = \dfrac{d}{R} \right)$

$$d_{er} = [13.347\,(1 - \cos\,(\beta)) + 31.624]^{1/2} \tag{2.20b}$$

also

$$d_{er} = 6.6235\,\frac{\sin(\beta)}{\cos(\eta)} \tag{2.20c}$$

Solar eclipses

The primary power of a geostationary satellite is derived from solar cells and therefore the cells must be continuously illuminated by rays from the Sun. However, satellites in a geostationary orbit undergo solar eclipses caused by the shadows of the Earth and the Moon. In order to avoid a disruption in services it is necessary to provide a secondary power source for satellite operation during such eclipses. These secondary sources, usually storage batteries, are heavy and consequently increase the spacecraft weight (and hence the launch cost). It is therefore preferable to provide only minimal service during an eclipse. Another effect of solar eclipses is the relatively rapid change in thermal conditions of a satellite. Effects of such thermal gradients must be considered in a spaceraft's thermal design.

Eclipse due to the Earth

The Sun appears to make a sinusoidal motion relative to the equatorial plane, with a period of a year (see figure 2.12a). This motion results from the difference between the inclinations of the equatorial and the ecliptic planes. The equatorial plane and the Sun appear along the same line on the Spring Equinox (March 21) and Autumn Equinox (September 21). Consequently geostationary satellites suffer an Earth eclipse around these periods. The duration of the eclipse is determined by the time taken by a satellite to traverse the shadowed region – a 17.4° orbital arc at the geostationary orbit at maximum. This is equivalent to a duration of ~69.4 minutes [(17.4/360) × orbital period]. An eclipse, in fact, begins several days before the equinoxes and lasts for an identical period after. As shown in figure 2.12(b), an eclipse begins on the day when the shadow of the Earth just grazes the geostationary orbit. The duration of the eclipse increases gradually as the Sun inclination approaches the equatorial plane until a maximum (69.4 minutes) is reached at the equinox (see left-hand side of figure 2.12b). The duration then gradually decreases identically until the Sun inclination exceeds 8.7°. Because of diffraction effects and the fact that the Sun is not a point source, the Earth's shadow is diffused and eclipse duration exceeds the theoretical value by ~2 minutes. The movement of the Sun relative to the equatorial plane (see figure 2.12a) can be approximated as (Maral and Bousquet, 1987)

$$\chi \approx 23\sin\left(\frac{2\pi t_d}{T_Y}\right) \tag{2.21}$$

where χ = Sun inclination in degrees
t_d = time in days
T_Y = 365 days.

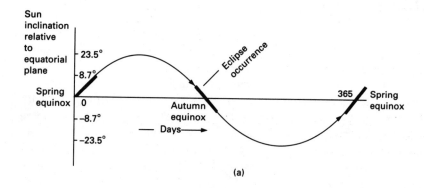

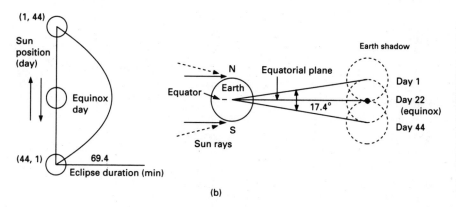

Figure 2.12   (a) Movement of Sun relative to equatorial plane; (b) Earth-
induced eclipse of geostationary satellite.

On substituting χ as 8.7° in equation (2.21) we see that the eclipse lasts for ~22 days before and after the equinoxes.

Eclipse occurs at local midnight of the sub-satellite point. The time of occurrence of an Earth eclipse in a satellite's coverage zone depends on the position of the satellite relative to the coverage area. Hence sometimes (e.g. in direct broadcasts systems) satellites are positioned so that eclipses occur late at night – a period of low demand – thereby reducing the requirements on the secondary power source of the satellite. The occurrence of the eclipse can be suitably timed by selecting the location of the satellite to the west of the coverage area according to the following equation (Mertens, 1976)

$$T_1 = 23.38 - (1/15)(\phi_s - \phi_{tz}) \tag{2.22}$$

where T_1 = local time at which the eclipse begins

ϕ_s = eastern limit of the orbital arc

ϕ_{tz} = longitude of the time zone of the coverage area.

Eclipse due to the Moon

Eclipse also occurs when the satellite is shadowed by the Moon. The total period of the eclipse caused by the Moon in a year is much less than that caused by the Earth. However, whereas Earth eclipses are predictable and may be designed to occur at an off-peak hour by a suitable choice of orbital location, Moon eclipses are more irregular and may occur when the traffic demand is high or at a peak viewing time for television broadcasts. Moreover, if these two types (i.e. Earth and Moon) of eclipses occur in quick succession, the on-board rechargeable batteries may not have had adequate charging time. It is therefore necessary for the system operators to be aware of the duration and time of occurrence of the Moon eclipses. A relatively simple method which makes use of a Nautical Almanac (a readily available publication) has been suggested for such use (Siocos, 1981; see also appendix B).

Solar interference

The reader should note that several terms used in this section are discussed further in chapter 4.

Solar interference occurs when the Sun appears within the beamwidth of an earth station antenna. As the Sun is a strong source of noise, the antenna noise temperature rises significantly, causing a disruption to communication in many cases. The increase in the antenna noise temperature (ΔT_a) during Sun transit depends on the beamwidth and efficiency of the earth station antenna together with the Sun's equivalent noise temperature at the frequency of operation. An accurate expression for ΔT_a would involve integration of the average Sun noise temperature over the earth station antenna's radiation pattern and is cumbersome. An approximate expression giving adequate accuracy for practical application can be obtained by considering the antenna pattern to have a uniform gain over the half-power beamwidth and zero gain outside (Mohamadi *et al.*, 1988)

$$\Delta T_a = pT_s\eta D_s^2 \tag{2.23}$$

where p = polarization factor to account for random polarization of Sun noise

= 0.5

T_s = Sun's equivalent noise temperature

= 120 000 $f^{-0.75}$ (where f is the frequency of operation in Gigahertz)

η = antenna efficiency

D_s = optical diameter of the Sun/ψ_{hp} (where ψ_{hp} is the half-power beamwidth of the antenna)

= 0.48/ψ_{hp} (for $\psi_{hp} > 0.48$)

D_s = 1 (for $\psi_{hp} < 0.48$).

A solar conjunction occurs within ± 22 days of the equinoxes when the declination angle of the Sun equals the angle that an earth station antenna makes with the equator. For an antenna of beamwidth $\psi_{hp}°$, solar interference occurs for a maximum number of days of

$$\frac{(\psi_{hp} + 0.48)}{0.4} \tag{2.24}$$

The maximum duration of Sun transit on the peak day is

$$\frac{(\psi_{hp} + 0.48)}{0.25} \tag{2.25}$$

Appendix B summarizes the technique for predicting the occurrence of Sun transit at any location on the Earth (Siocos, 1973).

Selection of orbital slot

The selection of an orbital location depends on the application and the associated radio regulations, the desired coverage area and the orbital crowding in the region of interest. Some of the basic points to consider are:

1. The service area should be served at as high an elevation angle as possible. This applies in particular to mobile satellite service. The reasons will become clear in chapters 3 and 4.
2. Satellite eclipse should occur as late at night as possible to minimize the need for storage batteries on the satellite (see 'Solar eclipses' earlier in this section).
3. The need to maintain a separation of 2–3° from the adjacent satellite to permit coexistence and to coordinate with other agencies who might be affected or might have an existing satellite.
4. Some services such as BSS have an existing plan which reserves an orbital slot to each country.

In practice the number of available slots may be very limited especially in regions of orbital overcrowding. For example, currently there is a great demand for orbital locations for the MSS throughout the world, and for the FSS in the Pacific area.

2.7 Launching of geostationary satellites

The total energy, U, of a satellite, for a two-body system, is given by

$$U = \frac{1}{2} mv^2 - \frac{GmM}{r} \qquad (2.26)$$

where m and v are the mass and the velocity of the satellite respectively, and r is distance from the geocentre, G is the gravitational constant and M the mass of the Earth. To achieve a geostationary orbit, the launch vehicle must be able to impart a velocity of 3070 m/s at the geostationary orbit height (42 165 km from the Earth's centre). In an equipotential field, the maximum velocity increment, Δv, which a launch vehicle of total mass, m_o, can impart may be given as (Collette and Herdan, 1977)

$$\Delta v = v_g \ln \left(\frac{1}{1 - \dfrac{m_f}{m_o}} \right) \qquad (2.27)$$

where v_g = effective exhaust velocity of the gas – a function of the type of fuel and the rocket nozzle design

m_f = mass of the expanded fuel.

In order to maximize Δv the ratio m_f/m_o should be maximized. Therefore it is usual to launch satellites by means of multiple stage rockets, each stage being jettisoned after imparting a given thrust. As m_o is progressively reduced, succeeding stages of rockets need to impart progressively lower thrust to achieve the desired orbit. The final velocity of the spacecraft is the sum of the velocity increments of all the stages.

Satellites may be launched directly into a geostationary orbit or via lower orbits, depending on the type of launcher. Most of the satellites launched today are initially launched into a low Earth 'parking orbit'. In the next phase, the satellite is injected into an elliptical transfer orbit which has an apogee at the height of the geostationary orbit and its line of apsides (perigee–apogee line) in the equatorial plane. Finally, the satellite is injected into the geostationary orbit. This is achieved by imparting a velocity increment at the apogee equal to the difference between the satellite velocity at the apogee of the transfer orbit and the velocity in the geostationary orbit. A transfer between two coplanar circular orbits via an elliptical transfer orbit requires the least velocity increment (and hence fuel). This principle was first recognized by Hohmann in 1925 and is therefore referred to as a *Hohmann transfer*. At this stage of the mission a geosynchronous altitude is achieved but the orbital inclination is determined by the latitude of the launch station and is given by

$$\cos(i) = \sin(\xi_l) \cos(\theta_l) \tag{2.28}$$

where i = inclination

ξ_l = azimuth of launch

θ_l = latitude of launch site.

The inclination is minimal for an easterly ($\xi_l = 90°$) launch and equals the latitude of the launch site. To achieve a geostationary orbit a further correction is therefore necessary. This is achieved by applying a velocity increment in a direction perpendicular to the orbital plane. The correction is usually performed at the same time as the circularization manoeuvre discussed above. The velocity increment is minimal if it is applied at the point of minimum velocity, i.e. at the apogee of the transfer orbit.

At present, geostationary satellites may be launched either by expendable launchers or by the reusable Space Transportation System (STS or space shuttle). The flight plan of a geostationary mission depends on the type of launcher involved. We will briefly examine the launch sequence of these two types of launch vehicles.

Launch from an expendable launcher

A typical launch sequence (zero inclination) is shown in figure 2.13. A satellite is launched in an easterly direction and as close to the equator as feasible to take maximal advantage of the Earth's rotational velocity (see equation 2.28). To minimize drag from the atmosphere a satellite is launched vertically. The vehicle is gradually tilted by its guidance system during the flight until, at the point of injection, it is tilted by 90°. After a few minutes the first-stage rockets are burnt and jettisoned. The second stage is ignited soon after. When the initial parking orbit – typically between 185 and 250 km – is reached the ignition of the second stage is cut off. The satellite together with the remaining second-stage rocket drift in the parking orbit. Shortly before reaching the equator the second-stage rocket is re-ignited. This stage is burnt to depletion, followed by a burn to depletion of the third stage (if any). This manoeuvre injects a satellite into the elliptical transfer orbit and the payload is separated from the launch vehicle.

The satellite trajectory is closely monitored by a network of tracking stations until an accurate set of orbital parameters is obtained. Several revolutions (lasting several hours) are usually necessary. Before performing the circularization manoeuvre the satellite attitude is corrected to obtain the desired orientation. An apogee-kick motor, which is a part of the spacecraft, is then fired at the apogee of the transfer orbit. The firing of the apogee-kick motor modifies the transfer orbit into a nearly circular geosynchronous orbit. The satellite begins to drift slowly with respect to the Earth and hence this phase is referred to as the drift phase. Small thrusters on-board the satellite

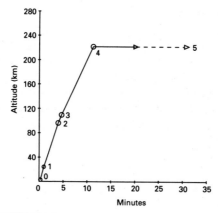

Event number	Event
0	Vertical lift-off
1	Guidance system begins tilting rocket towards east
2	First-stage drop-off
3	Second-stage ignition
4	Horizontal insertion into parking orbit 185 to 250 km
5	Second and third stages fired at equator to acquire transfer orbit

(a)

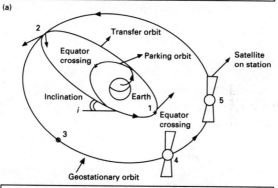

Event number	Event
1	Velocity increment to acquire transfer orbit; satellite spun for stabilization; attitude manoeuvres done before apogee kick-motor firing
2	Apogee kick-motor fired to give necessary velocity increment; orbit circularized and inclination reduced to near zero
3	Satellite despun
4	Three-axis stabilization acquired
5	Minor orbit corrections performed to correct residual orbital errors in orbit tests and position satellite on station

(b)

Figure 2.13 A typical launch sequence of a communication satellite by an expendable launcher, showing the sequence of major events: (a) acquiring a parking orbit; (b) injection into a transfer orbit and positioning of satellite at a specified location.

are fired for fine corrections until the satellite is positioned at the desired location. Other operations during this phase include (see figure 2.13) transition of stabilization of the satellite from the spin mode into a body-stabilized mode (for three-axis stabilized satellites), solar array deployment, and Sun and Earth acquisition. Once 'on-station' the performance of the satellite is checked using well-laid-down procedures before the actual operations begin.* These tests are called in-orbit tests. During the operational phase, drifts caused by various perturbations are corrected periodically to maintain the satellite within the specified limits.

Launch from space shuttle

The expendable launchers lose most of the expensive hardware during launch. Therefore one of the main design objectives of the space shuttle was to develop a reusable launch vehicle. In addition, the shuttle was designed with the capability to retrieve (and repair) satellites in low orbits.

The shuttle consists of a reusable orbiter which injects satellites to a low Earth orbit and re-enters the atmosphere, landing as an aircraft. The orbiter itself is launched vertically with the help of two recoverable solid rocket boosters. An expendable liquid-hydrogen/liquid-oxygen tank furnishes propellant to the three main engines. This tank is the only part of the shuttle that is not reused. Satellites can be launched in an easterly direction from the Kennedy Space Center. Such a launch gives an orbital plane inclination of 28°. The Vandenberg Airforce Base is used for launches into polar or near polar orbits. The shuttle can carry very large payloads into near Earth orbits – 29 500 kg into the eastern orbits and 14 500 kg into the polar orbits. The payload bay has a length of 18.3 m and a diameter of 4.6 m. The shuttle can only launch satellites in low Earth orbits and therefore additional propulsion is necessary to inject a satellite into the geostationary orbit. The heavy lift capability of the shuttle is effectively used for carrying shuttle upper stages for the extra propulsion. Various types of upper stage have been developed. These may be categorized as follows.

(a) *Perigee stages*
These stages provide the perigee thrust to inject a satellite into a geostationary transfer orbit. The geostationary orbit is attained by firing an apogee-kick motor which is usually incorporated into the satellite.

(b) *Integrated stages*
These types of upper stages combine the perigee and apogee motors into a single package.

---

* When the orbital position is already occupied by the 'current' operational satellite, pre-operational in-orbit tests on the new satelllite are performed at another convenient location to avoid causing interference to the operational system. After successful tests the satellite is moved to the designated location to replace the existing satellite.

The capabilities of such upper stages extend from PAM-D, which can launch 1250 kg into the geosynchronous orbit, to STS/Centaur capable of injecting up to 5910 kg into the geosynchronous orbit.

The spacecraft is placed by the shuttle into a parking orbit of ~290 km at an inclination of, say, 28°. At this stage the satellite is injected out of the shuttle by a suitable mechanism. The thrust to attain the geostationary transfer orbit is applied at the equator, using a suitable upper stage. The geosynchronous orbit is achieved by applying the required thrust at the apogee. The succeeding phases are similar to those of the expendable rockets.

Launch window

Before the launch of a satellite it is necessary to ensure that the launch time falls within a 'launch window'. This guarantees that the position of a satellite in respect of the Sun is favourable, thus ensuring adequate power supply and thermal control throughout the mission. Further, the launch must be so timed that the satellite is visible to the control station during all the critical manoeuvres. This set of conditions limits the launch time to certain specified intervals, designated the *launch window*.

The reader is referred to appendix B for a summary of useful associated topics which practicing engineers encounter from time to time. The topics included are:

1. effects of minor deviations of satellites from the geostationary orbit;
2. method of obtaining the coverage contour of a satellite;
3. prediction of Sun transit time at an earth station;
4. prediction of the duration and extent of solar eclipses on geostationary satellites;
5. conversion of Earth coordinates to the satelli-centric coordinate system, and vice versa;
6. estimating off-axis angle from earth stations and satellites;
7. estimating satellite position from orbital parameters;
8. optimization of a low Earth orbit satellite constellation coverage.

Problems

1. State Kepler's laws as applied to satellite communications. Briefly describe the orbital parameters with the help of a diagram.
2. Why does a satellite's orbit deviate from the predictions of Kepler's laws? Outline the principle of Sun-synchronous and Molniya orbits. Suggest some uses of the Molniya orbit.

What are the advantages and disadvantages of a geostationary orbit for satellite communications?

3. Determine the azimuth and elevation angles of a ground terminal located at Guildford, UK for communicating with geostationary satellites positioned at the following longitudes:
   (a) 335.5° east
   (b) 345.0° east
   (c) 10° west.
   Assume Guildford coordinates to be 51°N, 0.5°E.
   A transmission time delay of more than 400 ms causes difficulty in establishing a coherent telephonic conversation. Calculate the transmission time delay for the links given above, to examine if this delay criterion is satisfied. Two satellite hops are necessary to establish a communication link with a location outside the coverage area of a satellite (source: satellite 1 – intermediate earth station with common visibility to both satellites; satellite 2 – destination). Estimate the *order* of delay in such a communication link and comment on your results.

4. (a) A reliable link cannot be established with geostationary satellites which appear below about 5° elevation owing to unacceptable degradations in the received signal caused by ground reflections and atmospheric effects. Plot a contour of 5° elevation for a satellite located at 345°E.
   (b) Compute the geostationary arc within which communications can be established from London (approximate location 0°, 51.5°N) via a satellite. (*Hint*: Determine extreme longitudes which give an elevation of 5°.)

5. (a) Briefly discuss the phenomena of eclipse as applied to geostationary satellites and solar interference experienced at an earth station. How do these factors influence the system design?
   (b) Briefly discuss the sequence of events during the launch of a geostationary satellite.

References

Bate, R.R., Mueller, D.D. and White, J.E. (1971). *Fundamentals of Astrodynamics*, Dover Publications, New York.

Collette, R.C. and Herdan, B.L. (1977). 'Design problems of spacecraft for communication mission', *Proc. IEEE*, Vol. 65, March, pp 342–356.

Escobal, P. (1976). *Methods of Orbit Determination*, Wiley, New York.

King-Hele, D.J. (1978). 'Methods of predicting satellite orbital lifetimes', *Journal British Interplanetary Society*, Vol. 31, pp 181–196.

Maral, G. and Bousquet, M. (1987). *Satellite Communication Systems*, Wiley, New York.

Martin, R. and Davidoff, M.R. (1985). *The Satellite Experimenter's Handbook*, American Radio Relay League, Newington, Connecticut.

Mertens, H. (1976). *Satellite Broadcasting Design and Planning of 12 GHz Systems*, EBU Technical Review, Tech. 3220, March.

Mohamadi, F., Lyon, D.L. and Murrel, P.R. (1988). 'Effects of Solar-transit on K_u band VSAT systems', *International Journal of Satellite Communications*, Vol. 6, pp. 65–71.

Morgan, W.L. and Gordon, G.D. (1989). *Communications Satellite Handbook*, Wiley, New York.

Siocos, C.A. (1973). 'Broadcasting satellite coverage – geometrical considerations', *IEEE Trans. Broadcasting*, Vol. BC-19, No. 4, December, pp 84–87.

Siocos, C.A. (1981). 'Broadcasting satellites power blackouts from solar eclipses due to moon', *IEEE Trans. Broadcasting*, Vol. BC-27, No. 2, June, pp 25–28.

3 Frequency and Propagation Considerations

3.1 Introduction

In chapter 1 we learnt that a satellite communication network utilizes radio links for interconnections. The design of radio links is one of the most important considerations in implementing a satellite communication system. A large number of parameters must be considered and optimized. An important consideration in a link design is the selection of the operational frequency.

In this chapter we shall develop a basic understanding of some of the basic criteria in the selection of frequency. The choice is mainly governed by propagation considerations, the need for various services to coexist without causing undue interference to one another, radio regulations and the prevalent state of technology. Additional factors governing a radio link design will be discussed in chapter 4.

An important requirement in selecting the operational frequency is that radio signals suffer minimum degradation while propagating through the intervening medium. Degradations include attenuation, scintillation, noise contamination, etc. Other important considerations include the need to comply with the radio regulations to ensure coexistence of the various systems and to maximize utilization of the limited frequency resource; economic considerations which are related to the state of the technology; and the availability of the desired bandwidth in the preferred frequency band. An additional constraint, when expanding the bandwidth capacity of an operational system, is the need to ensure that the frequencies are chosen close to the operational frequency, in order to minimize the need to upgrade existing equipment.

In general, an attempt is made to minimize the transmitted power because the transmitter cost is significant and directly related to the power requirements, especially for a spacecraft; and the potential of interference to other radio systems is reduced with low transmitted power.

In the following sections we shall discuss the issues related to harmonizing the use of the radio spectrum, followed by a detailed discussion of propagation effects at frequencies used in satellite communications.

3.2 Equitable use of radio spectrum

The radio spectrum is a limited natural resource which should be shared by all types of radio services, be it terrestrial or via satellites. To avoid interference between the various radio systems, the International Telecommunications Union (ITU) allocates frequencies for each service on a global and regional basis. The use of frequencies for domestic applications is regulated by individual countries who assign frequencies according to radio regulations and ensure that radio transmissions originating in their respective countries do not cause interference either to domestic or international networks.

The ITU has categorized radio services according to their broad functions. Frequency allocations are made for each service. At present, about 35 radio services have been defined by the ITU. We are concerned here with the Fixed Satellite Service (FSS), the Broadcast Satellite Service (BSS) and the Mobile Satellite Service (MSS). The FSS applies to systems which interconnect fixed points such as international telephone exchanges. The BSS refers to broadcasts by satellite of television or radio programmes directly to the public. Finally, the MSS networks provide communication to mobile terminals.

The ITU has divided the world into three regions for the purpose of frequency allocation. These regions are:

Region 1:   Europe, Africa, the Middle-East and the Asian regions of the former USSR
Region 2:   The Americas
Region 3:   The remainder of Asia, plus Australasia

A frequency band can be allocated to one or more services either globally or by region. The allocations can have several 'types' of status. Each allocation type is governed by a specific set of regulations. The majority of allocations have a *primary* or *secondary* status. Departures from the international allocation are permitted within some countries according to individual requests, provided this arrangement is agreeable to all the ITU signatories. Such allocations are known as *footnote allocations*. Allocations may be *exclusive* to one service or *shared* between services. In shared allocation some services may have primary status and others secondary. When a frequency range has shared allocations the services with primary status are given preference and guaranteed interference-protection from services with secondary status. If two services, each with primary status, are planned to operate in the same frequency band (or one of the two systems is already operational), the agencies concerned coordinate and arrive at mutually acceptable solutions so that both systems can coexist. A service with secondary status is not guaranteed interference protection from services with primary status, nor is it permitted to cause interference to them. The ITU frequency allocations are published as tables in Article 8 of Radio Regulations.

Frequency allocations may also follow a regional or worldwide *plan.* In *planned* frequency allocation, each country is assigned frequencies according to its requirements. The main purpose of such an assigned plan is to ensure that technical parameters are retained for long periods (at least 10–20 years) and to guarantee that all countries can access the geostationary orbit when needed. Planned allocations are attractive for broadcast service where thousands of terminals may be used by the public who do not want frequent changes. In fact such a plan for satellite broadcast service has already been formulated by the ITU and is in force at the moment.

The allocation of frequencies by ITU is a complex process, Decisions are taken in international conferences known as World or Regional Administrative Radio Conferences (WARC or RARC), organized by the ITU. In addition to the technical considerations outlined above, political factors influence the allocations.

When an organization intends to expand an existing system, frequencies close to the existing are preferred in order to minimize both the impact of the planned expansion to the existing customers and also changes to the space segment. For example, a totally new ground and space segment would be necessary to change frequencies of a system operating on the 4/6 to 11/14 GHz bands.

We shall now outline the main considerations used in selecting frequency for a *new* satellite system. In some applications, such as direct-to-home satellite broadcast system, frequencies and other technical parameters have already been assigned by the ITU for each country in a plan, making the task straightforward. When such a frequency plan does not exist, the system designer selects frequencies from the ITU allocations and performs a trade-off analysis based on propagation, the state of the technology and economic considerations. When a frequency band has been selected, the frequency coordination procedure, recommended by the radio regulations, is invoked. In general the following stages are involved when a new system is to be introduced:

1. A frequency band allocated for the service is chosen from the ITU allocations, based on a trade-off analysis taking into account the technical and economic factors.
2. When the first stage is completed satisfactorily, steps are taken to resolve interference likely to be caused within the jurisdiction of the country.
3. At the same time the ITU is notified about the planned satellite location, type of service, bandwidth required, etc. This is done several years (typically 3–5 years) in advance of introducing the system.
4. The ITU notifies all its signatories regarding the planned system.
5. The agency proposing the new system coordinates with all system op-

Table 3.1   The main FSS frequency bands (below ~30 GHz).

Downlinks	*Uplinks*
3.4 – 4.2 GHz and 4.5–4.8 GHz	5.725– 7.075 GHz
7.25– 7.75 GHz	7.9 – 8.4 GHz
10.7 –11.7 GHz	
11.7 –12.2 GHz (Region 2 only)	12.75 –13.25 and 14.0–14.5 GHz
12.5 –12.75 GHz (Region 1 only)	
17.7 –21.2 GHz	27.5 –31.0 GHz

erators who might be affected, to resolve any potential problems (e.g. by reducing the satellite EIRP, or shaping the antenna footprint, etc.).

6. When the coordination activity is successfully completed, the ITU is notified. The ITU enters the allocation in its frequency register and the system can progress to the implementation stage.

7. Any changes to the assignment, which may be required at a later stage, are re-coordinated.

As an example, the main FSS allocation bands existing at present are summarized in table 3.1. The table is included only to provide a general idea and no attempt has been made to include all worldwide FSS allocations. Almost all the main bands are shared with terrestrial fixed services. The FSS has the maximum traffic demands and consequently the service has been allocated the largest bandwidth. The earliest systems used the 4/6 GHz (downlink/uplink) band with the result that this band is now congested in most regions of the developed world. Consequently, most new systems in these regions now prefer to use the 11/14 GHz band. Several experiments are being conducted into the 20/30 GHz band, to study the feasibility of utilizing this when the K_u band eventually becomes congested.

Two categories of BSS have been defined for the purpose of frequency allocation – *individual reception* and *community reception*. As the names imply, individual reception broadcasts are meant for simple, low-cost receivers in the homes of individuals, whereas community reception broadcasts are meant for larger and more complex receivers for community viewing or distribution of programmes over a limited area (e.g. by terrestrial rebroadcast or cable distribution). As mentioned above, a worldwide plan (WARC 77 and RARC 83 plan) has been formulated for individual reception. The plan allocates technical parameters such as satellite EIRP, coverage area contours, polarization, etc., and allows high power satellite transmissions enabling the use of small, inexpensive receivers. Allocations for BSS downlinks exist in the S, K_u and K_a bands. The uplinks of BSS, known as *feeder links*, generally share the same uplink frequencies as the

FSS. Additional allocations have therefore been made to the FSS to accommodate BSS transmissions. In addition to television broadcasts, these frequencies may be used for broadcasting sound channels. Satellite broadcast bands are used for radio broadcasts in the USA, Europe and India. There is growing interest in high-quality sound broadcasts via satellite directly to portable receivers. In WARC 92, specific allocations for such a service have been made in the L band.

The choice of frequencies for the civilian mobile satellite service (MSS) needs careful consideration. The suitability of a frequency band for the MSS is governed by the need to use small, affordable mobile terminals capable of operating in various types of environments. Therefore, satellites used in the MSS need to transmit at high power to offset the severely limited sensitivity of the mobile terminals and provide extra link margin in the presence of physical obstructions such as trees and buildings. These requirements make the sharing of the MSS frequencies with other services difficult, because of potentially large interference. Therefore primary and exclusive allocations are preferred for the MSS. The optimal range of frequencies from propagation, receiver cost and spacecraft size considerations is rather limited. When frequencies are below ~800 MHz, cosmic and man-made noise (e.g. from vehicles) is high, as is the cost of spot beam antennas for achieving high satellite transmitted power, remembering that the size of an antenna is directly related to the wavelength. Above ~2.5 GHz, the path loss begins to increase prohibitively making the received carrier level low, bearing in mind that the antenna gain of mobile terminals is intentionally kept low to minimize antenna complexity. Tracking antennas for mobiles tend to be complex and expensive, increasing the cost of a mobile terminal. Further, propagation degradation caused by shadowing begins to dominate at higher frequencies. Thus at present the range 800 MHz to 2.5 GHz is the favoured band for most civilian MSS applications. However, for military applications where cost is not a major consideration, higher-frequency bands (e.g. the X band) can be used.

As mentioned above, Article 8 of Radio Regulations gives a detailed list of frequency allocations, and the latest edition should be consulted for any specific need.

3.3 Propagation considerations

The intervening medium between earth stations and satellite – the *channel* – affects radio wave propagation in several ways. The parameters mainly influenced are path attenuation, polarization and noise. The factors to be considered are gaseous absorption in the atmosphere, absorption and scattering by clouds, fogs, precipitation, atmospheric turbulence and ionospheric effects. Several techniques have been developed and are being continually refined to quantify these effects in order to improve the reliability in system design. Because of the randomness associated with all such processes, statistical techniques are used in modelling their effects on radio wave propagation.

The most significant impairments of radio wave propagation occur in the troposphere and the ionosphere. The first few tens of kilometres of the atmosphere in which clouds and rain are formed is known as the *troposphere*, and the ionized region in space extending from about 80 to 1000 km constitutes the *ionosphere*. Ionization in this region is caused by the interaction of solar radiation with gas molecules. Therefore the electron content of the atmosphere is high during the day and also during periods of high solar activity.

At present, the frequencies being used or under consideration for satellite communication range between ~100 MHz and 30 GHz. The frequency window of ~3–10 GHz is least affected and therefore it is not surprising that most satellite systems operate within this band. Ionospheric effects are significant between ~30 MHz and ~7 GHz with major effects confined below ~3 GHz. Above 10 GHz, absorption in the troposphere begins to become significant. The main constituents which absorb RF energy in the troposphere are water and oxygen. Condensed water vapour existing in the atmosphere as rain, hail, ice, fog, cloud or snow produces the most significant impairment to radio wave propagation (see Ippolito, 1981 for details).

Mobile satellite communication systems require additional considerations. Here the earth–satellite path profile varies continuously whenever the mobiles (e.g. ships or land vehicles) are in motion. The environment in the vicinity of the mobile plays an important role in the propagation characteristics. For example, a mobile–satellite path may become shadowed by a building or may pick up scattered signals from trees, buildings, etc.

In the following sections we shall discuss the main effects of the troposphere and the ionosphere. This is followed by propagation considerations peculiar to mobile satellite communication systems.

A. Tropospheric effects

(i) Gaseous absorption

Figure 3.1 (CCIR Rec 390–4) shows the total estimated one-way attenuation at moderate, 0% and 100% relative humidity for a vertical earth–space path as a function of frequency between 1 and 200 GHz at 45°N latitude using US standard atmosphere. It may be noted that there are specific frequency bands where the absorption is high. The first absorption band, caused by water vapour, is centred around 22.2 GHz, while the second band, caused by oxygen, is centred around 60 GHz. Absorption at any frequency is a

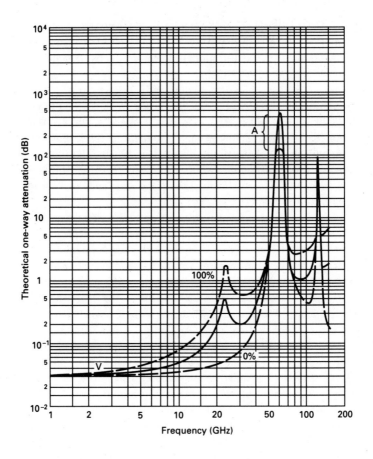

Figure 3.1 Theoretical one-way attenuation for vertical paths through the atmosphere (calculated using the United States standard atmosphere for July, at 45°N latitude). Solid curves are for a moderately humid atmosphere, dashed curves represent the limits for 0% and 100% relative humidity. V – vertical polarization; A – limits of uncertainty (CCIR Rec 390–4).

function of temperature, pressure, humidity of the atmosphere and the elevation angle of the satellite. Absorption increases as elevation angle is reduced. A variation given by $1/\sin(\eta)$ (where η = elevation angle) can be applied to transform the attenuation over a zenith path to any other elevation angle path. It is noticeable in the figure that, in the frequency range of current interest (1–18 GHz), the zenith one-way absorption is in the range ~0.03–0.2 dB. Applying the Cosec correction, this translates into an attenuation of ~0.35–2.3 dB at 5° elevation. The corresponding upper limit for 100% humidity is 0.7 dB at zenith, increasing to 8 dB at 5° elevation.

(ii) Attenuation due to hydrometers

Hydrometer is a general name for referring to condensed water vapours existing in the atmosphere. Thus rain, hail, ice, fog, cloud or snow are all examples of hydrometers. All types of hydrometers produce transmission impairments. However, raindrops produce by far the maximum attenuation by absorbing and scattering radio waves. In the following sections we shall discuss the mechanism and the extent of attenuation produced by rain, fog and clouds. We shall also discuss the techniques used in estimating the attenuation caused by rain.

Attenuation by rain

Attenuation of radio waves through rain (A_R), extending over length L of path, can be obtained from

$$A_R = \int_0^L \alpha \, d\alpha \tag{3.1}$$

where α = specific attenuation of rain (dB/km).

The most reliable estimates are obtained through measurements taken over several years in the geographical region, at frequency and elevation angle of interest.

In the following text, the classical development of specific attenuation is summarized to illustrate the significance of the various parameters that affect attenuation. The more practical techniques used in predicting attenuation due to rain are then discussed.

The following assumptions are applied in the classical development of specific attenuation due to rain:

1. a radio wave decays exponentially as it propagates through rain;
2. water drops are spherical;
3. attenuation due to each drop is independent and additive.

Specific attenuation is determined by integrating the attenuation cross-section, Q_t, over all drop sizes

$$\alpha = 4.343 \int Q_t(r, \lambda, m) \, n(r) \, dr \tag{3.2}$$

where $Q_t(r, \lambda, m)$, the attenuation cross-section of a drop, is a function of the drop radius, r, wavelength λ, and complex refractive index, m.

The attenuation cross-section is the sum of a scattering cross-section – implying loss due to scattering – and an absorption cross-section representing absorption loss. This parameter can be determined by employing the classical theory of Mie (Van De Hulst, 1957).

$n(r)$ is the drop size distribution and therefore $n(r)$ dr represents the number of drops per unit volume within the radius between r and $r + dr$. The drop size distribution is well represented by an exponential distribution. The applicable constants of exponential distribution have been determined by several investigators (e.g. Laws and Parsons, 1943).

Equation (3.2) shows the dependence of attenuation on the rain structure (e.g. light, heavy) and the frequency. By substituting the distribution into equations (3.2) and (3.1) we obtain the overall attenuation over the rain-affected path length L as

$$A_R = 4.343 \int_0^L [N_0 \int Q_t e^{-\Lambda r} \, dr] \, d\alpha \qquad (3.3)$$

where N_0 and Λ are empirical constants which have been determined by several investigators; Q_t is the attenuation cross-section which can be theoretically estimated at the operational frequency.

Calculations using the above equation show that rain attenuation increases with increase in frequency and rain-rate. It is also observed that rain attenuation becomes significant at frequencies above $\sim$10 GHz. These theoretical conclusions are well supported by measurements.

In practice, specific attenuation is estimated by the following relationship (Olsen *et al.*, 1978)

$$\alpha = aR^b \qquad (3.4)$$

where a and b are frequency- and temperature-dependent constants. R is the surface rain-rate at the given location.

The most reliable estimate of rain attenuation at a location is obtained by measurements taken over several years at that location. As noted earlier, rain attenuation becomes significant above 10 GHz and therefore when a satellite system is planned to operate above 10 GHz, extensive sets of measurements are obtained at several locations in the coverage area of the satellite system (e.g. Lin *et al.*, 1980).

The measured data are usually presented as a cumulative distribution because this form of presentation permits the *link margin* to be easily established. Link margin is the additional transmitter power necessary to compensate signal fades (see chapter 4 for further discussion of link margin). Figure 3.2 shows an example of such data at a frequency of 11.7 GHz for various USA locations (Ippolito, 1981). It can be observed that for link reliabilities of 99.9% and 99.99%, the required link margins are approximately 2.5±1 dB and 11±2 dB respectively. Table 3.2 shows the link margins necessary for the temperate continental climate regions of the USA, based on 30 station-years of data (Ippolito, 1981) for the 11 GHz, 20 GHz and 30 GHz bands.

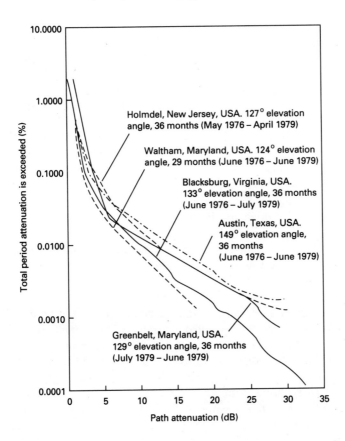

Figure 3.2 Attenuation distribution at 11.7 GHz measured at various locations in the USA (Ippolito, 1981).

Table 3.2

Link reliability (%)	Hours per year outage	Margin		
		11 GHz	20 GHz	30 GHz
99.5	44	1	3	6
99.9	8.8	3	10	20
99.95	4.4	5	20	30
99.99	0.88	15	>30	—

Several interesting observations can be made. For example, it can be seen that to provide 99.5% reliability, equivalent to traffic loss of 44 hours/year, a 6 dB margin is adequate up to 30 GHz. However, with a 6 dB margin only a 11 GHz link is capable of providing a link reliability of ~99.95%, equivalent to a traffic loss of 4.4 hours per year.

It has also been observed that in general the required link margin at a given location increases as the elevation angle to the satellite is reduced. This is a direct consequence of the increase in the extent of rain affecting the path (parameter L in equation 3.1) as the elevation angle is reduced.

At present, link margins of 3–4 dB can be provided without significant complexity and cost. When larger link margins are desired, special fade counter-measure techniques are needed. Some of the techniques which have been proposed are site diversity, the use of coding, and increasing satellite transmitted power by the use of spot beams. In a *site diversity* scheme the signal is received at different earth stations spaced sufficiently apart so that the probability of simultaneous fades at the two sites is minimized. The received signal with least fading can then be used for detection; alternatively the signals from the sites can be combined to obtain a higher carrier level. When *coding* is used, the information bits are protected by the use of redundant bits such that the specified bit rate is maintained even in the presence of a signal fade (see chapter 6 for a discussion of coding).

The use of a spot beam has the following advantages relative to a global beam. It permits the satellite power to be focused more effectively on the coverage area, thereby increasing the effective satellite transmitted power, and in addition has a better G/T while receiving than is possible with a global beam. Thus relative to a global beam, a spot beam can provide a larger link margin for the same satellite and earth station transmitter power.

Rain attenuation prediction techniques
To obtain measured data for all the locations within the coverage area is time-consuming and expensive. Therefore several rain attenuation prediction techniques have been developed (e.g. Lin, 1979; Crane, 1980). The techniques generally use the relationships given in equations (3.1) and (3.4), written in a modified form as

$$A = aR^bL(R) \tag{3.5}$$

where $L(R)$ = effective length of rain.

The difference in prediction techniques lies in the manner in which the effective length $L(R)$ of rain is estimated. Empirical relationships, obtained from measured data, are used for this purpose, and therefore the accuracy of a model depends on the applicability of the database for the location under consideration. CCIR study groups investigate various prevailing models and recommend a technique for use by system designers when measured data are not available (*CCIR Recommendations and Report*, Vols IV and V). The recommendations are updated regularly as the models are refined. One of the factors in the choice of a model by the CCIR is the applicability of the model to as wide an area as possible, preferably worldwide. Hence the estimates using some recommended models may not be as accurate as those obtained by using a 'local' model based on measurements specific to the region of interest.

Here we shall briefly described a simple method (Lin, 1979) to illustrate the approach. It should be emphasized that the main reason for selecting this model here is its simplicity, and that the model may not be applicable to all regions (see the next paragraph for a brief description of a more general model). Lin's model, developed in the USA, provides a simple empirical relationship between the attenuation of radio signals at a location and the 5-minute-averaged rain-rate. The rationale behind choosing 5-minute-averaged rain-rate is that long-term 5-minute-averaged rain-rate data are widely available in the USA. Further, it is assumed that a 5-minute-averaged rain-rate gathered at a single point represents a good average of the spatial variation of rain. The rain intensity varies along the radio path on an instantaneous basis. The model was developed from an attenuation database of terrestrial links in the USA and later extended to satellite paths. The attenuation is estimated by the relationship of equation (3.5)

$$A_{\mathrm{L}} = aR^{b}L(R)$$

Empirical relationships are used to calculate the effective path length, and the constants a and b are obtained theoretically. $L(R)$ is given as by

$$\frac{2636}{\left(R - 6.2 + \dfrac{2636\,\sin(\theta)}{4 - G}\right)} \tag{3.6}$$

where θ = elevation angle
R = 5–minute-averaged rain-rate (mm/h)
G = earth station height above mean sea level
a and b are frequency-dependent constants.
Table 3.3 lists the calculated values of a and b for the Marshall–Palmer drop distribution (Marshall and Palmer, 1948) at a rain temperature of 0°C and the calculated specific attenuation (A) at rain-rates of 10, 50 and 100 mm/h.

The predictions obtained from equation (3.6) have shown very good agreement with measured results from the USA.

A global model applicable worldwide has been proposed by Crane (Crane,

Table 3.3   Values of coefficients *a*, *b* and α for calculation of rain
attenuation.*

Frequency (GHz)	Coefficient		α (dB/km) for R specified in mm/h		
	a	*b*	R = 10 mm/h	50 mm/h	100 mm/h
2	3.45×10^{-4}	0.891	0.003	0.011	0.021
4	0.00147	1.016	0.015	0.078	0.158
6	0.00371	1.124	0.049	0.30	0.657
12	0.0215	1.136	0.29	1.83	4.02
15	0.0368	1.118	0.48	2.92	6.34
20	0.0719	1.097	0.90	5.25	11.24
30	0.186	1.043	2.05	11.0	22.7
40	0.362	0.972	3.39	16.2	31.8
94	1.402	0.744	7.78	25.8	43.1

* Marshall–Palmer drop distribution; rain temperature 0°C.

1980) on the basis of the climate of a region. The model relates the surface
rain-rate at a given point to a path-averaged rain-rate by means of an effective path average factor. To obtain the average rain-rate at any location, the
Earth has been divided into several zones on the basis of the climate, thereby
making it possible to estimate link margin throughout the world. The global
model has shown good agreement with measurements in the USA and is
currently recommended by the CCIR (see the latest Vol. V of *CCIR Green
Books*).

(iii) Attenuation from cloud and fog

Clouds and fog are suspended water droplets, usually less than 0.1 mm in
diameter. Attenuation to radio waves depends on the liquid water content of
the atmosphere along the propagation path. Clouds have liquid contents from
0.05 to 5 g/m^3. Calculations of specific attenuation show that for frequencies of current interests (i.e. <30 GHz) the largest attenuation from clouds
is equivalent to the attenuation caused by light rain (~10 mm/h).
    The liquid content of fog is of the order of 0.4 g/m^3 and typically fog
extends from 2 to 8 km. As a result, the attenuation from fog is negligible
for satellite communications.

Tropospheric scintillation
Scintillation is the rapid fluctuation in amplitude, phase or angle of arrival

of radio waves. It is produced by small-scale refractive index variations in the troposphere or the ionosphere (discussed later) causing the radio signals to arrive at a receiver via several paths. Hence scintillation is also sometimes called *atmospheric multipath*. In this section we shall discuss tropospheric scintillation. This type of scintillation, produced in the first few kilometres of the Earth's atmosphere, depends on the season, the local climate, the frequency and the elevation angle. Typical values of amplitude scintillation observed at high elevation (more than about 20° in temperate climates) have been ~1 dB peak-to-peak in clear conditions in summer, 0.2–0.3 dB in winter, and 2–6 dB in some types of cloud. Fading rates of 0.5 Hz to over 10 Hz have been observed. In general, the tropospheric amplitude scintillation magnitude increases with increase in radio frequency (unlike ionospheric scintillation, which reduces with increase in radio frequency); reduction in the elevation angle and antenna diameter; and increase in temperature, humidity and wind velocity. Tropical regions suffer higher scintillation than do temperate zones, with the severest events occurring in wet seasons.

Amplitude and phase fluctuations cause a deviation in the wavefront across the aperture of an antenna. As a result, there is a reduction in antenna gain in the presence of amplitude and phase scintillation. This degradation becomes higher as antenna aperture and frequency are increased and elevation angles reduced.

Depolarization

The atmosphere behaves as an anisotropic medium for radio propagation. Consequently power from one polarization is coupled to its orthogonal component, causing interference between the channels of a dual polarized system. Cross-polar discrimination (XPD) has been defined for antennas in section 4.2. A similar definition is applied for the atmospheric effects

$$\text{XPD} = 20 \log \frac{|E_{11}|}{|E_{12}|} \tag{3.7}$$

where E_{11} = received co-polarised electric field strength
E_{12} = electric field strength coupled to orthogonal polarization.

The main sources of depolarization on an Earth–space link are rain and ice. In the following sections we shall discuss the effects of each of these sources.

Depolarization caused by rain

The mechanism causing depolarization in a radio wave travelling through rain can be understood with the help of figure 3.3. The figure shows a model of a falling raindrop used in theoretical analysis, together with the

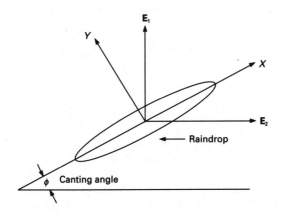

Figure 3.3　The model of a canting raindrop. $\mathbf{E}_1$ and $\mathbf{E}_2$ represent two orthogonally polarized waves.

two field vectors, $\mathbf{E}_1$ (polarization 1) and $\mathbf{E}_2$ (polarization 2), of a dual polarized system. An oblate shape is chosen because raindrops begin to take such a shape as their size increases. The angle that the falling raindrops make to the horizon is known as the *canting angle*. Such a medium causes coupling of signals between the field vectors. The vectors $\mathbf{E}_1$ and $\mathbf{E}_2$ can be resolved along the major (X) and minor (Y) axes of the ellipse formed by the raindrops. The resultant signal for each polarization is then

$$\mathbf{E}_{Rx} = T_1 \,(\mathbf{E}_{1x} + \mathbf{E}_{2x}) \qquad\qquad (3.8)$$
$$\mathbf{E}_{Ry} = T_2 \,(\mathbf{E}_{1y} + \mathbf{E}_{2y})$$

where　$\mathbf{E}_{Rx}$ and $\mathbf{E}_{Ry}$ are the resultant vectors at the output of the medium (i.e. a raindrop) for polarizations 1 and 2 respectively

T_1 and T_2 are the transmission coefficients in the X and Y directions respectively

$\mathbf{E}_{1x}$ and $\mathbf{E}_{2x}$ are the vector components of $\mathbf{E}_1$ and $\mathbf{E}_2$ respectively in the X direction

$\mathbf{E}_{1y}$ and $\mathbf{E}_{2y}$ are the vector components of $\mathbf{E}_1$ and $\mathbf{E}_2$ respectively in the Y direction.

From (3.8) we note that the two field vectors are cross-coupled in the anisotropic medium of a raindrop.

On assessing (3.8), the parameters which affect coupling between two polarizations of a dual polarized system caused by an oblate raindrop can be identified. The magnitude of depolarization is observed to be a function of: (a) orientation and shape of raindrop; (b) number of raindrops in the

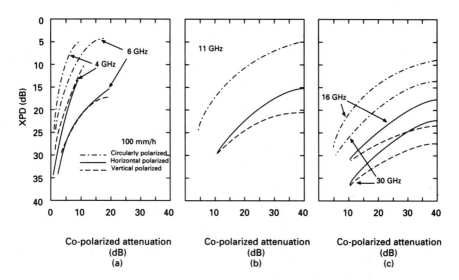

Figure 3.4   Cross-polar discrimination of horizontal, vertical and circularly polarized waves at frequencies of (a) 4 and 6 GHz, (b) 11 GHz and (c) 18 and 30 GHz at a canting angle of 25° (Chu, 1974).

propagation path; (c) polarization and frequency of radio wave.

Cross-polar discriminations have been estimated theoretically for a plane wave incident on an oblate spheroid (Oguchi, 1960). The results provide a good understanding of the behaviour of the cross-polar discrimination property as a function of the radio frequency and type of polarization. Figure 3.4 (Chu, 1974) shows the calculated values of cross-polar discrimination (XPD) at 4, 6, 11, 18 and 30 GHz for 100 mm/h rain and 25° canting angle for various types of polarization. Several interesting features can be noted from the figure:

1. Cross-polar discrimination at a given co-polar attenuation degrades with decreasing frequency.
2. Cross-polar discrimination reduces with increasing co-polar attenuation.
3. Cross-polar discrimination for the vertically polarized wave is better than that for horizontal polarization – except at 4 GHz.
4. Circularly polarized waves have ~10 dB lower cross-polar discrimination than horizontally polarized waves for the same co-polar attenuation.

These properties form a useful basis for the design of a dual polarized system. However, because of the random nature of rain, theoretical estimates can only serve as guidelines. The most accurate values are obtained by measurements and the use of simple empirical relationships developed from measured data. The following relationship between the co-polar attenuation A (in dB) and the cross-polar discrimination is useful when measured data are unavailable:

$$\text{XPD} = U - V \log(A) \qquad \text{dB} \qquad (3.9)$$

where U and V are empirical constants which depend on frequency, polarization, elevation angle and the canting angle.

The following semi-empirical relationship has been proposed by the CCIR for system design. (Note that CCIR reports and recommendations are updated regularly, therefore the latest issue must be used.)

$$U = 30 \log(f) - 40 \log(\cos \theta) - 20 \log \sin(2\tau) \qquad (3.10)$$
$$V = 20 \qquad 8 < f < 15 \text{ GHz}$$
$$ = 23 \qquad 15 < f < 35 \text{ GHz}$$

where f = frequency in GHz
θ = elevation in degrees
τ = polarization tilt angle with respect to local horizon
= 45° for circular polarization.

Table 3.4 summarizes the results of measurements performed in various countries (Kennedy, 1979; Bostian, 1980; Vogel, 1980). The results were consistent with the theoretical predictions summarized above. It was noted that the cross-polar discrimination (XPD) reduced as the frequency was increased (the exception being the 4 GHz Taipei result). It was also observed that the vertical polarization has better discrimination than the horizontal (see the 19 GHz result). Such data form a good basis for system design, especially at locations and frequencies for which measurements are performed. A practical design approach is to use measured data such as that given in table 3.4 when available, but the latest model recommended by the CCIR (or a 'local' model, if available) when measured data are not available.

One important feature in the design of a dual polarized system (see section 4.2) is the need to consider outage both for co-polar attenuation *and* cross-polar discrimination. A useful form of data representation is shown in figure 3.5 (Arnold *et al.*, 1980). This figure shows the outage probability as a function of co-polar attenuation and cross-polar discrimination. The figure provides permissible co-polar attenuation and cross-polar discrimination for a specified outage probability. For example, consider a system with the requirement of a 10 dB co-polar link margin and a 20 dB cross-polar discrimination: the achievable reliability from the figure is obtained as ~99.94%. When the cross-polar discrimination requirement is increased to 30 dB, the link reliability is reduced to ~99.6%. An important observation from the figure is that for a given cross-polar discrimination, increasing co-polar link

Table 3.4 Summary of cross-polarization discrimination measurements.

Location	Transmit polarization (tilt angle)	Elevation angle	Observation time period	XPD (in dB) for given percentage time						
				1	0.5	0.1	0.05	0.01	0.005	0.001
4 GHz										
Taipei, Taiwan	RHCP	20°	—	<35	31	22	20.5	18	17.5	16
Lario, Italy	RHCP	25°	3/77–6/78 (15 months)	41	37	30.5	28	24	22	20
Ibaraki, Japan	RHCP	30°	—	<35	32	30.5	28	24	22	20
11.7 GHz										
Blacksburg, Virginia, USA	RHCP	33°	1/78–12/78	24	23.5	22	21.5	20	18	14
			1/79–7/79	26	25.5	24.5	24	21	20	16
Austin, Texas, USA	RHCP	49°	6/76–6/79	<35	34	28	25.5	20.5	18	15
19 GHz										
Holmdel, New Jersey, USA	H LIN (21°)	38.6°	5/77–5/78	—	30	24	22	16	11	>10
	V LIN (78°)	38.6°	5/77–5/78	—	30	25	23	18	15	>10
28 GHz										
Holmdel, New Jersey, USA	V LIN (78°)	38.6°	5/77–5/78	—	28	22	20.5	12	>10	—
Blacksburg, Virginia, USA	V LIN (60°)	46°	1/78–12/78	18	17	15	14	13	12.5	9.5
			1/79–6/79	20	19	18	17	12.5	9	7

RHCP = right-hand circular polarization; H = horizontal; V = vertical; LIN = linear.

margin in excess of a specified value – the onset of flattening of curve – cannot improve the link reliability. Thus it is evident that the link reliability of a dual polarized system for a given link margin is lower than that for a system using single polarization.

Depolarization caused by ice
Depolarization is also caused by ice crystals existing in the atmosphere above the melting layer (the height at which the atmospheric temperature reduces to 0°C). Ice crystals produce a differential phase shift in the radio wave without differential attenuation. As a result, the depolarization caused by ice crystals occurs without co-polar attenuation. However, fortunately the depolarization due to ice is small and needs consideration only when the cross-polar discrimination requirement is of the order of 25 dB or more.

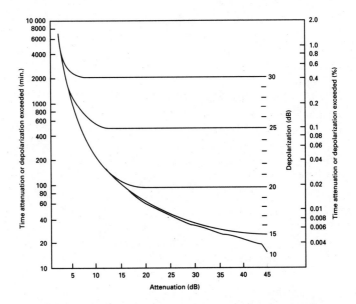

Figure 3.5   19 GHz linear polarized attenuation and depolarization for Holmdel (USA), elevation angle 38.6°, polarization 21° from vertical for the period 5/77 to 5/78 (Arnold *et al.*, 1980).

B. Ionospheric effects

The ionosphere is an ionized region in space extending from about 80 km to 1000 km. Ionization is caused by the interaction of solar radiation with gas molecules. Therefore the electron content (i.e. ionization) of the ionosphere is high during the day and also during periods of high solar activity. Radio waves propagating through the ionosphere are affected in a number of ways. The main effects of the ionosphere are *rotation in polarization* or the *Faraday effect*, and *scintillation*. Other effects such as absorption, propagation delay, dispersion and group delay are negligible at the frequencies of main interest for satellite communication, except for a small fraction of time under events such as solar flares. In general, most ionospheric effects reduce with frequency f, showing a $1/f^2$ dependence, and depend on location and the prevailing ionospheric conditions (see CCIR, Vol. VI for details).

(i) Faraday effect

The polarization angle of a linearly polarized wave is rotated in the ionosphere owing to interaction of the electromagnetic wave with the Earth's magnetic field. This phenomenon, known as the Faraday effect, is directly proportional to the total electron content (the integral of the electron content along the radio path) of the ionosphere. Faraday effect reduces as $1/f^2$

with an increase in the radio frequency (f), with significant effects limited to frequencies below around 2 GHz. Faraday rotation of up to 150° may sometimes be reached at 1 GHz. Applying frequency-dependent corrections, the Faraday rotations at 4 and 6 GHz are 9° and 4° respectively (CCIR, Vol. IV, Rep 205–2).

The Faraday effect is usually predictable (except under unusual atmospheric conditions such as a magnetic storm for a small percentage of time) and can be compensated by adjusting the polarization of the receive antenna. It should be noted that the directions of polarization rotation are opposite for transmit and receive signals. The impact of polarization rotation can be minimized by using circular polarization which remains virtually unaffected by the Faraday effect.

(ii) Ionospheric scintillation

Rapid fluctuation of signal amplitude, phase, polarization or angle of arrival is known as *scintillation*. In the ionosphere, scintillation occurs because of small-scale refractive index variations in the ionosphere caused by local concentrations of ionization. Ionospheric scintillation decreases as $1/f^2$ (where f = frequency) as the radio frequency is increased, the major scintillation being confined to frequencies below ~4 GHz. However in extreme conditions such as magnetic storm, scintillation can cause problems up to 7 GHz.

Ionospheric scintillation is produced in the F-region of the ionosphere (height of 200–400 km), which has the maximum electron content. Maximum levels of scintillation are observed in a region around the equator (~±25°) and during the period around the equinox. Scintillation also increases in conditions of high solar activity and has a diurnal variation with high levels occurring at night. Measurements show a strong correlation between sun spot numbers and scintillation. The peak levels occur approximately 1–2 hours after sunset. It is interesting to note that unlike tropospheric scintillation, ionospheric scintillation is independent of the elevation angle of the radio path.

When scintillation is expected at locations of interest, a link margin (i.e. additional transmitted power) is provided to mitigate the effect. Link margins of several dBs could be required to achieve high link reliabilities for unfavourably sited earth stations and operations below 3–4 GHz. Measurements at 4 GHz show that peak-to-peak attenuation may vary from 0.5 dB to ~10 dB for 0.1% of the time, depending on the sun spot number and the location.

C. Mobile communications channel

Propagation mechanisms discussed so far have assumed a fixed ground terminal. The received signal in applications using fixed terminals is essentially independent of reflections or scattering from the ground, except for a few exceptions such as operation at low elevation angles (<5°). For *mobile* ground terminals we need to take into account the following additional considerations:

1. Mobile–satellite path profile varies continuously while the mobile is in motion.
2. Mobile terminals use relatively broadbeam antenna which have only a limited discrimination against signals reflected and scattered from nearby objects such as buildings, trees, etc.

The resultant effect of these additional considerations are:

1. Signals suffer attenuation, whenever the mobile–satellite path is shadowed.
2. Signals fluctuate randomly because reflected and scattered random signal components arriving at the antenna are picked up. The effect is known as *multipath*.
3. The power spectral density of multipath noise is a function of the speed of the vehicle and the environment.

Depending on the environment in which a mobile operates, the mobile–satellite channels may be categorized as maritime, aeronautical or land based. Each category has its typical characteristics because of the differing environments.

Because of the inherent random nature of disturbances, signals are usually characterized statistically. In general, a signal arriving at the antenna of a mobile consists of the vector sum of a direct component and diffused components arising from multipath reflections. Figure 3.6 shows the geometry of a mobile channel propagation path. This geometry is applicable to all types of mobile environment, even though the figure shows a land-based mobile. The signal $r(t)$ received at the antenna consists of a direct component $a(t)$, a specular component $s(t)$ (i.e. a coherent reflected component such as from a metallic object) and diffused components $d(t)$ caused by scattering from nearby objects:

$$r(t) = \alpha(x) \, [a(t) + s(t) + d(t)] \tag{3.11}$$

where $\alpha(x)$ is an environment-dependent attenuation factor ($\alpha(x) = 1$ when there is no shadowing). In equation (3.11) $\alpha(x)$, $s(t)$ and $d(t)$ are random variables.

The probability distribution function of the received signal follows a Ricean distribution under the following assumptions:

1. received signal has a direct component;
2. in-phase and quadrature components of the resultant diffused signal are independent of each other, each component being normally distributed with zero mean; and
3. the phase of the received signal is uniformly distributed.

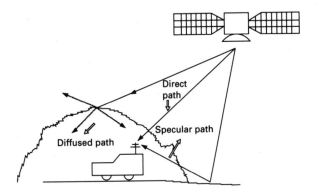

Figure 3.6   Geometry of mobile propagation channel.

When the direct ray is completed shadowed ($\alpha(x) = 0$), the Ricean amplitude distribution tends to a Rayleigh distribution.

The characteristics of the Ricean and Rayleigh distributions are well known and therefore the fade margin and time characterization of signals can be readily derived (Rice, 1944/1945). The cumulative distribution of a Ricean distribution can be expressed as a percentage of time (for example, see Sandarin and Fang, 1986). Multipath noise is sometimes expressed as carrier to multipath noise (C/M) ratio; this is termed the *Ricean factor*. Multipath noise increases at low elevation angles, when reflections can be easily picked up by an antenna and also when broadbeam antennas, which have lower discrimination against scattered energy, are used.

A practical approach for modelling a mobile communication channel is to include an element of environment dependence in the model. For example, in a land mobile communication system, a vehicle moving on an open motorway is less likely to be shadowed than a vehicle moving within a city with high-rise buildings. This environment-dependent approach has been widely used in the propagation modelling of terrestrial land mobile communication. Lutz *et al.* (1986) have proposed an environment-dependent approach for modelling land mobile satellite channel by considering the degree of shadowing as a variable. In this model, the probability distribution, $P(S)$, of the signal envelope at a mobile receiver is obtained by combining three types of probability distributions. When a direct path from the satellite is present the signal is assumed to follow a Ricean distribution, but when the direct path is shadowed the signal is assumed to follow a Rayleigh distribution possessing a log-normally distributed mean. (A signal is said to be *log-normally distributed* when the logarithmic value of the signal follows a normal distribution.) The log-normal distribution of the mean is attributed to slow variations of signals caused by gradual variations in the path profile. For example, reflections received off large distant objects such as hills, etc. are likely to change gradually, causing a slow variation in the mean of the received signal. The probability distribution, $P(S)$, of the resultant signal envelope is obtained as follows:

$$P(S) = (1 - A)P_{RC}(S) + A \int_0^\infty P_{RY} (S/S_0) \, P_{LN}(S_0)dS_0 \qquad (3.12)$$

where $P(S)$ is the resultant probability distribution of the signal
A is the fraction of time a mobile is shadowed
$P_{RC}(S)$, $P_{RY}(S/S_0)$ and $P_{LN} (S_0)$ are the Ricean, Rayleigh and log-normal probability distributions respectively
S is the instantaneous signal power and S_0 is the mean received power.

The parameters of the model depend on the type of environment in the vicinity of the mobile, e.g. urban, suburban, rural, etc. The model has been used successfully in the design of an experimental land mobile satellite system in Europe.

The time-dependent characteristics of the signal are also of interest. The motion of a mobile or the surrounding environment (e.g. motion of sea waves) results in frequency and phase variations of the radio signal. This effect is called the *Doppler effect* and affects the choice of modulation and coding (see chapter 2 for details of the Doppler effect). The parameters of interest to a system designer are multipath spectrum, fade duration, fade rate and inter-fade interval – all associated with the Doppler effect. The received signal spectrum at a mobile depends on the product of the receiver antenna gain function, $g(\theta)$, and the probability density function, $p(\theta)$, of the angle of arrival of the waves (Clarke, 1968). A ray arriving at an angle α with respect to the motion of a mobile has a Doppler shift, f:

$$f = f_m \cos(\alpha) \qquad (3.13)$$

where $f_m = V/(\lambda)$ is the maximum Doppler shift at a mobile velocity of V and carrier wavelength λ.

The Doppler effect mentioned here pertains only to the movement of the mobile. When considering the end-to-end link it is essential to include the Doppler shift caused by satellite motion, which depends on its orbital parameter. A simple technique for estimating the approximate Doppler shift on ground due to satellite motion is given in appendix B.

A detailed discussion of each type of mobile channel is outside the scope of the book. However the important features of each type of channel will be highlighted.

A *land mobile* satellite communication channel suffers the largest degradation. In fact, the propagation characteristics of land mobile satellite links are a major factor governing the viability of such a system for civilian use. The main problem arises because of the need to use small, simple and low-cost antennas/receivers in a hostile propagation environment. Such antennas

Table 3.5   Cumulative probability distributions of received power at L
band on a land mobile terminal in a European city and on a
highway using a 6 dBi toroidal antenna.*

Probability fade level (%) relative to mean dB level	*City*	*Highway*
99.75	< −30	−30
99	−30	−22
95	−22	−14
90	−18	− 7.5
80	−14	− 2
60	−10	0

* The satellite was at an elevation of 30°. City environment shows a much larger fading, indicating the strong influence of the environment (data from Lutz *et al.*, 1986).

are broadbeam with a low gain, resulting in a low sensitivity receiver. At the same time, physical environments consisting of obstructions such as buildings and trees cause severe signal attenuation and multipath, making the reception of good signals even more difficult. The solution to these problems is the transmission of high power signals from satellites and the use of robust modulation and coding techniques. Both solutions tend to increase the overall system costs. Therefore it is not surprising that the land mobile communication system was a late entrant into the mobile communication market, and at present the offered bit rate is severely limited in comparison with the maritime and aeronautical links.

Land mobile communication links need to operate in the greatest variety of physical environments. At one extreme, there are open stretches of road with direct visibility to the satellite, and at the other the heavily shadowed environments of cities with high-rise buildings. Hence, link margins are specified according to the type of environment, such as open, wooded, urban, etc. (see equation (3.12)). Table 3.5 shows the results of measurements taken in Europe, giving an indication of the order of link margins expected in various environments (Lutz *et al.*, 1986).

Another important variable in characterizing a land mobile channel is the elevation angle (Richharia, 1989). In general, the link quality improves with elevation angle because of a general reduction in:

1. the number of objects intercepting the slant path;
2. the path length through the intercepting objects;

3. the multipath noise picked up from antennas.

Thus the use of highly elliptical, inclined orbits has been proposed for operating land mobile satellite communications to serve high latitude locations in such regions as Europe, where geostationary satellites cannot provide high elevation angles. As noted in section 2.5, such orbits can be designed to provide high elevation angle coverage to such regions.

In a *maritime* environment, multipath is the dominant mode of signal fades. Multipath is caused by scattered signals picked up off the sea, and consequently depends on the sea state. Therefore, signal impairment increases at low elevation angle and with broad beamwidth antennas. On occasion, shadowing can be caused by a ship structure but this can be minimized by mounting the antenna so as to lessen the chance of shadowing by such a structure.

Experimental results show that the link is well represented by a Ricean model for 90–99% of the time (Sandarin and Fang, 1986). The Ricean model is valid under rough sea conditions for ship antennas which have a gain of more than 10 dBi and when the elevation angle is greater than 5°. Typically, a link margin of 3–4 dB provides a high link reliability. On rare occasions, when the sea surface is steady, specular reflections dominate and in such cases an estimate of the lower bound of fade depth can be obtained by developing geometrical relationship, assuming a perfectly smooth sea (e.g. General Electric, 1983; Karasawa and Shiokawa, 1984). Using such relationships the worst-case multipath fade depth for smooth sea for an omnidirectional antenna has been estimated as between 5 dB at 5° elevation angle and 1.25 dB at 15° elevation (General Electric, 1983). Experimental data on time/frequency characterization for maritime links are not well-behaved mathematically and therefore a useful practical approach is to extract the mean and maximum values for system design (Hagenauer *et al.*, 1984; Sandarin and Fang, 1986) (see Table 3.6).

In the *aeronautical* environment, observations taken over ocean show the validity of the Ricean model (Neul *et al.*, 1987). The Rice factor (C/M) depends on antenna beamwidth and elevation angle – improving for narrower beamwidth and higher elevation angle. Shadowing is mainly caused by aircraft structure when the orientation of an aircraft relative to the satellite is unfavourable. Therefore the radiation pattern characteristics and mounting of aircraft antennas require careful optimization. Mean fade duration caused by multipath decreases significantly with an increase in elevation angle. Measurements at the L band show maximum fade durations of 50–100 ms at a fade level of 5 dB (Neul *et al.*, 1987). Aeronautical channels require lower link margins than land mobile–satellite links. Usually a link margin of 3–4 dB is adequate for the L band for commercial applications. In applications such as automatic aircraft position reporting, where high link reliability is essential, propagation effects such as scintillation need consideration for existing L-band mobile satellite systems. The use of higher frequency

Table 3.6   Summary of measured fade duration times.

Fade threshold* (dB)	Mean fade length range (s)	Maximum fade length range (s)
−2	0.02–0.17	0.3 –3.5
−5	< 0.01–0.12	< 0.01–1.6
−8	< 0.01–0.08	< 0.01–0.9

* Relative to mean received power.

bands, such as the K_a band, may prove advantageous in such applications, but enough propagation data are not yet available to justify such an assumption.

Problems

1. Referring to International Telecommunication Unions (ITU) frequency allocation regulations, what do you understand by *primary, secondary* and *footnote frequency allocations*?

2. The ITU has divided the world into three regions for the purpose of frequency allocations. To which regions do the UK, USA and Singapore belong? For each of these countries list the primary FSS and MSS allocations in the frequency range 1–15 GHz, using Article 8 of the Radio Regulations.

3. What are the main considerations when selecting the operational frequency of a new satellite system?
   In practice, the communication demand in most satellite systems grows beyond the initially allocated bandwidth. What are the main considerations in selecting frequency when additional spectrum is required for an operational system?

4. Outline the main considerations in the selection of operational frequencies for the mobile satellite service.

5. (a) Figure 3.2 shows the cumulative distribution of attenuation at 11.7 GHz for various US locations. What are the factors which contribute to the wide location-dependent variation observed in the figure?
   (b) Use Lin's model to estimate the rain attenuation caused by 5-minute-averaged rain-rates of 10 mm/h and 50 mm/h at frequencies of 12 GHz and 20 GHz for a location situated at mean sea level. Assume the satellite to be located at an elevation of 30°.
   (c) A dual polarized system is affected by co-polar attenuation together with cross-polar degradation. Mention the advantage in presenting data as shown in figure 3.5.

From the figure, obtain the maximum useful co-polar margin for cross-polar discrimination requirements of 15 dB, 20 dB and 25 dB.
6. Compare the propagation effects in the maritime, aeronautical and land mobile satellite channels.

References

Arnold, H.W., Cox, D.C., Hoffman, H.H. and Leck R.P. (1980). 'Characteristics of rain and ice depolarisation for 19 and 28 GHz propagation paths from a COMSTAR satellite', *IEEE Transactions Antennas and Propagation*, Vol. AP-28, January, pp 22–28.
Bostian, C.W. (1980). 'A depolarisation experiment using COMSTAR and CTS satellites', *Final VPI and SU Report on Fourth Year of Work*, NASA Contract NASS-22577, March 25.
CCIR, ITU. *Recommendations* and *Reports*, updated regularly.
Chu, T.S. (1974). 'Rain-induced cross-polarisation at centimeter and millimeter wavelengths', *Bell Systems Technical Journal*, Vol. 53, No. 8, October, pp 1557–1579.
Clarke, R.H. (1968). 'A statistical theory of mobile reception', *Bell Systems Technical Journal*, July–August, pp 957–1000.
Crane, R.K. (1980). 'Prediction of attenuation by rain', *IEE Trans. Commun. Technology*, Vol. COM-28, September, pp 1717–1733.
General Electric (British) (1983). *Report on Standard-A Antenna Propagation Studies* (INMARSAT contract).
Hagenauer J. *et al.* (1984). 'Multipath fading effects and data transmission for small ship earth stations (Standard C)', DFVLR, *Final Report for ESA/ESTEC Contract 5323/82/NL/JS*, November 15.
Ippolito, L.J. (1981). 'Radio propagation for space communication systems', *Proc. IEEE*, Vol. 69, No. 6, June, pp 697–727.
Karasawa, Y. and Shiokawa, T. (1984). 'Characteristics of L-band multipath fading due to sea surface reflection', *IEEE Transactions Antennas and Propagation*, Vol. AP-32, No. 6, June, pp 618-623.
Kennedy, J. (1979). 'Rain depolarisation measurements at 4 GHz', *COMSAT Technical Review*, Vol. 9, Fall, pp 629–668.
Laws, J.O. and Parsons, D.A. (1943). 'The relationship of rain drops with size', *Transactions of American Geophysics Union*, Vol. 24, pp 452–460.
Lin, S.H. (1979). 'Empirical rain attenuation model for earth satellite paths', *IEE Trans. Commun.*, Vol. COM-27, May.
Lin, S.H., Bergman, H.J. and Pursley, M.V. (1980). 'Rain attenuation on earth satellite paths – summary of ten year experiments and studies', *Bell Systems Technical Journal*, Vol. 59, No. 2, February, pp 183–228.

Lutz, E., Papke, W. and Plöchinger, E. (1986). 'Land mobile satellite communications-channel model, modulation and error control', *Proc. ESA Workshop on Land Mobile Services*, ESTEC, 3–4 June, ESA SP-259, September.

Marshall, J.S. and Palmer, W.M. (1948). 'The distribution of rain drops with size', *Journal Meteorol.*, Vol. 5, August, pp 165–166.

Neul, A., Hagenauer, J., Papke, W., Dolansky, F. and Edbauner, F. (1987). *Propagation Measurements for the Aeronautical Channel*, IEEE, CH2429–9/87/0000–0090.

Oguchi, T. (1960). 'Attenuation of electromagnetic wave due to rain with distorted wave drops', *Journal Radio Research Lab.*, Vol. 7, No. 33, September.

Olsen, R.L., Rogers, D.V. and Hodge, D.B. (1978). 'The aR^b relation in the calculation of rain attenuation', *IEEE Transactions Antennas and Propagation*, Vol. AP-26, March, pp 318–329.

Rice, S.O. (1944/1945). 'Mathematical analysis of random noise', *Bell Systems Technical Journal*, Vol. 23, July 1944, pp 282–333; Vol. 24, January 1945, pp 96–157.

Richharia, M. (1989). 'Review of propagation aspects of highly elliptical orbits', *IEE Colloquium on Highly Elliptical Orbit Satellite Systems*, Digest No. 1989/86, 24 May.

Sandarin, W.A. and Fang, D.J. (1986). 'Multipath fading characteristics of L-band maritime mobile satellite links', *COMSAT Technical Review*, Vol. 16, No. 2, Fall, pp 319–338; Spring, pp 141–154.

Van De Hulst, H.C. (1957). *Light Scattering by Small Particles*, Wiley, New York.

Vogel, W.J. (1980). '*CTS Attenuation and Cross-polarisation Measurements at 11.7 GHz*, University of Texas at Austin, Austin, Texas, Rep. No. 22576–1, June.

4 Communication Link Design

4.1 Introduction

A satellite communication network consists of a number of earth stations interconnected via a satellite. The radio links used for interconnections are designed so as to deliver messages at the destination with acceptable fidelity. A compromise is exercised between the quality and quantity of delivered messages and practical constraints such as economics and the state of technology. To deliver a large amount of information at a very high quality may require unacceptably high cost under some conditions. Factors which need consideration in a link design include operational frequency, propagation effects, acceptable spacecraft/ground terminal complexity (hence cost), effects of noise and regulatory requirements. In chapter 3 we have discussed issues related to operational frequency. In this chapter we will develop an understanding of several other parameters and develop the necessary system design tools.

Figure 4.1 identifies the main elements of a network influencing the link design. The source-to-destination path can be partitioned for the purpose of radio link design as: the earth station–satellite link or *uplink*; the *satellite* path; and the satellite–earth station link or *downlink*. As we are only considering radio links, the word 'radio' will not generally be used in the following text.

Each component of the link has its individual characteristics. For example, when the destination earth station is a mobile terminal, the receive antenna size is small, resulting in low received carrier level. For such an application, therefore, the design of mobile terminal–satellite link is critical. A system designer attempts to optimize the overall link, taking into consideration the characteristics of each component of the link. During optimization it may turn out that the specified communication requirements are too stringent, leading to unwarranted demands on satellite or earth station size, cost and complexity. Accordingly a compromise is effected by accepting a lower message quality or quantity at the destination.

A basic element in a satellite communication link is the antenna. Therefore, the fundamental characteristics of an antenna that influence satellite communication link design are defined in the first section. A communication system design must include the effects of noise on communication performance. Satellite systems are susceptible to noise because inherently the carrier levels received are low. In the next section, the main sources of noise are identified and their combined effect on the system performance quantified. The final section of the chapter brings these concepts together and develops a methodology for a link design.

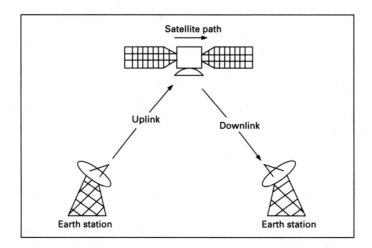

Figure 4.1 Partitioning of source-to-destination radio link for the purpose of end-to-end link design.

4.2 Antenna basics

Antenna characteristics play a vital role in the design of satellite communication systems. The subject of antenna has received considerable attention in the past and several books, covering the fundamental as well as implementation aspects, have been written on the subject (e.g. Silver, 1949; Jasik, 1961; Jordon and Balamain, 1968; Rudge *et al.*, 1982). The interested reader may refer to the literature for an in-depth treatment of the subject. Here, only the basic properties of antennas, affecting the design of satellite communication systems, are briefly reviewed. Chapters 9 and 10 discuss the implementation aspects of satellite and earth station antennas respectively.

The following must be borne in mind while reading the following text:

1. An antenna is a reciprocal device and therefore the receive and the transmit properties are identical at a given frequency.
2. An earth station antenna can be considered as representative for defining various parameters.
3. Most earth station antennas require movement along two axes to be able to follow a satellite.

Consider a satellite transmitting radio signals to an earth station. A plot of the received signal power level as a function of angle in each axis is known as the *radiation pattern* of the earth station antenna. Such a plot is

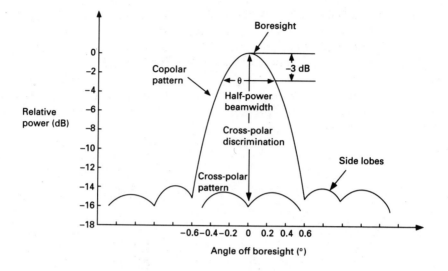

Figure 4.2   Antenna radiation pattern characteristics.

three dimensional. The direction in which the maximum power is received is known as the *boresight* of the antenna. When dealing with earth station antennas, it is usual to consider the radiation pattern of each axis separately, although in most instances the radiation patterns are identical.

The radiation pattern of an antenna can also be represented as constant gain contours. This form of representation is well suited for defining the coverage area of satellites.

Figure 4.2 illustrates a single-axis radiation pattern of an earth station antenna together with various useful parameters in satellite communication. The *half-power beamwidth* of an antenna is defined as the angular separation between the half-power (or −3 dB) signal points on the radiation pattern. For an antenna with a non-symmetric radiation pattern, each axis has its individual half-power beamwidth. Such *shaped* radiation patterns are often used on satellites.

In addition to the power received (or transmitted) by an antenna in the desired direction via its *main lobe*, some energy is also received (or transmitted) in unwanted directions through the *side lobes*. Power transmitted through the side lobes can cause interference in other radio systems and in turn may receive interfering signals. Such interference imposes a fundamental limit on the effective utilization of two important natural resources – the geostationary orbit and the radio spectrum. Therefore every effort is made to minimize the side lobe levels of antennas. The importance of the side lobe characteristics of antennas can be judged by the fact that International Radio Regulations recommend the use of specified radiation patterns for earth stations to permit coexistence of various radio systems (see sections 3.2 and 10.3).

Aperture antennas are commonly used for satellite communication applications because of their ability to focus transmissions within desired areas. The radiation pattern of aperture antennas depends on the distribution of the field pattern across the aperture. To minimize spill-over of energy the distribution is usually tapered across the aperture with the maximum at the centre. The half-power beamwidth, ψ_{hp}, depends on the aperture distribution, antenna diameter and operating frequency. A useful approximate relationship of ψ_{hp}, is

$$\psi_{hp} = \frac{N\lambda}{D} \tag{4.1}$$

where N is a constant dependent on aperture distribution
  $N \approx 58$ for uniform distribution
  $N \approx 70$ for a typical tapered distribution
  D and λ are antenna diameter and operational wavelength respectively.

The *radiation intensity*, $P(\theta, \phi)$, of an antenna in the direction (θ, ϕ) is defined as the power radiated from an antenna per unit solid angle in that direction.

Antenna directivity, $D(\theta, \phi)$, is a measure of the focusing property of an antenna, defined as

$$D(\theta, \phi) = \frac{P(\theta, \phi)}{P_{av}} \tag{4.2}$$

where $P(\theta, \phi)$ = radiation intensity in the direction (θ, ϕ)
  θ = elevation
  ϕ = azimuth
  P_{av} = average radiation intensity (or average power radiated from an antenna/unit solid angle)
  $= \dfrac{P_r}{4\pi}$
  P_r = total radiated power from an antenna (note that this is the power launched into the free space).

The definition of directivity does not take the efficiency of an antenna into account because P_{av} is related to the actual power launched into space.

In an antenna some power is lost as a result of energy spill-over, blockage of RF energy by sub-reflectors and supporting structures, manufacturing defects, and ohmic and reflection losses. Such losses reduce the antenna gain and are accounted for by associating an efficiency value to an antenna. Thus each antenna has an associated efficiency.

The *gain function*, $G(\theta, \phi)$, takes account of the antenna efficiency and is related to directivity through an efficiency factor:

$$G(\theta, \phi) = \eta \, D(\theta, \phi) \tag{4.3}$$

Hence

$$G(\theta, \phi) = \frac{P(\theta, \phi)}{P_i/4\pi} \tag{4.4}$$

where P_i = power fed into an antenna.

$P_i/4\pi$ can be viewed as the radiation intensity produced by a lossless radiator ($\eta = 1$) capable of transmitting uniformly in all directions, when fed with power P_i. Such a radiator is known as an *isotropic radiator*. An isotropic radiator is not realizable physically but is widely used as a reference because it has unity gain.

The maximum value of gain function is called the *gain* of an antenna. The gain is thus a measure of the increase in radiated power by an antenna relative to a lossless isotropic antenna emitting the same RF power.

Another useful fundamental relationship relates the antenna gain to its physical dimensions and the operating frequency:

$$G = \frac{4\pi A}{\lambda^2} \tag{4.5}$$

where A = aperture area of antenna
λ = wavelength of operational frequency.

Taking the antenna efficiency into consideration, the gain expression of (4.5) is modified to

$$G = \eta \, \frac{4\pi A}{\lambda^2} \tag{4.6}$$

where η is the antenna efficiency and ηA is called the *effective aperture* of the antenna. For typical parabolic antennas, the value of η is between 50% and 70%. For some types of horn antennas, the efficiency can be as high as 90%.

The *polarization* of an electromagnetic wave describes the orientation of the electric field vector in space. Polarization is determined by the manner in which RF is launched by the antenna into space. This function is performed by a polarizer which is part of the antenna's feed system. An antenna is capable of transmitting and receiving signals to which it is polarized.

A *linearly* polarized wave has an electric field vector oriented at a constant angle with respect to the horizontal or vertical axis, as it travels in

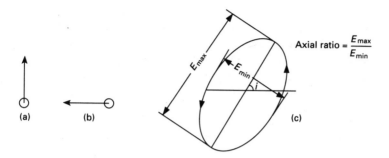

Figure 4.3   Linear and elliptically polarized waves: (a) vertical,
(b) horizontal, (c) elliptical. The wave travels into the plane
of the paper in each case.

space. Thus when the electric field vector is parallel to the horizon the
wave is *horizontally* polarized, and when the electric field is vertical the
wave is *vertically* polarized.

Figures 4.3(a) and (b) show the vertical and horizontal polarization of a
wave travelling into the plane of the paper. The electric field vector of a
circularly polarized wave describes a circle as the wave travels. The rotation is clockwise for a *right-hand circularly* polarized wave and counter-
clockwise for a *left-hand circularly* polarized wave. Rotation sense is usually
characterized as seen from antenna looking in the direction of propagation.
Distortion in a circularly polarized wave results in an *elliptical* polarization.
Figure 4.3(c) shows an elliptically polarized wave travelling into the plane
of the paper. The distortion is measured as the voltage axial ratio, A_R, defined as

$$A_R = \frac{E_{max}}{E_{min}} \tag{4.7}$$

where E_{max} and E_{min} are the major and minor axes of the elliptically polarized wave, as shown in figure 4.3(c).

Another important parameter of an elliptically polarized wave is the inclination angle i of the ellipse with respect to the reference axis (figure
4.3c). An elliptical polarization may be viewed as a generalized case because such a polarization approaches a linear polarization as $A_R \to \infty$ and a
circular polarization when $A_R = 1$.

Theoretically a dual polarized antenna can completely isolate waves travelling in orthogonal polarizations, permitting each polarization to be received
as a separate channel. A system utilizing this property of an antenna is known
as a *dual polarized* system. Note that the frequency of transmission for both
channels remains the same and therefore such a system doubles the band-

width utilization. However, in practice antennas cannot be matched perfectly to each polarization and some energy is always coupled into the orthogonal port. The coupled cross-polar energy appears as an intra-system interference and therefore great care is exercised in minimizing cross-polar coupling in antennas used in dual polarized systems. A dual polarized antenna is characterized by a co-polar radiation pattern (discussed above) and a *cross-polar pattern* (sometimes referred as an X-polar pattern). Figure 4.2 includes a plot of the cross-polar pattern.

The property of an antenna which discriminates against RF energy travelling in orthogonal polarization is termed the *cross-polar discrimination*. For a linearly polarized wave the cross-polar discrimination is the ratio of the co-polar signal to the cross-polar component. For an elliptically polarized wave, a cross-polarized wave has an opposite sense of rotation, the same axial ratio and an inclination angle 90° more than the original wave. The cross-polar discrimination, XPD (in dB), for such a wave is given by

$$XPD = 20 \log \frac{A_R + 1}{A_R - 1} \tag{4.8}$$

4.3 Transmission equation

The transmission equation is fundamental to the design of a radio communication link. The equation relates the received RF signal power at the destination to the RF power transmitted by the source, the transmission frequency and the transmitter-to-receiver distance. The level of received signal power governs the quality of message delivered to the destination, and its estimation is therefore vital to radio link design.

Consider an isotropic radiator (see section 4.2 for the definition). The power transmitted from such a source spreads uniformly outwards on an expanding sphere. At any distance D, the transmitted power is therefore uniformly spread over the area of a sphere of radius D. The received power flux density (i.e. power received/unit area) at a distance D from the source is given by

$$P_{FD} = \frac{P_s}{4\pi D^2} \quad W/m^2 \tag{4.9}$$

where P_s = transmitted power (watts) from the isotropic source
$4\pi D^2$ = surface area of a sphere of radius D.

When the isotropic antenna is replaced by an antenna of gain G_s, the power flux density in the direction of antenna boresight is increased by G_s:

$$P_{FD} = \frac{P_s G_s}{4\pi D^2} \ \text{W/m}^2 \tag{4.10}$$

The product $P_s G_s$ is known as the *effective isotropic radiated power* (EIRP) of the transmitter. Then

$$P_{FD} = \frac{\text{EIRP}}{4\pi D^2} \ \text{W/m}^2 \tag{4.11}$$

It is generally convenient to express the above expression in a logarithmic form because of the large numerical values involved (e.g. $D \approx 4 \times 10^7$ m). The expression for EIRP (in dB) is

$$\text{EIRP} = 10 \log (G_s) + 10 \log (P_s) \quad \text{dBW} \tag{4.12}$$

Similarly the power flux density (in dB) is given by

$$P_{FD} = \text{EIRP} - 10 \log (4\pi D^2) \ \text{dBW/m}^2 \tag{4.13}$$

With the expression for power flux density at any distance D established, the carrier power C received at the destination with an antenna of area A_d m^2 can be obtained as

$$C = P_{FD} A_d \quad \text{watts} \tag{4.14a}$$

As noted before, in practice some power is lost in the antenna as a result of miscellaneous losses inherent in an antenna system (see section 4.2), and therefore (4.14a) is modified to

$$C = \eta A_d P_{FD} \quad \text{watts} \tag{4.14b}$$

where η is the efficiency of the antenna.

Let us now express equation (4.14b) in terms of commonly used parameters of a communication link. From equation (4.6)

$$A_d = G_d \lambda^2 / 4\pi\eta$$

Substituting for A_d and P_{FD} (equation 4.10) in (4.14b) we obtain

$$C = P_s G_s G_d \left(\frac{\lambda}{4\pi D} \right)^2 \tag{4.15}$$

Again, expressing (4.15) logarithmically

$$C = P_s \text{(dB)} + G_s \text{(dB)} + G_d \text{(dB)} - 20 \log \left(\frac{4\pi D}{\lambda}\right) \qquad (4.16)$$

The term $20 \log (4\pi D/\lambda)$ is known as the *path loss*. Expression (4.15) (or 4.16) is commonly known as the *transmission* equation.

Example 4.1

Consider a satellite transmitting 25 W at a frequency of 4 GHz via an antenna of 18 dB gain. An earth station in the network uses an antenna of 12 m diameter with an efficiency of 65%. Determine:
(a) the gain of the earth station antenna;
(b) the path loss;
(c) the flux density at the earth station, assuming the satellite–earth station range to be 40 000 km;
(d) the power received at the output of the earth station antenna.
Repeat the exercise for a downlink frequency of 11.5 GHz. Compare the two sets of results and comment.

Solution

(a) Gain of antenna is given by equation (4.6):

$$G = \frac{4\pi A}{\lambda^2} \eta$$

Substituting the given values into this equation, the gain is obtained as 52.16 dB.
(b) Path loss is given as:

$$20 \log \left(\frac{4\pi D}{\lambda}\right)$$

Here $D = 40\,000$ km, so path loss $= 196.53$, dB.
(c) The flux density is given by equation (4.13):

$$P_{FD} = \text{(EIRP)} - 10 \log (4\pi D^2) \quad \text{dBW/m}^2$$

Here

EIRP $= 10 \log (25) + 18$

$\qquad = 31.98 \text{ dBW}$

Substituting the given values in the expression for P_{FD} gives $P_{FD} = -131.05$ dBW/m².

(d) The received signal level at the output of the antenna is given by equation (4.16). Substituting the given values into this equation gives

$P_d = 31.98 + 52.16 - 196.53 \text{ dBW}$

$\qquad = -112.39 \text{ dBW}$

Repeating the exercise with a downlink frequency of 11.5 GHz gives
(a) Antenna gain $= 61.33$ dB
(b) Path loss $= 205.7$ dB
(c) Received power flux density $= -131.05$ dBW/m²
(d) Received power $= 31.98 + 61.33 - 205.7$
$\qquad\qquad\qquad\qquad = -112.39 \text{ dBW}$.

A comparison of the results at these two frequencies reveals some interesting characteristics of satellite communication radio links. We note that the antenna gain increases with an increase in radio frequency but the path loss reduces. As a consequence, the received power at the output of the antenna remains the same even though the frequency increased from 4 to 11.5 GHz. This can be explained by comparing the expressions for the free space path loss and the antenna gain. The wavelength relationships (equation 4.6 and path loss component in equation 4.15) are the inverse of each other and therefore an increase in path loss compensates the corresponding increase in the gain of the antenna. It would appear that a satellite link is insensitive to changes in frequency. However, there are technical and economic factors which favour a certain range of frequency for a given application. As an example, consider the use of a satellite link for communications between very small earth stations or for direct television broadcasts to homes. For such applications, earth stations must be low cost with very small antennas mountable directly on to customers' premises. We observed that for a given antenna diameter the gain of an antenna increases at high frequencies. When we restrict the diameter of antennas to meet technical and cost constraints, higher frequencies appear favourable because larger gain is possible for both satellite and ground antennas for a given diameter. Thus it is possible to transmit high EIRP from the satellite using relatively small and simple antennas and at the same time use small high-gain antennas on the ground. This is one reason for the extensive use of the frequency range between 11 and 14 GHz for broadcast and business applications, and a growing interest in the use of 20/30 GHz band. We conclude that even though the satellite link appears to be insensitive to a variation in frequency, other factors influence the choice of frequency and usually there is an optimal frequency range for each type of service and application. However, as we have seen in chapter 3, the selection of frequency for satellite communication is a complex process and governed by a wide range of factors.

4.4 Noise considerations

Noise introduces a fundamental limit on the capacity and performance of any communication system. Satellite communication systems are particularly susceptible to noise because of their inherent low received power. Several sources introduce noise into a satellite communication link. Natural and man-made noise are picked up from antennas; thermal noise is introduced by the first stages of receivers; intra-system noise is generated by inherent non-linearities of devices used in the system, adjacent channel interference cross-polar coupling, etc. In this section we shall discuss and quantify the main types of noise. This knowledge can then be used to minimize the introduction of noise by judicious choice of system components and to quantify system performance.

A. Thermal noise

All active devices used in a communication system introduce thermal noise which affects system performance to various degrees. To understand the effect, consider a resistor in an electrical circuit. The energy generated by the random motion of electrons in such a resistor appears as a randomly fluctuating voltage or *noise* across the terminals of the resistor. The mean square voltage of noise, $e_n^2(t)$ is given by

$$e_n^2(t) = 4kTB_nR \quad \text{(volt)}^2 \tag{4.17}$$

where k = Boltzmann's constant
T = absolute temperature of resistance
B_n = measurement bandwidth
R = resistance.

The resistor can therefore be represented as a noise generator. Maximum power from the noise generator, P_n, is transferred when a load is matched to the generator, and is given by

$$P_n = e_n^2(t)/4R$$

$$= kTB_n \tag{4.18}$$

The noise behaviour of an electronic device such as an amplifier can be represented by an equivalent resistor using equation (4.17) to quantify the noise power. It is useful to note at this stage that the noise power increases with increase in physical temperature of a device and the bandwidth under consideration. We can use these characteristics to minimize the effects of noise.

The noise generated in a device is characterized by the *noise figure* and the *noise temperature*.

Noise figure

The noise figure of a device is defined as the ratio of the signal-to-noise power ratio at the input to the output:

$$F = \frac{P_i/N_i}{P_o/N_o} \tag{4.19}$$

$$= \frac{N_o}{GkT_iB_n}$$

where P_i = available input signal power
N_i = available input noise power
$\quad = kT_iB_n$
T_i = ambient temperature in Kelvin
P_o = available output signal power
N_o = available output noise power in a noise-free amplifier
G = average power gain over the specified frequency band.

An actual amplifier adds noise, ΔN, generated within itself to the available output noise power.

Noting that

$$P_o = GP_i$$
$$N_o = GkT_iB_n$$

equation (4.19) can be written as

$$F = \frac{GkT_iB_n + \Delta N}{GkT_iB_n}$$

Therefore the noise figure can also be expressed as

$$F = 1 + \frac{\Delta N}{GkT_iB_n} \tag{4.20}$$

Noise temperature

The noise temperature of an amplifier is defined as the temperature, T_e, of a resistance which provides the same noise power at the output of an ideal (i.e. noise-free) amplifier as that given by an actual amplifier which has its input terminated at a noise-free resistance (i.e. $T_i = 0$). The noise power, ΔN, generated within the actual amplifier is therefore GkT_eB_n. Substituting in equation (4.20) gives

$$F = (1 + T_e/T_i) \tag{4.21}$$

When the ambient temperature is 290 K

$$F = 1 + T_e/290 \tag{4.22a}$$

or

$$T_e = 290(F - 1) \tag{4.22b}$$

Noise figure and noise temperature of a lossy network

The treatment of noise figure and temperature in the previous section dealt with active devices such as amplifiers. Signals received by an antenna are coupled to the first stage of a receiver through lossy components such as a waveguide or coaxial cables, diplexers and switches. Such components, in addition to attenuating the RF signals, also introduce thermal noise. In this section we will apply the equations developed above to lossy networks.

First we define a loss factor L:

$$L = P_i/P_o$$

The noise at the output of the attenuator is

$$P_o = kT_iB_n/L + P_a \tag{4.23}$$

The first term represents the attenuated input noise power, and the second the noise introduced by the attenuator.

From the definition of noise temperature, the equivalent noise temperature of the attenuator can be expressed as

$$T_e = T_i(1 - 1/L) \tag{4.24}$$

and when $T_i = 290$ K

$$T_e = 290(1 - 1/L) \qquad (4.25)$$

The noise figure can be obtained using equation (4.21).

Noise figure and noise temperature of networks in series

For system design we need to specify the noise characteristic of a complete receiver consisting of various components connected in series. Here we develop the noise temperature of a cascaded network. Consider a network of two amplifiers in series. The effective noise temperature, T_o, at the output of the network is

$$T_o = G_1 G_2 T_1 + G_2 T_2 \qquad (4.26)$$

where T_1 and T_2 are respectively the noise temperatures of the first and second stage. Also, when the noise temperature of the network is T_e:

$$T_o = G_1 G_2 T_e \qquad (4.27)$$

From equations (4.26) and (4.27):

$$T_e = T_1 + T_2/G_1 \qquad (4.28)$$

Extending the analysis to n amplifiers in cascade, the equivalent noise temperature for n cascaded amplifiers can be shown to be

$$T_e = T_1 + T_2/G_1 + T_3/(G_1 G_2) + \ldots + T_n/(G_1 G_2 \ldots G_{n-1}) \qquad (4.29)$$

where T_n and G_n are respectively the effective noise temperature and the gain of the nth stage.

Substituting the equivalent noise figures in above expression (4.29), the noise figure of n cascaded amplifier is given as

$$F_n = F_1 + (F_2 - 1)/G_1 + \ldots + (F_n - 1)/(G_1 G_2 \ldots G_n) \qquad (4.30)$$

where F_n is the noise temperature of the nth stage amplifier.

Equations (4.29) and (4.30) provide an important insight into the system noise behaviour of the system. Notice that the noise contribution of the first stage is the largest, whereas the contributions from the succeeding stages are reduced successively as G_1, $G_1 G_2$, etc. Therefore every effort should be made to minimize the noise of the first stage of a receiver.

B. Antenna noise temperature

An antenna receives noise emanating from various external sources together with the useful satellite signal. The antenna noise temperature is a measure of the noise entering a receiver via the antenna. The antenna noise temperature is obtained by integration of the noise components from all the external noise sources:

$$T_a = \frac{1}{4\pi} \int_0^{2\pi} \int_0^{\pi} G(\theta, \phi) T_b(\theta, \phi) d\Omega \qquad (4.31)$$

where $d\Omega$ is an element solid angle and refers to the solid angle subtended
         by a source at the antenna
         $G(\theta, \phi)$ is the gain function of the antenna in the direction θ, ϕ
         $T_b(\theta, \phi)$ is the brightness temperature in the direction θ, ϕ
         θ, ϕ are respectively the elevation and the azimuth angle of the
         antenna.

Noise sources can be natural or man-made. The main natural sources are cosmic noise, noise introduced by the Sun, the Moon, the Earth, atmospheric absorption and rain attenuation. Such noises are collectively known as *sky noise*. Man-made noises emanate from sources such as vehicles, industrial machinery, etc. and from other terrestrial and satellite systems operating at the same frequency. This latter type of noise, referred to as *inter-system noise*, is treated separately in a system design and although received via the antenna, is not considered as part of antenna noise. Each of these noise sources are discussed in more detail in the following paragraphs.

Cosmic noise originates in outer space and is radiated by the hot gases of stars and inter-stellar matter. The average cosmic noise power decreases with frequency, the noise power being negligible above about 1000 MHz (<30 K). In certain parts of the sky the noise power is very low (sometimes called 'cold sky') and in others it is relatively high ('hot sky'). There are also a number of discrete high-intensity point sources in the sky. The angles subtended by such sources are of the order of a few minutes of arc, consequently their contribution to the average noise temperature is low unless an antenna is highly directional and pointed directly at such a source. The locations of such radio stars are well known and their flux densities on earth predictable. Therefore such sources are often utilized for calibration of the G/T of large earth stations (see sections 4.5 and 10.2 for definition of G/T).

The Sun is another major source of sky noise. The noise temperature of the Sun depends on the frequency. The average noise temperature can be approximated as $12\,000 f^{-0.75}$ K (Shimbukuro and Tracey, 1968; see also section 2.6). The solar temperature depends on the solar conditions, increasing by 10^2 to 10^4 during periods of high sun-spot activity. Earth stations are intentionally never pointed at the Sun but solar noise enters earth station antennas through side lobes. However, unavoidably the Sun appears within the main lobe of all earth station antennas during the equinox periods, causing disruption in service (see section 2.6 and appendix B). Fortunately such events occur for only a few minutes each year and are predictable.

The Moon acts as a black body radiator for microwaves. The average brightness temperature of the Moon is between 200 and 300 K. Hence the noise contributions of the Moon are small unless an antenna is pointed directly at it.

The propagation medium introduces noise when the radio energy is absorbed by the medium. The equivalent noise temperature can be determined by using equation (4.24) with loss L set to the attenuation introduced by absorption. Noise is introduced by absorption due to oxygen, water vapour and various forms of hydrometer such as rain. The physical temperature of rain can be assumed to be 300 K for estimating the rain noise temperature. At frequencies above 10 GHz the noise introduced by rain attenuation dominates all other types of noise.

Figure 4.4 (Hogg, 1960) shows the antenna noise due to oxygen, water vapour and galactic noise as a function of elevation angle and frequency for a typical earth station antenna. It can be seen that noise reduces as elevation angle is increased. This is caused by a reduction in the length of the radio path through the atmosphere, which lessens as the elevation angle is increased. It can also be observed that there is a low noise window between approximately 1 and 10 GHz. This is one reason why this frequency band was preferred for the first satellite communication systems (the other was the state of the technology).

While the methodology for computing the antenna noise temperature is simple, the difficulty lies in estimating the brightness temperature $T_b(\theta, \phi)$ for the various sizes and orientations of antenna used in a typical satellite communication system. Therefore, system designers prefer to use a general curve such as that shown in figure 4.5 (CCIR) for estimating the sky noise temperature. The sky noise temperature contributions for medium and small earth stations are relatively low in comparison with other system noise sources such as receiver front-end, and therefore such an approximation in antenna noise estimation provides acceptable accuracy in practice. At frequencies above 10 GHz, rain becomes a dominant contributor to antenna noise and therefore the system noise temperature becomes a statistical parameter and also depends on the geographical region under consideration (see CCIR Report 868).

The discussion above is valid for an earth station antenna pointing towards the sky. When estimating the noise temperature of an antenna onboard a satellite, the conditions are different. In such a configuration the antenna points towards the Earth, which has a physical temperature of about 300 K, and therefore the noise temperature of a satellite antenna approaches this value. In practice, the noise temperature has been observed to be less

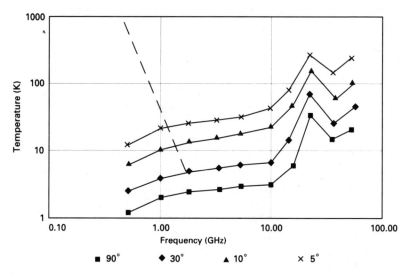

Figure 4.4   Estimated antenna temperature due to oxygen and water
vapour absorption (solid line) and galaxy noise (dashed line),
as a function of frequency and antenna elevation for standard
summer conditions. The peak is caused by water vapour
absorption. The frequency window between about 1 and 10 GHz
has the lowest noise temperature (data from Hogg, 1960).

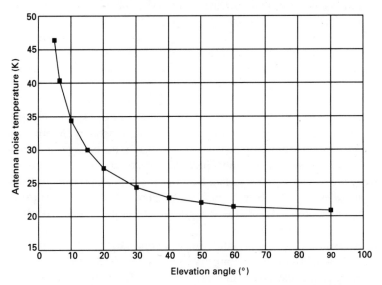

Figure 4.5   Total noise temperatures of a typical earth station antenna at
4 GHz under clear sky conditions without considering
miscellaneous antenna losses.

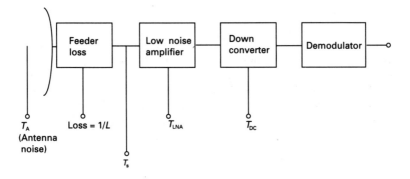

Figure 4.6   Noise components considered in evaluating a receiver.
T_{LNA} = noise temperature of low noise amplifiers;
T_{DC} = temperature of down converter.

than 300 K, because part of the antenna beam looks at the relatively cold space behind the Earth.

C. System noise temperature

All major noise components which are considered in specifying an earth station and satellite receivers have been identified in the preceding sections. The parameter for quantifying the integrated effect of noise in a receiver is the *receiver noise temperature*. Figure 4.6 shows the main elements of a receiver used for obtaining the receiver noise temperature. The antenna noise power is attenuated by $1/L$ in the passive components such as the feed and the transmission line leading up to the front-end amplifier. At the same time, these dissipative elements introduce noise. The receiver consisting of front-end amplifiers and down-converter adds further noise. The system noise temperature is the summation of all noise components and is given by

$$T_s = T_A/L + T_i(1 - 1/L) + T_r \tag{4.32}$$

where T_i = ambient noise temperature (usually assumed to be 290 K)
   T_r = effective noise temperature of the receiver (obtained by using equation 4.29).
Note that in equation (4.32) the noise is referred to the input of the low noise amplifier. If necessary, other reference points may be taken.

D. Inter-modulation noise

A major source of noise in a satellite communication system is the inter-modulation noise caused by non-linearites in the high power amplifier stages of spacecraft and earth stations. Two types of non-linearites exist – amplitude non-linearity which causes AM–AM conversion, and phase non-linearity which causes AM–PM conversion. (Amplitude modulation (AM) and Phase modulation (PM) are discussed in chapter 5.) Further, out-of-band inter-modulation products transmitted from earth stations result in interference to other systems. To minimize such harmful emissions, radio regulations restrict such out-of-band transmissions from earth stations to very low levels.

Typical transfer characteristics for a travelling tube amplifier (TWT) commonly used in satellites is shown in figure 4.7. The characteristic is linear at low input drive level but becomes increasingly non-linear as the output power approaches saturation. To minimize inter-modulation noise, an attempt is made to operate the TWT in the linear region by reducing or 'backing off' the input power level. Referring to figure 4.7, the input back-off, B_i, is given by $(P_{is} - P_{ic})$ and the output back-off, B_o, by $(P_{os} - P_{oc})$.

The frequency division multiplexed access (FDMA) scheme (see chapter 8) used for accessing a satellite transponder is particularly susceptible to inter-modulation noise because many carriers use the transponder simultaneously, leading to the requirement of backing off the drive level. Therefore spacecraft power amplifiers remain under-utilized. Typically, the back-off for a TWT to be operated with a FDMA system is in the range 3–10 dB. For example, a TWT capable of transmitting 20 W can transmit only 2 W with a 10 dB back-off. Another associated problem in a frequency division multiplex access system is the need for earth stations to maintain accurate control of transmit power in order to minimize overdrive of the satellite transponder and the consequent increase in inter-modulation noise. Inter-modulation considerations also apply to *earth stations* which transmit multiple carrier. In this case the earth station high power amplifiers remain under-utilized. The Time Division Multiple Access (TDMA) mode of accessing was developed to eliminate this inefficiency of the FDMA scheme. In TDMA the inter-modulation noise is minimized but AM–PM conversion caused by the TWT non-linearity needs to be considered. It is however possible to operate to within 1–2 dB of the TWT saturation (see chapter 8 for a detailed description of the FDMA and TDMA schemes).

We have noted that the satellite link constitutes one component in the link design. Degradation caused by transponder inter-modulation noise is an important consideration when quantifying the satellite part of the link. Therefore quantitative analysis of the non-linear effects is essential for system design. The analysis is highly involved and the details are outside the scope of this book. A simple technique to quantify the inter-modulation noise from measurements performed on communication satellites is included later in the sec-

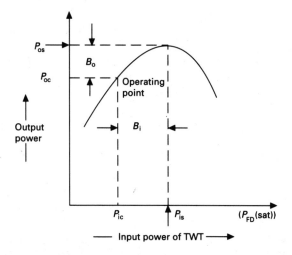

Figure 4.7 Typical transfer characteristic of a TWT used on a communication satellite.

tion. Obviously, the use of *measured* inter-modulation noise provides a more accurate assessment of the overall signal quality and often the link calculations, based on theoretical analysis, performed during the planning stage are revised to reflect the actual operating environment.

Consider a FDMA system with x unmodulated carriers accessing a satellite transponder. The TWT non-linearity is then said to produce inter-modulation products of order n at a frequency f_n given by

$$f_n = m_1 f_1 + m_2 f_2 + \ldots + m_x f_x \tag{4.33}$$

where $f_1 \ldots f_x$ are input carrier frequencies and $m_1 \ldots m_x$ are integers.

The order n of an inter-modulation product is

$$n = |m_1| + |m_2| + \ldots + |m_n| \tag{4.34}$$

For frequencies used in satellite communications only the odd-order inter-modulation products fall within the useful frequency band. Further, it can be shown that the power of the inter-modulation products decreases very rapidly as the order of the inter-modulation product is increased. In practice, the third-order products contribute the major proportion of the inter-modulation noise power. There are two types of third-order product: the first type takes the form $(2f_i - f_j)$ and the second the form $(\pm f_i \pm f_j \pm f_k)$ with only one minus sign being permitted (i, j, k are integers). When N carriers are present there are $N(N - 1)$ products of the first type and $N(N - 1)(N - 2)/2$ products of the second type.

Several methods have been developed for estimating inter-modulation noise (e.g. Fuenzalida *et al.*, 1973). Numerical techniques are best suited to solve the complex expressions obtained.

Generally the inter-modulation characteristics of a satellite TWT are available from measurements. One commonly used method is to drive the satellite up to the operating point with symmetric equi-spaced carriers such that all the significant inter-modulation products fall on the centre frequency of the transponder. The power measured at the centre frequency P_{IM} is therefore the sum of all the significant inter-modulation products. We can assume that under operational conditions P_{IM} is uniformly distributed across the complete transponder band B_T. With this assumption the inter-modulation power density is (P_{IM} [dBW] $-$ 10 log (B_T) [dBW/Hz]). This power spectral density can be used to approximate C/N_0 (IM) in the link design. This approach, however, gives optimistic results.

It is implicit from the above discussions that the effect of inter-modulation noise can be minimized by exercising care in positioning the carriers in the transponders. Good frequency planning is particularly important for Single Carrier per Channel (see chapter 8) systems where a large number of small carriers (e.g. several hundred) share a single transponder. Several methods have been proposed for frequency planning (e.g. Fang and Sandarin, 1977; Hirata, 1978).

E. Interference

There are various sources of interference in a satellite communication system. These may be broadly classified as intra-system and inter-system. We shall discuss each separately.

Intra-system interference

A number of sources can cause intra-system interference. Interference can be caused by coupling of orthogonally polarized signals in a dual polarized system. A horizontally polarized signal can interfere with a vertically polarized signal and vice versa. This type of interference occurs in earth station and satellite antenna and feed systems, and is also caused during propagation through rain and ice. Such interference can be minimized by using well-designed earth station and spacecraft antennas. Typical values of cross-polar discrimination in well-designed antennas are of the order of 25–30 dB. However, depolarization caused by rain and ice imposes a fundamental limitation on this type of system (see section 3.5).

Interference can occur in satellites when the filters used for isolating adjacent channels do not have sufficient roll-off characteristic. Such interference can be minimized by using adequate guard bands between adjacent channels and the use of well-designed channel separation filters. There are other miscellaneous sources of intra-system interference such as multiple path in satellites. The magnitude of these interferences is difficult to predict analytically but with the use of well-designed equipment and by exercising care in the system design a link margin of less than 1 dB is adequate to compensate for intra-system interference.

Radio regulations and inter-system interference

Care must be exercised to minimize interference from (and to) other satellite and terrestrial systems during initial system design. Interference may occur between a satellite system and terrestrial systems or between two satellite networks whenever the same frequencies are shared. Procedures laid down by the ITU are used to minimize inter-system interference. The procedures are complex and depend on the type of service. No attempt will be made here to describe these procedures in detail but only a basic guideline will be laid out (see also section 3.2). The interested reader may refer to the prevailing radio regulations, published by the ITU regularly, for details. The procedures involve the use of sharing constraints, frequency coordination and the use of regional or global pre-assigned plans.

Sharing constraints impose limits on transmission levels from terrestrial transmitters, earth stations and satellites in certain bands, and services which permit sharing.

Coordination involves the use of procedures which have been developed to permit coexistence of two networks sharing the same frequencies by, among other steps, mutual agreement (e.g. avoiding high power transmissions in a specific section of the band) and allowing for entry of a certain amount of interference.

The use of a *plan* involves using worldwide or regional plans which have been developed for certain services. Such plans assign to each participating country an orbital position, frequency, bandwidth and satellite power for the duration of the plan, thus greatly simplifying the implementation of the system and at the same time making efficient use of orbital locations and RF spectrum. Worldwide plans are in use for the direct broadcast service and fixed satellite service. The use of a plan makes the system parameters invariant for long periods, which is an attribute well suited to applications such as direct broadcasts by satellite where the general public is involved. Also, developing countries are guaranteed access to the geostationary orbit according to their needs.

In general, the procedures for the fixed and mobile satellite services are the most complex, involving an elaborate set of sharing constraints and frequency coordination procedures.

In summary, the regulations which affect the planning and design of satellite communication systems pertain to:

1. frequency allocation for various satellite communication services (Article 8 of Radio Regulations);
2. constraints on the maximum permissible RF power spectral density from earth stations (CCIR Rec 524);
3. antenna pattern of earth stations (CCIR Recs 465 and 580);
4. constraints on the maximum permissible transmission levels from satellites (CCIR Rec 358);
5. orbital position, station-keeping requirements and antenna-pointing tolerance of satellites (Article 29 of Radio Regulations);
6. permissible interference from other networks (e.g. CCIR Rec 466, Rep 867, Rec 523 and Rec 483).

Noise is introduced by other satellite and terrestrial systems in both the uplinks and downlinks. Typically, the maximum single entry noise from external sources is limited to 4%. A total interference noise budget of 20% is adequate for most purposes (i.e. interference noise = 20% of total noise). Although the contribution of carrier-to-interference ratio noise is small in an operational system, this is achieved during the planning stage through negotiations and mutual adjustments of technical parameters with those satellite/terrestrial system operators sharing the band. Often this process takes several years and must not be taken for granted.

In systems used for military communications where intentional interference is possible, special techniques are used for making the transmissions resistant to interference (e.g. the use of spread spectrum communication – see chapter 8).

4.5 Link design

General

The main purpose of a satellite communication system is to deliver a message of the desired quality to the destination. A link design endeavours to achieve this objective by the judicious choice of various link parameters. The figure of merit to measure the signal quality is the carrier-to-noise ratio measured at the input of the destination earth station demodulator. The carrier-to-noise ratio is related to the baseband signal quality, the message, through a modulation-dependent factor. The parameters affecting the link design may be categorized according to the system element to which they relate (refer to figure 4.1), i.e *earth station, satellite* or *propagation channel*, as follows.

Earth station related

- Geographical location provides an estimate of rain fades, satellite look angle, satellite EIRP in the direction of the earth station and earth station–satellite path loss.
- Transmit antenna gain and transmitted power provide the earth station EIRP.
- Receive antenna gain is related to the G/T of the earth station.
- System noise temperature gives the sensitivity of the earth station and is related to the G/T.
- Inter-modulation noise affects the total carrier-to-noise ratio.
- Equipment characteristics (e.g. demodulator implementation margin, cross-polar discrimination, filter characteristics) dictate the additional link margins.

Satellite related

- Location of satellite relates to the coverage region and earth station look angle.
- Transmit antenna gain and radiation pattern provide the EIRP and coverage area.
- Receive antenna gain and radiation pattern are related to the G/T and coverage area.
- Transmitted power is related to the satellite EIRP.
- Transponder gain and noise characteristics are related to EIRP and G/T.
- Inter-modulation noise affects the total carrier-to-noise power at the earth station receiver.

Channel related

- Operating frequency is related to path loss and link margin.
- Modulation/coding characteristics govern the required carrier-to-noise ratio.
- Propagation characteristics govern the link margin and the choice of modulation and coding.
- Inter-system noise.

In the following sections all these parameters are brought together and equations used in satellite communication system link design developed.

Total carrier-to-noise ratio

As noted above, the main objective in the link design is to establish the desired carrier-to-noise ratio at the input of an earth station demodulator, within all practical constraints. The carrier-to-noise ratio at the demodulator input is a function of the uplink and downlink EIRP; the noise introduced in the earth station receiver and the satellite link; and the amount of interference. The received message quality is related to the carrier-to-noise ratio at the demodulator input as follows:

$$S/N \text{ (or bit error rate)} = f\frac{C}{N_T} \tag{4.35}$$

The relationship for obtaining the total carrier-to-noise ratio is developed as follows.

The total noise N_T at the receiver is the summation of noise from all sources:

$$N_T = N_U + N_D + N_I + N_i$$

where N_U = uplink noise, measured at the satellite
        N_D = downlink noise, measured at the receiving earth station
        N_I = inter-modulation noise in the satellite link
        N_i = Interference noise.

The total carrier-to-noise ratio is then

$$\frac{C}{N_T} = \frac{C}{N_U + N_D + N_I + N_i} \tag{4.36a}$$

By rearranging, C/N_T can be expressed as

$$\frac{C}{N_T} = \left[\left(\frac{C}{N_U}\right)^{-1} + \left(\frac{C}{N_D}\right)^{-1} + \left(\frac{C}{N_I}\right)^{-1} + \left(\frac{C}{N_i}\right)^{-1} \right]^{-1} \tag{4.36b}$$

For digital systems it is usual to measure the system performance as bit energy-to-noise power density E_b/N_o. Then E_b/N_o directly replaces C/N_T in equations (4.36) and noise power is expressed in terms of power density. Equations (4.36) form the basis for apportioning the carrier-to-noise ratio in various sections of the link. For example, when the receiver is small (e.g. a VSAT) the downlink (C/N_D) makes the most significant contribution and this can be offset by reducing the contribution from the uplink noise.

In the following sections the relationship for obtaining carrier-to-noise ratio of individual sections of the link is developed. We begin by developing the equation for the received carrier power in a practical environment, followed by discussions of the carrier-to-noise ratio of each part of the link.

Received carrier power

The transmission equation (4.16) is the basis for the derivation of the received carrier level in terms of the effective isotropic radiated power (EIRP), the path loss and the receiver antenna gain. The transmission equation as given by (4.16) assumes an ideal condition with path loss as the only loss factor. However, in practice additional losses and link degradations must be considered. Such degradations are compensated by transmitting additional power, termed *link margin*. Thereby it is ensured that the desired quality objective is met under the worst possible conditions. The definition of *worst condition* is governed by the acceptable reliability of the link. For example, assuming a 11.7 GHz system and 99.9% link reliability, we note from figure 3.2 that the link is likely to suffer a fade of ~2.5 dB whereas for 99.99% reliability the expected fade level increases to ~11 dB. Thus link margins of ~2.5 dB and 11 dB provide reliabilities of 99.9% and 99.99% respectively. It should be noted that degradations are usually present for only a small fraction of the time and therefore the link quality is in effect better than specified for most of the time. It is worth mentioning here that the overall system cost increases as the link reliability is improved because the transmitter power or receiver sensitivity or both must be increased, tending to increase the system cost. Hence a compromise is struck between the link reliability and system cost.

The main components considered in obtaining the downlink margin are:

- antenna tracking loss;
- atmospheric absorption;
- a statistical loss parameter due to hydrometers (mainly rain);
- a statistical loss parameter due to shadowing and multipath when a mobile system is considered;
- a statistical loss parameter associated with scintillation;
- intra- and inter-system interference;
- miscellaneous losses, e.g. wet radome, equipment ageing, demodulator inefficiencies.

A simple arithmetic addition of the statistical loss parameters gives a pessimistic fading scenario. Hence a root sum square of the standard deviations of various losses is often used in practice.

The link margin depends on the frequency of operation and service (e.g. mobile, fixed, direct-to-home broadcasts). For example, a satellite link from a mobile terminal is susceptible to shadowing and multipath whereas a link from a large fixed station suffers no such degradation. Therefore the transmission equation (4.16) can be modified to

$$C = P_e - L_p - L_m + G_d \quad \text{dB} \qquad (4.37)$$

where P_e = EIRP (dBW) [earth station or satellite]
        L_p = uplink path loss (dB)
        L_m = link margin (dB)
        G_d = destination antenna gain (dB) in the direction of the transmitter.
The range of the earth station to the satellite for obtaining the path loss can be determined by equation (2.19) given the longitude of the satellite and the latitude/longitude of the earth station (see section 2.6). Another consideration is the satellite antenna gain, G_s, towards an earth station. An earth station can be located anywhere in the coverage area of a satellite and therefore satellite antenna gains towards an earth station need to be considered on an individual basis when possible. The antenna pattern of a satellite often takes complex shapes and therefore G_s can be obtained by superimposing the earth station position on the coverage contour of the antenna. The position of all earth stations to operate in the network may not always be known at the planning stage, and in fact for a mobile communication system the position could change continuously. A more general approach is therefore to assume the location of an earth station at the edge of service coverage. This provides the worst-case estimate of the link quality/transmitted power. Using this approach, the desired link quality is guaranteed for all the earth stations in the network.

In the following sections, equations (4.36) and (4.37) are applied to obtain the carrier-to-noise ratio of each link component.

Uplink carrier-to-noise ratio

Considering the *uplink* only (earth station–satellite), the main source of noise power at the destination is the thermal noise power at the satellite receiver. From section 4.4 we know that

$$N = kT_sB$$

where k is the Boltzman constant
        T_s the noise temperature of the receiver
        B the pre-detection bandwidth.
The carrier-to-noise ratio (C_u/N_u) in dB at the satellite is then

$$C_u/N_u = (P_e - L_{pu} - L_{mu} + G_s) - 10 \log (kT_sB)$$

or

$$C_u/N_u = P_e - L_{pu} - L_{mu} + G_s - 10 \log (T_s) - 10 \log (kB)$$

The above equation can be rearranged as

$$C_u/N_u = P_e - L_{pu} - L_{mu} + G_s/T_s - 10 \log (k) - 10 \log (B) \quad (4.38)$$

where P_e = earth station EIRP
      L_{pu} = uplink path loss
      L_{mu} = uplink margin
      G_s/T_s = G/T of the satellite in the direction of the earth station
      $10 \log (k)$ = -228.6 dBW/K.

The carrier-to-noise ratio, when expressed in per Hertz of bandwidth, is known as the carrier-to-noise power spectral density and obtained from (4.38) by setting B to unity. The term G_s/T_s is the figure of merit for measuring the receiver sensitivity. The sensitivity increases with an increase in G_s/T_s (see also section 10.2).

The uplink carrier-to-noise ratio is not a critical factor in applications when large earth stations are in use because the uplink power can be quite readily generated on the ground. When considering extremely small terminals, such as used in mobile earth stations which are EIRP limited, the uplink carrier-to-noise ratio needs careful optimization.

Downlink carrier-to-noise ratio

Similarly the *downlink carrier-to-noise ratio* C_d/N_d is

$$C_d/N_d = P_s - L_{pd} - L_{md} + G_e/T_e - 10 \log (k) - 10 \log (B) \quad (4.39)$$

where P_s = EIRP from satellite in the direction of the earth station
      L_{pd} = downlink path loss
      L_{md} = downlink margin
      G_e/T_e = G/T of destination earth station.

Satellite path

The noise in the *satellite component* of the link is dominated by the noise power introduced by the receiver and the inter-modulation noise introduced by the final stages of the amplifier. We shall discuss each of these components.

The lowest possible noise temperature of the satellite receiver is limited by the noise temperature of the satellite antenna which points in the direction of the 'hot' earth having a temperature of about 300 K. Therefore the G/T of the satellite receiver can only be improved by increasing the satellite antenna gain. The consequence is that the antenna beamwidth (and hence the coverage area) is reduced. The antenna gain and coverage area optimization is a complex process, mainly governed by a trade-off analysis between antenna complexity, advantages gained in terms of increased satellite

EIRP and frequency reusability by spatial isolation, and by the need to have a low G/T on the satellite. Note that generally the low sensitivity of satellite receivers can be easily compensated by transmitting higher power from the ground except when earth stations are very small.

The inter-modulation noise introduced in the high power stages of the satellite transmitter can be detrimental to the overall carrier-to-noise ratio unless care is exercised to minimize the noise. With due care, this contribution can be made insignificant.

Sometimes, the satellite can be represented as a block with a *gain G*. Then the satellite EIRP is obtained as

$$P_s = C_i + G \text{ dBW} \qquad (4.40)$$

where C_i = received carrier level (dBW) at the satellite, assuming an isotropic receiver antenna (gain = 0 dB)

P_s = satellite EIRP (dBW)

G = satellite gain (dB).

From equation (4.37)

$$C_i = P_u - L_{pu} - L_{mu} \qquad (4.41)$$

This is a useful representation of the satellite for determining the available EIRP from the satellite for a given input power (or vice versa). However, it should be borne in mind that setting of satellite gain is not arbitrary. It is based on meeting the overall carrier-to-noise ratio by optimally apportioning the carrier-to-noise ratio of each part of the link. For example, setting a high value of satellite gain enables the use of low uplink EIRP but to compensate for the resulting degradation in the uplink C/N a higher downlink C/N is necessary. Thus the trade-off is between minimizing the EIRP of the ground terminal or of the satellite. When the transponder is shared by several types of carrier, the gain is obtained after analysis of all the carriers.

Equation (4.40) is useful for performing link calculations on an operational satellite which has already been assigned a specific gain.

It should be mentioned here that the maximum EIRP of a satellite can be given as either a single or a multiple carrier value. The latter provides a more accurate estimate of the available EIRP from the satellite, as it is a closer representation of an operational satellite.

Some satellites use an automatic level control circuit which maintains the output EIRP constant irrespective of the input drive level. An example is ESA's MARECS B2 satellite currently being used by INMARSAT to provide service in the Atlantic (East) Ocean region. By maintaining the same output level, this type of transponder maintains a similar carrier to inter-modulation performance over a broad range of drive levels. However, the output carrier level is a function of the number and size of carriers sharing the transponder. The desired signal quality is maintained by ensuring that the carrier EIRP is adequate under fully loaded conditions when the available EIRP is shared among all the users. The difficulty with this type of transponder is the need to 'load' the satellite under light traffic conditions (i.e. when the number of carriers is a small fraction of the total) so as to avoid large EIRPs for the few carriers in use resulting in large inter-modulation spikes or transmission of wideband noise in the absence of any traffic carriers. This situation can be avoided by using an *ALC* circuit which only becomes operational after the satellite has been loaded to a specified threshold. This type of transponder is used in INMARSAT's second-generation satellites.

Link design considerations

Now that we have developed a basic understanding of individual components of a satellite communication link, all the concepts must be tied together. In this section we shall briefly review the steps involved in planning a satellite communication system and follow it by illustrating the link design of a VSAT system.

The following broad phases are used when considering the implementation of a satellite network:

1. A need for the use of satellite communication is identified.
2. A comparative analysis with other types of transmission media is performed (i.e. satellite versus microwave radio relay or optical fibre, etc.). In some applications the choice of satellite communications may be obvious. In others, a satellite system may be favoured in a complementary role.
3. When the outcome of step (2) favours a satellite system implementation, the basic satellite system model developed in (2) is refined further. Probably the most important aspect at this stage is the arrangement of capital to finance the project. The running/maintenance costs of satellite communication project are high, and this must also be considered in costing (chapter 9 includes an example cost model for a space segment cost sensitivity study). It is also useful to discuss with manufacturers for realistic cost estimates.
4. Implementation aspects such as regulatory process, equipment procurement and site selection are initiated.
5. The system is implemented and a mechanism set in place to operate and maintain the network to ensure reliable service and guarantee continuity.

The study mentioned in step (2) is important in the planning process, requiring a preliminary definition of the satellite system. Some of the parameters considered include communication traffic forecast, existing infrastructure, the state of technology, the necessary investment and revenue potential.

Several approaches are taken in developing a system model. In a 'no-risk' approach only existing and proven technologies are considered. In some instances leasing of transponders from operational satellites is also considered, thereby eliminating all risk and complexity of owning and operating a satellite. Constraints are applied to the model, where necessary. For example, in a VSAT network, a limit is imposed on the G/T and EIRP of VSATs to enable the use of off-the-shelf systems. A constraint is imposed on the upper limit of satellite EIRP, based on the available state of the art (or available EIRP, if a transponder is leased). With all constraints applied, the capacity of the system is derived. The system is considered viable if the capacity is adequate for the given application and the economics are favourable.

Another planning approach is to include a certain degree of risk by opting for some innovation. The risk is balanced against advantages such as increased satellite lifetime, improvement in system capacity resulting in reduction in cost/circuit, etc. Such an evolutionary approach is generally favoured by established system operators – both international and domestic. For example, INTELSAT has always included a significant amount of innovation in each generation of spacecraft, thereby maximizing the use of frequency and vastly increasing the circuits offered. A similar approach has been taken by INMARSAT. Domestic systems such as the Indian domestic satellite system have also followed an evolutionary approach.

Another route is to take a totally innovative approach in system design. Although innovative satellite technology has generally been funded by governments because of the expense and risk involved, service-providers and investors are willing to assume a great degree of risk by opting for unproven and innovative technology when there is a large potential market. An example is the current interest in the provision of service to hand-held mobile terminals. Some of the proposed solutions are novel and imaginative, and yet there is a great deal of commercial interest in funding and implementing such systems.

Detailed discussion of each aspect mentioned above is outside the scope of this book. In the remaining part of the chapter we shall discuss a link design example of FSS to illustrate the practical application of some of the concepts developed so far.

VSAT link design example

Very small apertutre terminals or VSATs are small and inexpensive terminals capable of being mounted directly on customers' premises, thereby providing a direct communication between a central point and a large number of remote points. VSATs are *fixed* terminals and are therefore considered to be a fixed satellite service. Because of the advantages offered by such networks, their use is increasing throughout the world. The system architecture consists of a large central earth station, called the hub, communicating with small terminals. The hub-to-VSAT link is called the outbound (or forward) link and the VSAT-to-hub link is called the inbound (or return) link. At present, communication between VSATs is only possible via the hub in two satellite hops.

Vuong *et al.* (1988) provide a useful summary of design considerations for VSATs.

Choice of frequencies

Being a FSS application, VSAT network frequencies are chosen in the FSS band. Allocations exist in the C (6/4 GHz), K_u (14/11 GHz) and K_a (30/20 GHz) bands. As VSATs are small terminals, their sensitivity (i.e. G/T) is low and it is desirable to be able to have as high a satellite EIRP in the outbound direction as possible. In many countries, C-band FSS allocations are shared with terrestrial systems. To facilitate sharing, regulatory authorities in such countries have imposed an upper limit on transmitted power flux density from satellites. To achieve the regulatory power spectral density limits, the radio spectrum has to be spread. It is possible to spread the radio signals by using modulation schemes which spread the radio energy. For example, BPSK modulation can be used instead of QPSK (see chapter 5). However the use of spread spectrum modulation which spreads the signal over a wide bandwidth gives real advantages. Therefore spread spectrum modulation has been used for C-band VSATs where sharing constraints are more severe. The sharing problem is less severe at the K_u band. This advantage, coupled with the fact that VSAT antenna sizes can be smaller, has made the K_u band very popular for VSAT applications. The use of the K_a band is also under active consideration.

We shall consider a K_u band system in the numerical example given later in this chapter.

Availability of link

Availability refers to the percentage of time that the link guarantees a specified link quality. Availability is an important consideration in the design of VSAT links. The factor which mainly affects link availability in the K_u and K_a band is rain fade. The affect of rain fades can be mitigated by providing a suitable link margin, which can be obtained by using the techniques discussed in chapter 3. Another commonly used technique is channel coding where the information signal is digital. Here redundant bits are added to the information signal for protection (see chapter 6). The net result is a reduction in power requirement of the radio link, thereby increasing the link margin for the same transmitted power. Further improvements can be achieved by varying the degree of coding according to channel conditions. Advanced techniques such as the use of regenerative transponders and very narrow spot beams are likely to emerge towards the end of the century.

Other sources causing link quality degradation to a lesser extent are outage caused by sun transit through the VSAT antenna beam – typically causing a total outage of about an hour in the outbound link and about half an hour in the inbound link (see chapter 2); and hardware failure including satellite transponder failure which together have a reliability of the order of 99.9%.

Channel quality

The main aim of any communication network is to provide an information signal of an acceptable quality to the user. From the discussions above, the parameter closest to the end user is the total carrier-to-noise density ratio. In chapter 5 we shall see that the signal quality of the information signal (or baseband signal) is related to the carrier-to-noise ratio (or energy bit-to-noise density ratio) by a modulation-dependent factor. The parameter for specifying the baseband signal quality is signal-to-noise ratio for analog signals and bit error rate for digital signals.

VSATs are generally used for carrying digital signals. Binary Phase Shift Keying (BPSK) or four-phase (QPSK) modulation schemes with forward error correction using Viterbi or sequential decoding technique are often used (see chapter 6). In the example here we shall assume that a typical energy bit-to-noise power density ratio (E_b/N_o) is required for VSATs, without going into details of the modulation/coding scheme. Readers will be in a position to select an appropriate modulation and coding scheme for their intended application after studying chapters 5 and 6.

Noise sources

In the preceding section we have identified several sources of noise. These include thermal noise in the front-ends of satellite and earth station receivers; inter-modulation noise in the satellite transponder; and various types of interference.

Thermal noise contribution is included in the G/T specifications. Note that the G/T of the K_u and K_a band system depends on atmospheric conditions. The G/T degrades in the presence of rain or clouds because of noise contributions from water vapours. In the numerical example here we shall assume clear sky G/T and account for degradation by adding a link margin. Inter-modulation (IM) noise depends on the drive level of the satellite TWT. IM noise can be estimated by using measured C/IM data (as discussed in section 4.4) or by theoretical analysis. We will use typical C/I_m values. Finally, all interference sources must be identified and their contribution quantified. The various interference sources are listed below.

Adjacent satellite interference

Adjacent satellites sharing the same band contribute interference. Interference from adjacent satellites is kept at acceptable levels by means of radio regulations and mutual agreement between system operators. This type of interference depends on the EIRP and antenna pattern of the satellites and terminals sharing the band. CCIR has recommended antenna patterns to be used for such calculations for the FSS. In countries such as the USA the regulatory bodies have specified a limit on the power flux density from satellites.

Cross-polar coupling

Cross-polar coupling in a dual polarized system occurs at earth station antennas (both transmitter and receiver), satellite input and output antenna and is induced in the atmosphere by rain and ice. Typically, antennas can provide cross-polar discrimination of the order of 25–30 dB. Depolarization caused by rain and ice can be estimated by the techniques proposed in section 3.3. The net effect of cross-polar coupling is obtained by summing the individual components of cross-polar discrimination.

To minimize interference from strong cross-polarized carriers, VSAT carriers are assigned frequencies well away from the centre frequency of cross-polarized carriers such as television.

Adjacent channel interference

The level of adjacent channel interference depends on the guard band between adjacent carriers sharing the same transponder. Ideally, the entire transponder bandwidth should be utilized. However this would cause excessive adjacent channel and inter-modulation noise. Therefore a compromise is made between carrier spacing and bandwidth utilization. There are two components of adjacent channel interference – uplink and downlink. The uplink component is dependent on spectral overlap between carriers whereas the downlink component is caused by spectral spread occurring in the satellite transponder as a result of non-linearity. A typical value of total adjacent carrier interference is of the order of >75 dB.

Adjacent transponder interference

Interference can arise from carriers occupying adjacent transponders. In the uplink, adjacent transponder interference occurs when the spectral components or IM products of carriers occupying adjacent transponders fall within the wanted VSAT carrier bandwidth. These components are minimized by imposing a constraint on out-of-band transmissions from earth stations. Typically such interference (per Hertz) is kept 75–80 dB below VSAT carriers. In the downlink adjacent transponder, interference occurs as a result of spectral spreading of carriers in the adjacent transponder caused by transponder non-linearity and IM products generated in the transponder. In many transponder designs, adjacent transponders are well isolated with filters. However, if transponder outputs share the same high power amplifiers, this type of interference needs to be minimized by operating the output stage in a linear region. The per Hertz magnitude of this interference is of the order of >100 dB below the VSAT carrier.

Interference from terrestrial systems

This type of interference is generally governed by spectrum regulations. As most of the terrestrial systems arise within the same country operating the VSAT, the regulatory bodies of individual countries are involved except when there is a potential of cross-country interference, when the affected countries have to coordinate the spectrum sharing. In the USA the problem of terrestrial interference in the K_u band has been minimized by allocating a secondary status to terrestrial transmissions in a specific part of the band reserved for VSAT application.

Numerical example

A typical set of earth and space segment parameters, which may be used in the outbound link of a K_u band VSAT system, are presented below. Determine the bit rate that such a system can offer to a VSAT user.

Uplink
Frequency: 14.5 GHz
Maximum hub earth
station EIRP: 55 dBW
Fade margin: 6 dB
Total carrier-to-interference
noise density ratio: 70 dBHz
Access: TDM

Modulation: BPSK
Coding: 1/2 FEC Convolution Code

Space segment
Satellite location: 70°E
Elevation angle at edge
of coverage: 5°
Satellite G/T: 0 dB/K
EIRP/VSAT carrier: 20 dBW
Transponder carrier-to-IM
noise power density ratio: 70 dB

Downlink
Frequency: 12 GHz
Fade margin: 5 dB
Total carrier-to-interference
noise density ratio: 69 dBHz
VSAT G/T: 20 dB/K
VSAT antenna diameter: 1.5 m
E_b/N_o (BER rate = 10^{-7}): 6.5 dBHz

Solution

We will assume that the hub earth station and VSAT are both located along the edge of the satellite foot print. Knowing the geometry of the satellite–edge of coverage, the range can be calculated using equations (2.17a) and (2.20a). Assume that the range is 41 000 km. The path loss may be calculated using the path loss formula given in equation (4.16).
Uplink path loss: 207.95 dB
Downlink path loss: 206.31 dB
Using (4.39) we obtain the downlink carrier-to-thermal noise density ratio (C/N_o) as 57.29 dBHz. Note that to obtain C/N_o, bandwidth is set to 1 Hz. Using (4.36) the total carrier-to-downlink noise density ratio is obtained as 57.01 dBHz. (Note that only two components are used in summation – N_D and N_i.)
Similarly the uplink total carrier-to-noise density ratio is obtained as 66.81 dBHz.
Finally, the total carrier-to-noise power density ratio at the receiver is obtained by summing the uplink, satellite inter-modulation and the downlink carrier-to-noise power densities, according to (4.36b). The magnitude of the total carrier-to-noise power density is obtained as 56.38 dBHz.

E_b/N_o is related to C/N_o total as follows:

$$C/N_o = E_b/N_o + 10 \log (R) \qquad (4.42)$$

where R = data bit rate.

Substituting the values of E_b and C/N_o in (4.42), the transmission bit rate possible with the above link parameter is calculated as 97.275 kbps. The use of 1/2 rate FEC coding doubles the transmission bit rate. The user bit rate is therefore half the channel rate or 48.637 kbps.

Problems

1. Develop the transmission equation. If you were to apply this equation to a satellite communications link design, what additional losses must be considered?
   Briefly discuss various noise sources which affect a satellite communications link design. Hence develop the relationship for the carrier-to-total noise ratio at the input of an earth station demodulator.
2. The earth to satellite path affects radio waves in several ways. Briefly discuss the role of propagation effects on the design of a satellite communication link.
3. A satellite communication system is expected to provide 17 000 two-way digital telephone circuits. Individual satellite transponders have the capacity to carry 120 Mbps of traffic. Using the link parameters given below determine:
   (a) uplink EIRP
   (b) energy per bit to noise power density (E_b/N_o) at the satellite input
   (c) power flux density at the earth station (downlink)
   (d) receive antenna gain
   (e) earth station G/T
   (f) carrier-to-noise power density at the receiver
   (g) E_b/N_o at the receiver.
   Available link parameters:
   Uplink
   Transmitter output power at saturation = 2 kW
   Backoff and combining loss = 7 dB
   Transmit antenna diameter = 15 m
   Antenna efficiency = 55%
   Frequency of transmission = 14 GHz
   Atmospheric loss = 0.6 dB
   Satellite G/T = −5.3 dB/K

Downlink
Satellite EIRP at beam edge = 40.2 dBW
Free space loss = 205.6 dB
Downlink frequency = 11.7 GHz
Atmospheric loss = 0.4 dB
Receive system noise temperature = 270 K
Feeder loss = 0.7 dB
Assume noise from other sources as negligible and that link margin is adequate for propagation losses due to rain etc.

4. An earth station has the following characteristics:
Antenna diameter = 30 m
Efficiency = 70%
Feeder loss = 1 dB
Physical temperature = 300 K
Low noise amplifier (LNA) gain = 30 dB
LNA noise temperature = 350 K
Gain of following stage = 30 dB
Noise figure = 3 dB
What is the *G/T* of the earth station if the effective antenna temperature is 40 K?

5. The following parameters are typical of a mobile satellite service for maritime communications. Use these parameters to determine the *G/T* of a ship earth station.
Frequency = 1535–1543.5 MHz
Range = 38 000 km
Number of channels = 40
Channel RF bandwidth = 30 kHz
Total transmitted power from satellite = 10 W
Satellite antenna gain at edge of coverage = 17 dBi
Required downline carrier-to-noise ratio = 5.1 dB

References

CCIR Recommendation and Reports, updated regularly.
CCIR (1980). *Handbook of Satellite Communications*, ITU, Geneva.
Fang, R.J.F. and Sandarin, W.A. (1977). 'Carrier frequency assignment for non-linear repeaters', *COMSAT Technical Review*, Vol. 7, No. 1, Spring, pp 227–244.
Fuenzalida, J.C., Shimbo, O. and Cook, W.L. (1973). 'Time domain analysis of intermodulation effects caused by non-linear amplifiers', *COMSAT Technical Review*, Vol. 3, No. 1, Spring, pp 89–141.

Hirata, Y. (1978). 'A bound on the relationship between intermodulation noise and carrier frequency assignment', *COMSAT Technical Review*, Vol. 8, No. 1, Spring, pp 141–154.

Hogg, D.C. (1960). 'Problems in low noise reception of microwaves', *IRE Transactions, Fifth National Symposium Space Electronics and Telemetry*, pp 1–11.

Ippolito, L.J. (1981). 'Radio propagation for space communication systems', *Proc. IEEE*, Vol. 69, No. 6, June, pp 697–727.

Jasik, H. (ed.) (1961). *Antenna Engineering Handbook*, McGraw-Hill, New York.

Jordon, E.C. and Balamain, K.G. (1968). *Electromagnetic Waves and Radiating Systems*, Prentice-Hall, Englewood Cliffs, New Jersey.

Rudge, A.W., Milne K., Olver, A.D. and Knight, P. (1982). *The Handbook of Antenna Design, Vols I and II*, Peter Peregrinus, Hitchin, Herts.

Shimbukuro, F.J. and Tracey, J.M. (1968). 'Brightness temperature of the quiet sun at centimeter and millimeter wavelengths', *Astrophysics Journal*, Vol. 6, pp 134–139.

Silver, S. (1949). *Microwave Antenna Theory and Design*, McGraw-Hill, New York.

Vuong, X.T., Zimmermann, F.S. and Shimabukuro, T.M. (1988). 'Performance analysis of Ku-band VSAT networks', *IEEE Commun. Magazine*, Vol. 26, No. 5, May, pp 25–33.

5 Modulation

In chapter 4 we learnt the basics of radio link design. The design goal was to achieve a specified carrier-to-noise ratio at the demodulator input. It was mentioned that the carrier-to-noise ratio is related to the demodulated signal quality by a modulation-dependent factor – without in any way attempting to quantify the factor or explain the modulation process.

In this chapter the topic of modulation is discussed. Modulation is the process of imparting baseband information to a carrier. The subject of modulation has received extensive coverage in the literature (e.g. see Taub and Schilling, 1986; Schwartz, 1987) and is a subject in its own right. Here we briefly review at a system level only those aspects of modulation relevant to satellite communications. The basic characteristics of various types of modulation used in satellite communication systems are covered.

5.1 Introduction

A student new to the field of radio communication may be curious to appreciate the need to modulate a signal. Therefore, to begin with, let us examine the fundamental reason for modulating and up-converting baseband signals to RF for transmission. A fundamental requirement for transmitting a signal efficiently into space is that the antenna size must approach the wavelength of the signal. Consider a 10 kHz baseband signal – its wavelength is 30 km! To couple and transmit such a signal directly into space would require an antenna size approaching 30 km – an obviously impractical solution. Hence it becomes necessary to translate the baseband signal to an adequately high-frequency signal where it becomes practical to construct antennas. This process of translating a baseband to a high-frequency signal includes the process of modulation. From a practical viewpoint, it is easier to perform modulation at an intermediate frequency – typically 70 MHz. Satellite communications usually operate at frequencies in the Gigahertz range and therefore the modulated signal is subsequently up-converted to the transmission frequency and amplified before transmission.

Modulation is a process in which some characteristic of a high-frequency carrier is varied in accordance with the baseband signal. Consider a sinusoidal wave $f(t)$:

$$f(t) = A \cos\{2\pi f_c(t) + \theta\} \tag{5.1}$$

Modulation can be achieved by altering the amplitude, A, frequency f_c or phase θ of the wave in accordance with the information. When its amplitude A

117

is varied, the wave is said to be *amplitude modulated*; similarly when the frequency f_c or phase θ is varied, the signal is said to be *frequency* or *phase modulated* respectively. Note that in general a sinusoid is used as the high-frequency wave and hence this type of modulation is also called *continuous wave* or *sinusoidal* modulation. The reverse process of recovering the information signal from the modulated signal is known as *demodulation*.

The term *modulation* is also used in relation to certain types of baseband processing – pulse amplitude modulation and pulse code modulation (see chapter 7). In a more general sense, modulation has been defined as a process in which the information is altered into a form which is more efficient for transmission.

5.2 System consideration

It has already been mentioned that the relationship between carrier-to-noise ratio and baseband signal-to-noise ratio is vital in any system design. A system designer uses the relationship to determine the desired carrier-to-noise ratio for achieving a specified baseband quality.

There are a number of additional system level considerations when studying the applicability of modulation to a given application. These are discussed below.

(a) *Spectral occupancy of the signal*
For efficient use of spectrum, it is essential that the bandwidth of the modulated signal be as small as possible.

(b) *Sensitivity of modulation scheme to signal impairments*
Depending on the type of channel, signal impairments are caused by a number of factors such as thermal noise, inter-modulation noise, multipath noise and signal fading. It is essential that the modulation be robust in the presence of such degradations and at the same time not require transmission of high carrier power. The resistance of the signal to degradation can be increased further by the use of channel *coding* (see chapter 6).

Here we shall discuss the sensitivity of modulation schemes to thermal noise, because satellite communication links are generally limited by this type of noise.

(c) *Hardware complexity*
It is essential that hardware required for modulation and, in particular, demodulation be compatible with the permitted cost and complexity. In general, realization of demodulators is more complex than modulators because of the difficulty in implementing demodulation in the presence of noise and

often the cost constraint imposed by a need for low-cost receivers.

The impact of a modulation scheme on these parameters helps develop basic guidelines and the rationale for selecting a modulation scheme for a specified application. These issues are therefore addressed for each modulation scheme considered in the following sections. Detailed derivation of relationships has not been included unless it is essential for understanding.

The discussion begins with a description of various categories of modulation systems – linear modulation schemes, followed by a non-linear modulation scheme known as frequency modulation, and finally various types of digital modulation schemes. A section discussing issues related to the selection of modulation schemes concludes the chapter.

5.3 Linear modulation schemes

We shall begin by introducing a category of modulation scheme in which the baseband signal is linearly related to the modulated signal. These schemes are sometimes known as *linear modulation*. To date, linear modulation schemes have been only of limited use in satellite communication RF links. However, at least one of the schemes, the single side band (SSB), has a potential application in future satellite systems. Three schemes – amplitude modulation (AM), double side band suppressed carrier (DSB-SC) and SSB – are discussed here. Though not of great practical use for satellite communication at present, an understanding of the AM and DSB-SC schemes is useful to appreciate the principles of SSB systems fully.

(a) Amplitude modulation

As mentioned in the introductory section, a carrier is said to be amplitude modulated when the amplitude of the carrier varies in accordance with the message signal. An amplitude modulated signal may be represented as

$$v(t) = A\{1 + m(t)\}\cos(\omega_c t) \tag{5.2}$$

where $m(t)$ is the message signal and ω_c is the angular frequency of the carrier.

An examination of equation (5.2) shows that the amplitude of the carrier [the term $A\{1 + m(t)\}$] varies in accordance with the signal. Assume $m(t)$ is a sinusoid with amplitude A_m then

$$v(t) = A\{1 + A_m \sin \omega_m(t)\}\cos(\omega_c t) \tag{5.3}$$

The spectral characteristics of $v(t)$ can be obtained by expanding (5.3) and applying trigonometric identity. It can be shown that the spectrum consists of the carrier (f_c), an upper side band ($f_c + f_m$) and a lower side band ($f_c - f_m$). In a more general representation, the upper and lower side bands have the same spectral shape as that of the message signal $m(t)$.

An amplitude modulated signal may be generated either by the use of a non-linear device or by switching the modulating signal via the carrier. One of the main advantages of amplitude modulation is its simple demodulation capacity. An amplitude modulated carrier can be detected simply by the use of an envelope detector. The rectified signal contains message components, together with other undesirable high-frequency components. Such undesirable components, being at a significantly higher frequency than the carrier frequency, are easily filtered out with a low pass filter. A simple *RC* circuit with a large time constant is commonly used.

There is a major limitation in using amplitude modulation for satellite communications. An amplitude modulated system is susceptible to signal fluctuations because the message is superimposed as a variation in amplitude. From our discussions on propagation in chapter 3 we know that RF signals used in satellite communication links suffer various types of amplitude perturbations caused either during propagation or by inherent limitations of equipment.

The consequent degradation to an amplitude signal is unacceptable. Another major difficulty is the requirement of a very high carrier-to-noise ratio to achieve acceptable received message quality. Consequently, amplitude modulation is not in general use in Earth–space satellite links. However, amplitude modulation in the form of 'on–off' keying has been considered advantageous in laser inter-satellite link (ISL) because the scheme can be readily implemented with lasers. Further, in the absence of atmosphere, signal fluctuations in ISLs are minimal.

(b) Double side band suppressed carrier (DSB-SC)

In amplitude modulation the information is carried only in the *side bands* and therefore power in the carrier remains un-utilized. In a *double side band suppressed carrier (DSB-SC)* modulation scheme the carrier is suppressed and only side bands are transmitted. The amplitude of such a wave does not follow the signal amplitude and consequently the inherent simplicity of using envelope detection is lost. A different type of demodulation, known as *synchronous detection*, is used for DSB-SC.

In a synchronous detection scheme the modulated signal is multiplied with a carrier which is synchronized in frequency and phase to the transmitted carrier. Figure 5.1 shows the main elements of this scheme. Consider the demodulation of a DSB-SC modulated carrier. Such a carrier is represented as $m(t)\cos(\omega_c)t$. Referring to figure 5.1, the output of the multiplier is

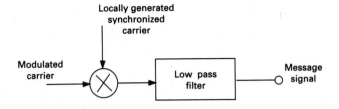

Figure 5.1 Principle of synchronous detection

$$e_o = m(t) \cos^2(\omega_c t) \tag{5.4a}$$

$$= m(t) [\{1 + \cos(2\omega_c t)\}/2]$$

$$= m(t)/2 + m(t)\cos(2\omega_c t)/2 \tag{5.4b}$$

The second term, at twice the carrier frequency, can be easily filtered out to give the desired information signal.

One of the main problems in using this scheme is the difficulty of generating a synchronized carrier at a receiver thousands of kilometres away. An error of $\Delta\theta$ in phase results in a demodulated signal of $m(t)\cos(\Delta\theta(t))$, and a frequency error of Δf results in a demodulated signal of $m(t)\cos(\Delta f)$.

DSB-SC modulation is not used for satellite communications. However, the concept is useful in understanding single side band modulation, discussed next.

(c) Single side band modulation

A variant of amplitude modulation called *single side band (SSB)* modulation exhibits characteristics which make the scheme a potential candidate for applications where bandwidth is at a premium, such as mobile satellite services. In an amplitude modulation system the baseband information is contained in *both* side bands, showing that the transmission contains redundancy. In SSB this redundancy is removed by filtering out one of the side bands at the modulator. The bandwidth of the RF carrier is therefore the same as that of the baseband signal. For example, telephone signals can be transmitted in a bandwidth of 5 kHz. It will be shown later that other types of modulation scheme require considerably more bandwidth and the spectral advantage offered by SSB makes it a potential candidate for frequency-limited applications. The SSB scheme is called a single side band suppressed carrier (SSB-SC) scheme when the carrier is suppressed. The most common application of SSB modulation in satellite communication is to multiplex voice signals into a composite baseband (see chapter 7).

A SSB signal can represented by

$$e_o(t) = m(t) \cos(\omega_c t) - \bar{m}(t)\sin(\omega_c t) \qquad (5.5)$$

where $\bar{m}(t)$ is a signal derived by shifting all the frequency components of $m(t)$ by 90°. The term $\bar{m}(t)$ is also called the *Hilbert transform* of $m(t)$. Equation (5.5) forms a basis for generating SSB signals.

As the amplitude of a SSB modulated carrier is constant, the inherent simplicity of the process of demodulating amplitude modulated signals is not possible in the case of SSB modulated signals. A SSB modulated signal is recovered by using *synchronous detection*.

The local carrier for synchronous demodulation of SSB systems can be generated either by an open-loop or a closed-loop scheme. In an open-loop system, receivers use a highly stable local oscillator such that the maximum frequency error is <10–30 Hz. In a closed-loop system, a low level pilot is transmitted together with the signal. The pilot is recovered at the receiver. Errors in phase ($\Delta\theta$) and frequency ($\Delta\omega$) of the recovered carrier give the demodulated signal terms $\cos(\omega t - \Delta\theta)$ and $\cos(\omega t + \Delta\omega)$ respectively. The human ear is rather insensitive to phase error but sensitive to frequency error. Therefore when using SSB transmission for music or voice it is essential to retain frequency errors <30 Hz, beyond which the errors become discernible.

SSB for satellite communication

There are two aspects considered in assessing the suitability of SSB for satellite communication – the required carrier-to-noise ratio and the occupied bandwidth. We shall consider each in the following section, and use the knowledge to assess the feasibility of SSB for satellite communication.

Effects of noise

In a real environment the received signal is always corrupted by noise. It is therefore necessary to develop a suitable model for estimating the effects of noise on a demodulated signal.

When the input signal to the multiplier (see figure 5.1) is contaminated by gaussian noise with a power spectral density of N_0, the signal-to-noise ratio at the output of a SSB-SC demodulator for a sinusoidal modulating signal is given by

$$S/N = (C_i/N_0 f_m) \qquad (5.6)$$

where C_i is input carrier power
        N_0 is input noise power spectral density
        f_m is maximum frequency of the modulating signal.
Signal-to-noise ratio of the order of 30–40 dB is necessary to provide ad-

equate signal quality for good-quality telephony. From equation (5.6) it is noted that the input carrier-to-noise ratio should also be of the same magnitude. However a typical satellite communication link can economically provide carrier-to-noise ratios of the order of 10–12 dB, making SSB transmissions inefficient from power considerations. This disadvantage can be offset to a large extent by the use of compandors (see chapter 7) which offer a signal-to-noise ratio advantage of ~15–20 dB. With this advantage the SSB transmissions begin to appear attractive. Such a scheme is called Amplitude Companded Single Side Band (ACSSB) transmission.

Occupied bandwidth

The occupied RF bandwidth of a SSB transmission is the same as the baseband bandwidth. Hence a 4–5 kHz RF bandwidth is adequate for a single telephone channel transmission. Compare this with the typical bandwidth necessary for other schemes – 30 kHz for frequency modulation or 20 kHz using offset quadrature phase shift (O-QPSK) keying modulation. Both the FM and the O-QPSK schemes are widely used in satellite communications and are discussed later in this section. A significant bandwidth advantage is evident in using SSB.

Bandwidth efficiency is essential for mobile satellite service and for this reason ACSSB has been considered favourably for such applications.

5.4 Frequency modulation

Frequency modulation (FM) systems are extensively used in satellite communications. Examples of FM applications are multiplexed telephony, single channel per carrier systems and television broadcast/distribution. Frequency modulation systems are well suited for those cases where the baseband signal is in analog form. The scheme also offers advantages for transmission of digital data in applications where simple receivers are essential. An example application is digital transmissions of INMARSAT's paging system. An important requirement of a paging system is the need for simple, low-cost and rugged receivers.

In a frequency modulation system, the frequency of a carrier is altered in accordance with the information signal. Frequency modulation belongs to a more general class of modulation called *angle modulation* represented by

$$e_c(t) = A \cos[\omega_c t + \theta(t)] \tag{5.7}$$

When the modulating signal $m(t)$ changes the phase $\theta(t)$ of the signal, the wave is said to be *phase modulated*. Recall that the derivative of phase, $d\theta/dt$, is the angular frequency of the wave. Therefore a *frequency modulated* wave is produced when the derivative of the phase angle $\theta(t)$ in (5.7) is changed in accordance with $m(t)$.

Consider $m(t)$ as a sinusoid, $\beta\sin(\omega_m t)$. Differenting $m(t)$ and substituting in (5.7) we obtain the equation of a frequency modulated wave as

$$e_c(t) = A \cos[\omega_c t + \beta\omega_m\cos(\omega_m t)] \tag{5.8}$$

Instantaneous frequency of the wave represented by (5.8) is

$$f_i = \underset{Carrier\ term}{f_c} + \underset{Modulation\ term}{\beta f_m\cos(\omega_m t)} \tag{5.9}$$

The maximum variation in frequency Δf_m is βf_m

$$\Delta f_m = \beta f_m \tag{5.10a}$$

or

$$\beta = \frac{\Delta f_m}{f_m} \tag{5.10b}$$

β is known as the modulation index of the frequency modulated carrier.

The non-linearity prevalent in the process (the component $\beta\omega_m\cos(\omega_m t)$ in equation (5.9)) gives rise to a number of side bands spaced symmetrically around the carrier. This is unlike the linear modulation system which has only two symmetrically spaced side bands. On expanding equation (5.8) we obtain

$$v(t) = J_0(\beta) \cos(\omega_c t) - J_1(\beta)[\cos(\omega_c - \omega_m)t -$$
$$\cos(\omega_c + \omega_m)t] + J_2(\beta)[\cos(\omega_c - 2\omega_m)t +$$
$$\cos(\omega_c + 2\omega_m)t] - \ldots \tag{5.11}$$

where $J_n(\beta)$ is a Bessel function of the first kind and order n.

Let us examine equation (5.11) more closely as it provides valuable insight into the frequency modulation process. It is noted that magnitude of each side band (say the nth) is weighted by a Bessel function of order $n - J_n(\beta)$. The magnitude of Bessel functions as a function of β have the following characteristics:

1. For low values of β the power is contained in only a limited number of side bands – the number of side bands increases with modulation index. For example when $\beta = 1$, only the first two side bands are significant whereas for $\beta = 5$, up to 8 side bands may have to be considered.

2. The total power of the carrier remains constant and therefore as the modulation index is increased the power is distributed between the carrier and the side bands.
3. An interesting property is that the carrier disappears when the modulation index is 2.4. This is known as the *first Bessel zero*. This property of FM can be utilized to adjust the carrier deviation in a FM modulator.

Example 5.1

Consider a television system with a peak deviation of 6 MHz and the maximum modulating frequency of 4.5 MHz. Determine the frequency of the first Bessel zero.

Solution

The relationship between modulation index, peak deviation and the carrier frequency is given by (5.10). Substituting into this equation

$$2.4 = 6/(f_B)$$

where f_B = frequency at which the first Bessel zero occurs
$f_B = 2.5$ MHz.
Now assume that the nominal voltage input to the modulator at 2.5 MHz is known, then the deviation of the modulator can be adjusted so that the carrier at the output of the modulator disappears when a 2.5 MHz baseband signal is applied to the modulator at the nominal level. A spectrum analyser or a selective level meter is convenient for such an adjustment.

Finally we observe (from equation 5.11) that although the theoretical bandwidth (i.e. the number of side bands) extends up to infinity, most of the power in a modulated signal is contained within a finite bandwidth.

Bandwidth of a frequency modulated wave

For studying the performance of FM systems a quantitative definition of the occupied bandwidth becomes essential. For practical purposes the received signal distortion is acceptable provided that 98% of the energy is contained in the modulated spectrum. It can be shown that for a sinusoidal baseband, this condition is satisfied when

$$B = 2(\Delta f + f_m) \qquad (5.12)$$

where Δf is the peak carrier deviation and f_m is the frequency of the sinusoid.

The bandwidth obtained by using (5.12) is called *Carson's bandwidth.*
Thus in the television system of example 5.1 the required transmission
bandwidth using Carson's formula is 2(6 + 4.5) or 21 MHz.

Equation (5.12) gives the bandwidth of a signal modulated with a sinusoid
but in practice the message signal is more complex. The power spectral density
of many message signals encountered in practice can be approximated as
gaussian-shaped. The FM waveform of such a signal is also gaussian and the
bandwidth B containing 98% of the side band is (Taub and Schilling, 1986)

$$B = 4.6(\Delta f_{rms}) \qquad (5.13)$$

where Δf_{rms} is the standard deviation of the power spectral density.

When the spectral shape of the signal is unknown, the occupied bandwidth can be estimated by using the largest frequency of significant power
in the message signal as f_m, when applying Carsons's rule.

Effects of noise

In this section, effects of noise on frequency modulated signals and the related system issues are discussed.

Noise characteristics at the output of a FM demodulator

To begin with, we shall examine the characteristics of noise at the output of
a FM demodulator. The power spectral density of noise at this point is given by

$$G(f) = \frac{\alpha^2 \omega^2}{A^2} N_0 \qquad (5.14)$$

where A = amplitude of RF carrier
α = a constant
ω = angular frequency of the baseband
N_0 = noise spectral density at the input.

From (5.14) it can be seen that the noise power spectral density (PSD) at
the output of a FM detector is proportional to f^2 (where f = frequency). A
PSD plot of such a noise is *parabolic* in shape and therefore $G(f)$ is often
referred to as parabolic noise. If the response of the post-detection filter is
chosen as inversely shaped (i.e. its amplitude response varies as $1/f^2$) the
mean value of the received noise at the output decreases. However, this
type of response also causes attenuation to higher-frequency components of
the message waveform. Now consider that the spectral density of the message signal at the transmitter is shaped as f^2 and an inverse parabolic filter is

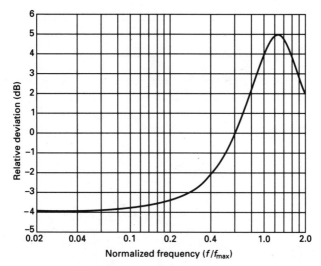

Figure 5.2   Pre-emphasis characteristic recommended by the CCIR for frequency division multiplexed telephony. f_{max} is the highest telephone channel baseband frequency of the system (CCIR Rec 464–1).

used at the output of the demodulator. With such an arrangement the frequency components of the output message signals are unaffected and at the same time the mean value of the noise is reduced. This concept is universally used in FM satellite communication systems. The process of shaping the transmitted waveform is called *pre-emphasis* and its inverse operation at the receiver, *de-emphasis*. To facilitate interworking between national and international systems the CCIR recommends (CCIR, 1982) specific pre-emphasis/de-emphasis filter characteristics for television and frequency division multiplexed carriers. As an example, figure 5.2 (CCIR Rec 464–1) shows the pre-emphasis characteristics for a frequency division multiplexed (FDM) telephone channel (see chapter 7). Note that the attenuation (Y-axis) of the pre-emphasis curve is given relative to a reference frequency of $0.6132f_{max}$ (f_{max} = highest frequency). The cross-over frequency can therefore be used as the test tone frequency for setting the FM deviation, since a test tone remains unaffected by the pre-emphasis circuit. To use any other frequency as a test tone for modulator adjustments, the pre-emphasis circuit must be removed from the path to avoid attenuation introduced by the circuit.

   In a FDM system, several telephone channels are multiplexed (i.e. stacked in frequency) and the telephone channel at the highest frequency suffers maximal noise contamination. By using the pre-emphasis/de-emphasis network of figure 5.2, the magnitude of noise in the highest channel is reduced significantly. An advantage of about 4 dB is achieved in this channel by using a pre-emphasis/de-emphasis circuit.

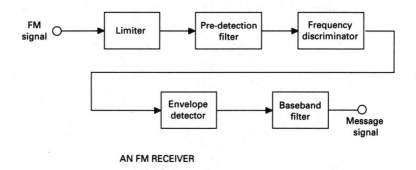

Figure 5.3   A block diagram of the FM detection process.

FM equation

Figure 5.3 shows the block diagram of the FM detection process.

The relationship between the input and output signal-to-noise ratio of such a receiver is known as the FM equation. For a condition $C_i \gg N_i$ and an arbitrary message signal $m(t)$ the relationship is (Taub and Schilling, 1986)

$$\frac{S}{N} = \left(\frac{3k^2}{4\pi^2}\right)\left(\frac{m^2(t)}{f_m^2}\right)\left(\frac{A^2/2}{N_0 f_m}\right)$$

$$= \alpha\left(\frac{m^2(t)}{f_m^2}\right)\left(\frac{C}{N}\right) \qquad\qquad (5.15)$$

where C = carrier power
N_0 = noise power density
A = carrier amplitude
f_m = bandwidth of the *message* signal
k = constants associated with detection hardware
α = constant.

Notice that here N is the noise power in the message bandwidth, however it is usual to specify C/N in the occupied RF bandwidth when dealing with satellite communication system.

The relationship is shown in figure 5.4. Notice that the output signal-to-noise ratio shows a rather abrupt degradation beyond a threshold. This sudden loss of output signal quality is known as the *threshold effect* and is discussed in detail in a later section. Equation (5.15) applies to operation above the threshold.

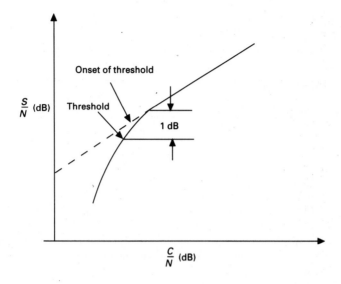

Figure 5.4   The relationship between the input carrier-to-noise ratio (C/N) and the output signal-to-noise ratio (S/N) of a FM demodulator.

For a sinusoidal modulating signal equation with a peak deviation of Δf, (5.15) can be simplified into the form

$$\frac{S}{N} = \frac{3}{2}\left(\frac{\Delta f}{f_m}\right)^2 \frac{C}{N_0 f_m} = \frac{3}{2}\frac{\Delta f^2}{f_m^3} \cdot \frac{C}{N_0}$$

$$= \frac{3}{2}\beta^2\frac{C}{N} \tag{5.16a}$$

and when baseband extends from f_1 to f_2

$$\frac{S}{N} = \frac{3}{2}\frac{\Delta f^2}{(f_2^3 - f_1^3)} \cdot \frac{C}{N_0} \tag{5.16b}$$

Notice from (5.16a) that the output signal-to-noise ratio improves by a factor $\frac{3}{2}\beta^2$ over the input carrier-to-noise ratio. This improvement is called *FM improvement*. Examining equation (5.16), it can further be observed that it is possible to trade off bandwidth with carrier power in achieving a specific signal-to-noise ratio. This trade-off is of great utility in satellite links. For example, in power limited satellite links the received C/N is low and to achieve acceptable signal-to-noise ratio it is customery to use a large-frequency deviation. A further improvement in signal-to-noise ratio is made possible using pre-emphasis and de-emphasis. The perceived signal-to-noise ratio to a user is improved a step further, because of the non-uniform perception of noise by human ears (for voice transmission) or eyes (for television transmission). This perceived improvement is called *weighting* advantage. Several types of weighting characteristic have been proposed. The weighting characteristic recommended by the CCIR is called *psophometric weighting*. After incorporating these advantages, the FM equation modifies to

$$(S/N)p.w$$

where p is the pre-emphasis advantage and w the weighting advantage.

Equation set (5.16) gives a general form of the FM equation. Depending on the application, this equation can be expressed in several ways. In each case the equation is simplified according to the specific characteristic of the baseband and approximations applied to the equation where possible. Consequently, the resulting equations all appear different. We shall consider television, FDM and SCPC systems, all of which use FM.

For television, the signal-to-noise ratio is given in the form of the peak-to-peak amplitude of the luminance signal-to-rms noise voltage. Equation set (5.16) is accordingly modified to

$$\frac{S}{N} = \frac{3}{2}\left(r\frac{\Delta f}{f_m}\right)^2 \frac{C}{N}\frac{B_N}{f_m} W_T P_T \qquad (5.17a)$$

or expressed in logarithmic form

$$\left(\frac{S}{N}\right)dB = \frac{C}{N} + 20\log\left(r\frac{\Delta f}{f_m}\right) + 10\log\left(\frac{B_N}{f_m}\right) +$$

$$W_T\ (dB) + P_T\ (dB) + 1.76 \qquad (5.17b)$$

where $\quad r\ = V_{pl}/V_{pc}$
V_{pl} = peak-to-peak luminance signal
V_{pc} = composite television signal including synchronization pulse
B_N = RF bandwidth
P_T and W_T are pre-emphasis and weighting advantages respectively.

The pre-emphasis/de-emphasis advantage in television transmission depends on the television standard. The advantage is specified at a given frequency (15 kHz in CCIR recommendations) varying from 2 to 3.1 dB for various television systems (CCIR Rec 568; CCIR, 1982). The advantage for B, C, G and H (used in UK) systems is 2 dB. Weighting advantage for television also depends on the standard used, varying from 9.9 to 15.8 dB. For example, the weighting advantage of 625 lines B, C, G and H systems is 14.3 dB.

Now consider a frequency division multiplexed telephony baseband. In the example considered above, it has been mentioned that the highest FDM telephony channel (i.e. the telephony channel stacked on top of the multiplexed baseband) suffers the largest noise contamination because of the parabolic noise characteristics of FM. Equation (5.16) is accordingly approximated and simplified to give the S/N ratio in the worst channel as

$$\frac{S}{N} = \frac{C}{N} \cdot \frac{B_N}{b} \left(\frac{\Delta f_{rms}}{f_m}\right)^2 W_t P_t \tag{5.18a}$$

or expressed logarithmically

$$\left(\frac{S}{N}\right) dB = \frac{C}{N} + 20 \log \left(\frac{\Delta f_{rms}}{f_m}\right) + 10 \log \left(\frac{B_N}{b}\right) + W_t + P_t \tag{5.18b}$$

where b = bandwidth of the top telephone channel
f_m = maximal frequency of the multiplexed baseband.
Notice that deviation is given here as the rms value.

The pre-emphasis/de-emphasis advantage in the worst affected channel in FDM telephony is 4 dB and the weighting advantage is 2.5 dB.

In single channel per carrier systems the baseband consists only of a single channel extending from f_1 to f_2. The FM equation is therefore modified accordingly and the resulting FM equation for SCPC is

$$\frac{S}{N} = \frac{3}{2} \frac{\Delta f^2}{(f_2^3 - f_1^3)} \frac{C}{N} B_N W_s P_s C_a \tag{5.19a}$$

where W_s = weighting advantage
P_s = pre-emphasis advantage
C_a = companding advantage.
Expressing logarithmically

$$\left(\frac{S}{N}\right) dB = \frac{C}{N} + 20 \log \left(\frac{\Delta f}{f_2^3 - f_1^3}\right) + 10 \log (B_N) +$$

$$W_s + P_s + C_a + 1.76 \tag{5.19b}$$

Companding is often used with SCPC systems and therefore we obtain a companding advantage C_a in addition to the pre-emphasis and weighting advantages (see section 7.2 for an explanation of companding). Companding advantages range from 15 to 20 dB. Pre-emphasis improvement is taken as 6.3 dB and psophometric weighting advantage as 2.5 dB.

Channel loading with voice baseband

The design of a FM link involves the adjustment of the C/N and modulation index to achieve an optimum trade-off. A major problem when considering a voice channel is its noise-like amplitude characteristics. Such a characteristic makes it difficult to estimate the baseband speech power needed to obtain the frequency deviation and the occupied bandwidth of the RF signal; and adjustments of frequency deviation of modulators so as to achieve the desired deviation under operational condition. If the speech signals are under-deviated the signal-to-noise ratio will be lower than expected, whereas an over-deviation causes a distortion of speech signal. The problem of determining channel loading in a FDM system will therefore be considered in more detail.

To resolve the difficulty of adjusting the frequency deviation of a FM modulator, a deviation corresponding to a defined level (usually 0 dBm) sinusoidal test tone is specified. Obviously, the level of the test tone must have a relationship with the level of the actual speech, otherwise speech signals may over (or under) deviate the carrier. To convert the test-tone deviation into a multichannel deviation, formulas, based on extensive measurements, have been developed by several national (e.g. Bell Systems in the USA) and international bodies (CCIR). The CCIR defines a parameter called noise loading ratio (NLR). The noise loading ratio depends on the number of baseband speech channels as follows:

$$20 \log (\text{NLR}) = [-1 + 4\log(N)] \quad \text{dB}, \quad N < 240 \text{ channels} \quad (5.20a)$$

$$20 \log (\text{NLR}) = [-15 + 10\log(N)] \quad \text{dB}, \quad N > 240 \text{ channels} \quad (5.20b)$$

The multichannel rms deviation Δf_m is obtained by multiplying the test-tone rms deviation Δf_t by the noise loading ratio, NLR:

$$\Delta f_m = \text{NLR} \cdot \Delta f_t$$

To obtain the occupied bandwidth of the FDM signal we need to convert the multichannel rms deviation to the peak deviation. The multichannel rms deviation, Δf_m, is related to the peak deviation Δf_p by a *multichannel peak factor*, G, which is a measure of the 'peakiness' of the FDM signal. For a given number of active channels at a constant volume, the multichannel peak factor is defined as the ratio of overload voltage to rms voltage. The peak deviation of a multichannel FDM is then

$$\Delta f_p = G \, \Delta f_m \quad (5.21a)$$

$$= G \cdot \text{NLR} \cdot \Delta f_t \quad (5.21b)$$

Using Carson's rule, we finally obtain the multichannel bandwidth as

$$B_{rf} = 2[\Delta f_p + f_m]$$
$$= 2[G \Delta f_m + f_m]$$
$$B_{rf} = 2[G . NLR . \Delta f_t + f_m] \tag{5.22}$$

When the number of telephony channels is large (typically $N > 24$) the channel peak factor is about 3.16 – around 99.8% of signals being within 3.16 of the rms signal level. However the peak factor increases as the number of channels reduces; values of 6.5–8.5 have been used. The highest peak factor is observed in single channel per carrier (SCPC) systems. The peakiness in SCPC systems is reduced by *companding* (see chapter 7). Note that when the signal is peaky, the required RF bandwidth increases whereas the *mean deviation* of the signal remains low, resulting in a lower average signal-to-noise ratio.

Example 5.2

Determine the multichannel rms deviation and the RF bandwidth in a satellite transponder for a 252-channel frequency division multiplexed system given

rms deviation of 0 dBm0 test tone = 358 kHz
Multichannel peak factor = 3.2
Maximum baseband frequency = 1052 kHz
RF guard band = 10% of the occupied bandwidth.

Solution

Noting that number of channels is >250, we use equation (5.20b). Substituting the value of N:

$$20 \log(NLR) = -15 + 10 \log(252);$$

or

$$NLR = 2.82$$

The multichannel rms deviation is then 2.82×358 kHz or 1.01 MHz.
To convert the rms deviation to the peak deviation the value of $G = 3.2$ is substituted into equation (5.21a), giving the peak deviation value as 3230.6 kHz.

The occupied bandwidth can be obtained by using (5.22). Substituting values:

$$B_{rf} = 2(3230.6 + 1052)$$
$$= 8.565 \text{ MHz}$$

Adding the 10% guard band, the RF bandwidth is finally obtained as 9.42 MHz.

We could also begin by allocating the RF bandwidth for a specified number of channels and thereby obtain the required test-tone deviation using (5.22). The signal-to-noise ratio in the worst channel can then be obtained by using any one of equation set (5.18). However, re-adjustment of test-tone deviation/RF bandwidth would be necessary if the signal-to-noise ratio is unacceptable.

It should be noted that in calculating the *S/N* of a FDM signal (equation set 5.18) we use the frequency deviation of the test tone, but the RF bandwidth is obtained under speech loading, following the technique outlined above.

Threshold effect in FM

Recall that the FM equation is valid for $C_i \gg N_i$. A plot of carrier-to-noise ratio against output signal-to-noise ratio is shown in figure 5.4. We have already remarked that the performance of a FM demodualator deviates rapidly from the linear relationship (in dB) below a specific carrier-to-noise ratio. The *FM threshold* is defined arbitrarily as the magnitude of input carrier-to-noise ratio where the deviation between the extrapolated and the actual output signal-to-noise ratios is 1 dB (see figure 5.4). This point is rather abrupt and in practice can quite easily be estimated subjectively. It manifests as clicks in voice signals or as dots moving randomly on television pictures, often resembling 'flying pigeons'. An experienced user can subjectively estimate the threshold quite accurately by viewing the television picture.

The mechanism of spike generation can be explained as follows. At high carrier-to-noise ratio the amplitude of noise follows a gaussian distribution which implies an 'even' distribution of noise. As the carrier level is reduced, its magnitude and phase become increasingly influenced by noise, causing large random 'uneven' fluctuations in the amplitude but more importantly in the *phase*. Recall that the FM demodulator produces a voltage proportional to $d\theta/dt$ at its output. It can be shown mathematically that whenever a carrier vector rotates by more than 2π radians as a result of noise, the demodulator produces a noise spike of considerable energy (Taub and Schilling, 1986). As the carrier-to-noise ratio is progressively reduced, rotation of carrier vector due to noise increases until eventually the effect becomes dominant causing the threshold effect. It can be shown that the onset of threshold occurs at larger *C/N* as the modulation index of the carrier is increased.

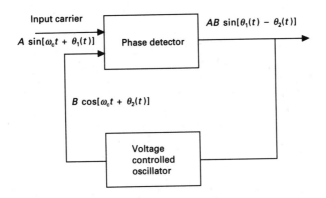

Figure 5.5 The main elements of a phase lock loop.

In an earlier section we noted that one of the main attributes of FM is the large improvement in signal-to-noise ratio made possible by using a large modulation index. However, the threshold phenomenon limits the potential of FM precisely when needed most, i.e. at low carrier-to-noise ratio. Hence a number of techniques have been developed for improving the FM threshold to obtain the maximum benefit of FM. The *phase locked loop* and FM demodulator using *feedback* are the two commonly used techniques.

Figure 5.5 shows the main elements of a phase locked loop. It consists of a phase detector and a voltage controlled oscillator (VCO). The phase detector produces a voltage proportional to the phase difference between the incoming carrier and a carrier generated by a VCO. The frequency of the VCO is controlled by the output voltage of the phase detector fed back to the VCO. When used as a FM demodulator, the phase detector is adjusted to output zero volt when the received carrier is unmodulated. When the incoming carrier is frequency modulated, the phase detector produces a voltage proportional to the difference in phase between the VCO and the carrier. This voltage is in fact $d\theta/dt$ or $m(t)$, the demodulated message signal. The effect of impulse noise is reduced by this circuit because of a dampening caused by the feedback. It is possible to extend the FM threshold by at least 2–3 dB by using this technique.

The principle of *FM demodulator using feedback* is shown in figure 5.6. The technique uses a VCO and a frequency multiplier together with a FM demodulator. The input to the frequency multiplier consists of the incoming FM signal and a carrier generated by the VCO. The frequency of the VCO carrier is controlled by a feedback from the output of the FM detector. The bandpass filter at the output of the frequency multiplier is centred to pass the message signal. This signal is then fed into a conventional FM discriminator. The outputs of the discriminator are fed to the baseband filter to retrieve the message signal in addition to the VCO. Here the feedback loop

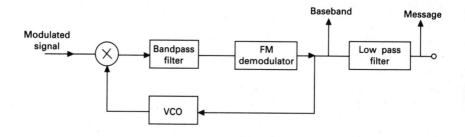

Figure 5.6   A block diagram of a FM demodulator using feedback.

to the VCO compresses the modulation index of the FM signal by a factor of $1/(1 + \alpha G)$, where α and G are constants depending on the discriminator and the VCO characteristics. Consequently the effective bandwidth of the FM signal is compressed and the bandwidth of the bandpass filter can therefore be made less than the bandwidth of the actual carrier. The carrier-to-noise ratio at the output of the bandpass filter remains the same as before, because carrier and noise are both affected equally. However it can be shown that the circuit reduces the total average number of noise spikes which are the basic cause of the threshold phenomena, thus extending the threshold. Using this technique, several dB of threshold extension can be achieved.

Effect of group delay on FM signals

In an ideal environment all the frequency components of a signal arrive at the destination at the same time. In a real environment the RF signal goes through non-ideal components such as filters and as a result a frequency-dependent transmission delay is introduced. This type of delay is measured in terms of group delay. Group delay is defined as $d\theta/d\omega$ where θ and ω are the phase and angular frequency of the wave respectively. Envelope delay is the steady-state delay of the envelope of a modulated wave, given as $d\theta(\omega)/d\omega$ where $\theta(\omega)$ is the phase characteristic. Most group delay measurements in fact produce the envelope delay. Group delay introduces distortion into a frequency modulated wave because the frequency components of the wave are delayed asymmetrically with respect to each other.

The effect of group delay can be minimized by introducing group delay equalizers before demodulation. Equalizers introduce an opposite group delay to that on the link, and thereby nullify the net group delay effect (see Calvert, 1984a, b for more details).

Energy dispersal

Most satellite systems share frequency bands either with other satellite systems or with terrestrial systems. For a given transmitted power, the power spectral density of RF signals increases as bandwidth is reduced. Therefore under lightly loaded or no-load conditions, the power spectral density of RF signals is bound to increase, causing a higher interference to other services sharing the band. To minimize such effects the CCIR recommends that under low traffic load conditions the energy of the RF signal be spread. Services in the FSS and BSS use *energy dispersal* baseband signals, which are introduced under lightly loaded traffic. Sawtooth waveform is the most commonly used. The aim is to match the transmitted spectral density under lightly loaded conditions to the fully loaded condition.

5.5 Digital modulation schemes

Digital signals use the same principle to modulate a carrier as do analog signals. Amplitude, phase and frequency modulation schemes are all applicable to digital modulation. The digital equivalents of these modulation schemes are known as Amplitude Shift Keying (ASK), Frequency Shift Keying (FSK) and Phase Shift Keying (PSK). Additionally, some modulation schemes have been developed specifically to optimize digital modulation. These are hybrid phase/amplitude schemes called Quadrature Amplitude Modulation (QAM). In satellite communications, PSK is the most commonly used modulation scheme. FSK has also been used in certain applications where receiver simplicity is essential. ASK schemes are not generally used in Earth–space links because of the uncertainties associated with the amplitude of the received signal. Here we will confine the discussions to PSK and FSK.

(a) Phase shift keying

Basic concepts

In a phase shift keying (PSK) system the phase of the carrier is changed in accordance with the baseband digital stream. Digital data are characterized by discrete level changes and therefore the phase of the modulated wave also changes in discrete steps. A general form of expression of a PSK scheme is

$$v(t) = A \cos(\omega_0 t + \phi_m) \tag{5.23}$$

where ϕ_m is the phase angle varied in accordance with the information signal. ϕ_m is defined as

$$\phi_m = (2m + 1)\pi/M \qquad (5.24)$$

where m assumes values from 0 to $(M - 1)$.

It is possible to group several bits of information as a *symbol*; when N information bits are combined together, there is a possibility of $M = 2^N$ phase states. When $N = 1$, giving $M = 2$: m takes values of 0 and 1. Substituting these values of M and m into equation (5.24), ϕ_m may assume values of $\pi/2$ and $3\pi/2$, i.e. the carrier phase is changed by 180° for each bit change. This type of modulation is known as *bi-phase shift keying* (or *BPSK*).

Extending this concept further, if two bits are combined as a symbol then there are four possible phase states corresponding to 0°, 90°, 180° and 270°. Such a scheme is a four or *quadrature phase shift keying* (or *QPSK*) modulation scheme. Thus a carrier with 0° change in phase relative to the unmodulated carrier may correspond to a bit pattern of 0,0; 90° phase to 0,1 and so on.

In a more general sense, when N baseband *bits* are combined to give M carrier phase states, such a scheme is called an *M-ary PSK*.

For a general case of an *M*-ary PSK, the relationship between symbol rate R_s *bauds* (unit for symbols/s) and the baseband bit rate R_b *bits per second* is

$$R_s = R_b/\log_2 M \text{ bauds} \qquad (5.25)$$

For example, in a BPSK system $M = 2$, so the baseband bit rate and the symbol rates are the same. Whereas in a QPSK system $M = 4$ and therefore the symbol rate is *half* the baseband bit rate. Similarly, if the baseband bit duration is T_b, the symbol duration is *increased* to $(T_b \log_2 M)$.

Equation (5.23) is an expression of an *M*-ary PSK. Expanding (5.23) using standard trignometric identity we get

$$V(t) = (A \cos \phi_m) \cos \omega_0 t - (\beta \sin \phi_m) \sin \omega_0 t \qquad (5.26)$$

From equation (5.26) it can be assessed that Mth-order PSK can be generated by substituting the appropriate value of ϕ_m into equation (5.26). This principle is used in generating PSK modulation.

It is also possible to vary carrier phase *and* amplitude compositely in accordance with the baseband signal. The resulting modulation is called a *quadrature amplitude modulation* (or *QAM*) scheme. Thus a M-QAM scheme has M possible states in a vector *I–Q* plane (discussed later).

Note that QAM schemes carry useful information in the signal phase *and* amplitude. Thus a QAM signal is susceptible to amplitude fluctuations caused by noise (e.g. signal fade caused by rain etc. or obstructions such as trees etc. in MSS) and amplitude non-linearities in the system. RF signals in a satellite system generally suffer amplitude fluctuations and pass through several non-linear stages. Therefore, QAM systems are not yet considered favourably for satellite communication systems. We shall therefore not consider such schemes in detail here.

PSK modulation schemes can be conveniently represented in an *I–Q* plane which consists of *I* and *Q* axes normal to each other. A signal state can be represented as a point in such a coordinate system. The *I*-axis may be considered to represent the channel which is in phase with the modulating carrier and the *Q* axis the channel which is in quadrature to the modulating carrier (i.e. the modulating carrier phase shifted by 90°). The transmitted carrier vector can be obtained by joining the origin to the specified point on the plane. Figures 5.7(a) to (d) show respectively the BPSK, QPSK, 8-phase-PSK and 16-QAM schemes in the *I–Q* plane. A BPSK signal consists of two phase states, 0° and 180°, on the *I* (or *Q*) channel and a QPSK signal consists of four phase states corresponding to bit patterns 00, 01, 10 and 11. Note that the carrier vector has a constant amplitude in BPSK schemes. In a 16-QAM scheme, in addition to the phase change, the amplitude change is also permitted. Thus its *I–Q* plane representation shows a 16-state grid which permits both carrier amplitude and phase change.

In general, as the order *M* of the phase modulation increases, the required RF bandwidth decreases but the signal becomes more susceptible to noise because the phase difference between adjacent permitted phase states decreases. For example, in a BPSK system the minimum phase shift $\Delta\phi$ between state change (0 or 1) is 180° whereas for QPSK, $\Delta\phi$ reduces to 90°. Therefore a QPSK scheme is more likely to be affected by noise spikes. One way to mitigate the effect of noise is to increase the transmitted carrier level. However, it is not always sufficient to increase the transmitted power. An additional consideration is the feasibility of implementing demodulators. Generally demodulator complexity increases as the order of modulation increases. A compromise has to be exercised between an increase in the transmitted power, demodulator complexity and the RF bandwidth reduction. Satellite communication systems to date have inherently been power limited – operating at low carrier-to-noise ratio – and therefore higher-order PSK schemes, even though offering higher spectral efficiency, have not been considered favourably. At present QPSK is the highest-order phase modulation scheme capable of providing a favourable trade-off between satellite EIRP and bandwidth. However, with an increasing scarcity of RF spectrum, there is growing interest in the use of higher-order PSK schemes.

Spectral efficiency

In the above discussions, we have noted that the occupied RF bandwidth reduces as the order of PSK is increased, without attempting to quantify the advantage. In this section the spectral efficiency of various schemes will be compared more precisely. For this purpose, we define an *efficiency factor* (η) as the number of user bits transmitted/Hz:

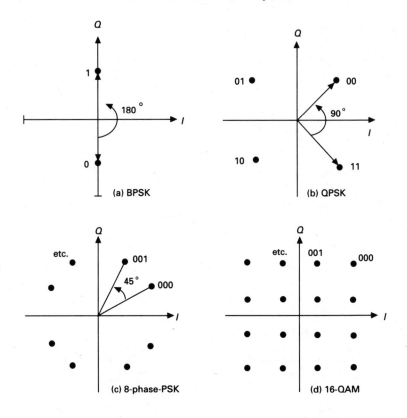

Figure 5.7   Representation of PSK signals in the *I–Q* plane.

$$\eta = R_b/B_{rf} \tag{5.27}$$

where R_b = user bit rate
       B_{rf} = RF bandwidth.
Expressed in terms of number of symbol rate R_s and phase state *M*:

$$\eta = R_s(\log_2 M)/B_{rf}$$
$$= (\log_2 M)/B_{rf} T_s \tag{5.28}$$

Theoretically, the maximum efficiency is achieved by making $B_{rf} T_s = 1$. Let us then compare the theoretical spectral efficiency of various PSK schemes. Substituting values of *R* and *M*, we obtain the following efficiencies:

BPSK: 1 bit/s/Hz
QPSK: 2 bit/s/Hz
8-phase-PSK: 3 bit/s/Hz

In practice the product $B_{rf}T_s$ is $(1 + r)$, where r is the 'roll-off' factor discussed in chapter 7 – typically 0.4–0.5 in practice. Assuming α to be 0.4, the corresponding efficiencies for BPSK and QPSK reduce to 0.7 and 1.4 respectively.

Modulators for PSK schemes

All the PSK schemes may be generated using balanced amplitude modulators which consist of the configuration shown in figure 5.8. The input digital stream is split into two channels called an in-phase (I) and quadrature phase (Q) channel. The digital-to-analog converter (D/A) and low pass filter (LPF) convert the digital stream into an analog form which is fed into a multiplier, the other input of which consists of the in-phase or quadrature carrier. The multiplier output contains the desired term ($A \cos \phi_m \cos \omega_c t$) in the I channel and ($B \sin \phi_m \sin \omega_c t$) in the Q channel. Additionally, there are some undesired terms (carrier term and message term). The I and Q signals are then summed, giving the modulated signal as shown in equation (5.26). The summing process also cancels the undesired components.

Demodulation of PSK signals

PSK modulated signals are recovered by *coherent demodulation*. The principle of coherent demodulation is illustrated in figure 5.9 for a BPSK scheme. The circuit consists of a pre-detection bandpass filter, a carrier recovery loop, a multiplier and the bit extractor. This is followed by a timing recovery circuit and an analog-to-digital converter.

The received signal can be represented by

$$V_{bpsk}(t) = Aa(t) \cos(\omega_c t + \phi) + n(t) \tag{5.29}$$

where $a(t)$ is the user bit
          ϕ is an arbitrary phase component introduced by the propagation channel delay and phase delay introduced by various components
          $n(t)$ is the system noise
          A is an arbitrary constant depending on the characteristics of the receiving hardware.

We shall ignore the noise component in the present discussion. If we multiply a locally generated carrier by the received signal we obtain a signal given by

$$Aa(t) \cos^2(\omega_c t + \phi) = Aa(t)(1/2)[1 + \cos 2(\omega_c t + \phi)] \tag{5.30}$$

The first term in the expression is the desired BPSK signal whereas the

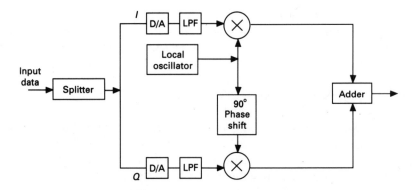

Figure 5.8   A general scheme for generating PSK.

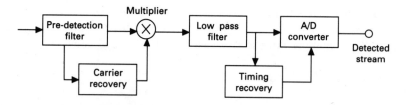

Figure 5.9   A block diagram representing the principle of coherent demodulation of a BPSK signal

second term is a factor at twice the carrier frequency. We can now recover the message signal by filtering out the second term. However, it is also necessary to convert the analog voltage at this point to a digital stream, and also to synchronize the timing to the transmitted bit stream. This is achieved by timing recovery and analog-to-digital converter circuits. One commonly used circuit for this purpose is the *integrate and dump* circuit. In this technique the signal is integrated for the duration of a bit and, at the end of the bit period, the accumulated charge is transferred (or 'dumped') to the following circuit stage. The integrator is then reset to receive the next bit. For correct detection of bits it is necessary to synchronize the start of each bit. The synchronization is achieved by the use of a *bit-synchronizer* circuit.

Any phase deviation between the locally generated carrier and the received signal introduces noise and consequently an error into the recovered bit stream. To minimize the effects of random phase changes occurring in the received signal, the locally generated carrier used for demodulation is extracted from the received signal by means of a *carrier recovery circuit*. One technique for carrier recovery is as follows. The received signal is squared to give $r_c(t)$:

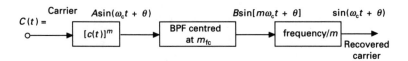

Figure 5.10   A block schematic of a clock recovery circuit for demodulating an *M*-ary PSK.

$$r_c(t) = \cos^2(\omega_c t + \phi)$$
$$= 1/2[1 + \cos 2(\omega_c t + \phi)] \qquad (5.31)$$

The first term being a DC is filtered out. The second term is a signal at twice the carrier frequency, retaining the instantaneous phase of the received carrier. The carrier can be recovered by dividing the second term by 2. In general, the carrier of an *M*-ary PSK scheme can be extracted by performing the operation (carrier)^M which provides the carrier at a frequency of Mf_c, followed by a division-by-*M* operation. Figure 5.10 shows a block diagram for the carrier recovery of an *M*-ary PSK scheme.

Notice that in the BPSK scheme (figure 5.9), the carrier recovery circuit consists of a squaring circuit, therefore the output of the carrier recovery circuit remains the same irrespective of the sign of the received carrier causing an ambiguity in the received bit stream, i.e. '0's and '1's can reverse. This ambiguity can be removed by using modified forms of PSK – *differential phase shift keying (DPSK)* or *differentially encoded phase shift keying (DEPSK)*. Both schemes combine two bits of information logically for transmission and reception.

In a DPSK modulation scheme the transmitted bit is coded in such a way that whenever the user bit is '1' the carrier phase does not change. However when the user bit is '0' the carrier undergoes a phase change of 180°. An exclusive-OR gate having the user bit as one input and a 1-bit delayed version of the gate output can provide this form of modulating signal. A DPSK demodulator does not require a synchronized carrier for demodulation. Demodulation is achieved by multiplying the received carrier and a 1-bit delayed version of the carrier. However, the DPSK scheme is more susceptible to noise than the coherent BPSK, as errors tend to occur in pairs although occasionally only single errors may occur.

A similar type of bit coding is used in the DEPSK scheme. A DEPSK scheme eliminates the need for a 1-bit delay device at the carrier frequency, by performing the required coding at the baseband. But the DEPSK scheme requires a coherent demodulator. The error performance of a DEPSK scheme is slightly worse than that of a DPSK scheme, because in the DEPSK scheme errors *always* occur in pairs but in a DPSK scheme single errors may also occur.

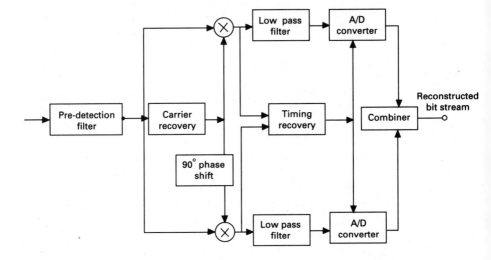

Figure 5.11   A block diagram of a QPSK demodulator.

The principle of coherent demodulation can be extended to the QPSK scheme. Figure 5.11 shows the principle of a QPSK demodulator. The incoming signal is divided into *I* and *Q* channels, each of which contains a coherent demodulator. A part of the received signal is also fed into a carrier recovery circuit consisting of a 'raise-to-power-*four*' circuit followed by a 'divide-by-*four*' circuit. The extracted carrier is divided into an in-phase and a quadrature-phase component which are each fed respectively into *I* and *Q* channel multipliers. Each multiplier is followed by an integrate and dump circuit which provides the recovered bit in each channel. Finally the outputs of each channel are combined with the help of a switch synchronized to the bit stream to give the recovered bit stream.

In a QPSK scheme, a phase change of 180° occurs whenever bits in the *I* and *Q* channels change simultaneously. During a bit transition the phase change occurs in a finite time, causing a variation in carrier amplitude. The amplitude variation increases as the required phase change increases. Such amplitude variations are exaggerated when the modulated signal is filtered, and amplified in a satellite channel containing non-linearity. The amplitude variations caused by a 180° phase shift can be reduced by using an *offset QPSK* (or *O-QPSK*) scheme. In an O-QPSK scheme the maximum permitted phase change is limited to 90°. Therefore O-QPSK offers advantages for satellite communication channels and is used in satellite links.

The principle of QPSK demodulation can be further extended to an *M*-ary PSK scheme, although the physical implementation increases with an increase in *M*. The recovered carrier is used for coherent demodulation of the *I* and *Q* channels. The coherently demodulated signal has an analog magnitude proportional to the digital signal. An analog-to-digital converter, a timing recovery and a combiner are then used to give the recovered bit stream.

Effects of noise

In chapter 4 we have seen that the main goal of a system designer is to provide a specified carrier-to-noise ratio (C/N) at the input of a demodulator by optimizing various segments of the link. The C/N is related to the bit error rate through a modulation-dependent factor. The target bit error probability of a system is defined at an early stage in the system design. For example, if only digital voice communication is required, bit error rate is not critical, as human ears are quite insensitive to errors. Whereas in an application which requires transferring computer data files, it is necessary to have virtually error-free transmission.

The error rate is best specified in terms of bit error rate *probability* because of the random nature of the errors. Thus the bit error rate of a transmission which has an average of n errors/10 000 bits is $n \times 10^{-4}$.

There are a number of sources of impairments which cause errors in digital modulation systems. These are, thermal noise in the front-ends of satellite and earth station receivers, phase noise introduced by imperfect synchronization between the received carrier and the regenerated carrier at the receiver, and noise caused by imperfect bit synchronization. The main sources of error considered in a satellite system design are caused by thermal noise. Errors caused by other sources of noise are minimized by adding an extra margin into the link design.

In a mobile communication system, signals sometimes suffer other types of degradation such as Ricean or Rayleigh fading or even total signal loss. Under such conditions, a careful evaluation of various candidate modulation (and coding) schemes becomes essential. Analytical methods and simulation are used to examine such a link in order to derive the best suited modulation scheme. Such channels are not discussed here and the reader, if interested, should refer to the literature (e.g. Lutz *et al.*, 1986). It may be of interest to note here that, to date, the most commonly used digital modulation schemes in mobile satellite communications have been BPSK and variants of QPSK. This is not surprising because, in general, higher-order modulations are more sensitive to channel impairements.

Effect of thermal noise

In this section, expressions of bit error rate probability as a function of carrier-to-noise ratio for BPSK, QPSK and M-ary PSK schemes caused by gaussian noise are presented. It may be recalled that thermal noise exhibits a gaussian distribution. This assumption is used for estimating the required carrier-to-noise ratio to achieve any specified bit error rate probability. Expressions of bit error rate probability for gaussian noise may be derived in a number of ways. Here we shall not attempt to derive the expressions.

Let us first define an integral called *error function* or *erf* which is commonly encountered in specifying bit error rates of PSK systems. For a variable x, erf(x) is defined as

$$\text{erf}(x) = \frac{2}{\sqrt{\pi}} \int_0^x e^{-x^2} dx \qquad (5.32)$$

The *complementary* error function is defined as $[1 - \text{erf}(x)]$.

The integral is commonly encountered in communication system theory and is tabulated in books of mathematical tables (e.g. Abramowitz and Stegun, 1972). The error function has a value of 0 for $x = 0$ and 1 for $x = \infty$.

The probability of error P_e(BPSK) for a BPSK is given by

$$P_e(\text{BPSK}) = 1/2 \text{ erfc } \sqrt{\frac{E_s}{N_0}} \qquad (5.33)$$

where E_s = energy/bit (note: In BPSK, symbol rate = bit rate)
       N_0 = noise power spectral density.

The probability of error P_e(DPSK) for a DPSK system is given by

$$P_e(\text{DPSK}) = \frac{1}{2} e^{-E_s/N_0} \qquad (5.34)$$

It has already been mentioned that the DPSK system is more error prone than is the BPSK system.

The error rate of the QPSK system is derived by noting that the in-phase and quadrature channels can individually be treated as a BPSK system. The bit error rate in each channel is therefore given by equation (5.33). Noting that each channel component of the QPSK scheme carries one bit of baseband data, we observe that the *bit* error probability of QPSK is the same as that of a BPSK scheme when energy per bit is the same in both cases:

$$P_e(\text{QPSK}) = P_e(\text{BPSK}) \qquad (5.35)$$

The symbol error rate of QPSK, P_{es}(QPSK) is

$$P_{es} = \frac{1}{2} \text{ erfc } \sqrt{\frac{E_s}{2N_0}} \qquad (5.36)$$

where E_s = energy/symbol.
When the bit rates, RF bandwidth and satellite EIRP are the same, QPSK

has a higher bit error rate. However it has twice the capacity of a BPSK. More generally, the bit error probability of an *M*-ary PSK is given by Taub and Schilling (1986) as

$$P_e(\text{MPSK}) = \text{erfc} \left[\frac{NE_b}{N_0} \sin^2 \left(\frac{\pi}{m} \right) \right]^{1/2}$$ (5.37)

where N = number of bits/symbol
  m = number of symbols.

A matter of interest is the relationship between bit error rate and symbol error rate. In BPSK, bit errors and symbol errors are the same. In a QPSK scheme, two bits are combined into a symbol and therefore a symbol error occurs whenever either of the two bits is in error or both bits are in error. The weighted average probability of these errors provides an upper bound of bit error rate. For an *M*-ary PSK assuming equi-likely transmission of bits and *Grey coding* (in Grey coding, adjacent symbols differ by a single bit), for a given symbol rate the bit rate P_{eb} is bounded by (Taub and Schilling, 1986)

$$\frac{P_e}{N} \leqslant P_{eb} \leqslant \frac{M/2}{M-1} P_e$$ (5.38)

For a QPSK the bit error rate is between 1/2 and 2/3 of the symbol error rate.

Figure 5.12 shows a carrier-to-noise ratio versus bit error probability of various digital schemes including BPSK and QPSK. For QPSK, symbol rate obtained using (5.36) is given. *Bit* error rates of BPSK and QPSK are the same. The error rate requirements for satellite communication systems generally lie in the range 10^{-3} to 10^{-7}. For BPSK and QPSK the corresponding E_b/N_0 required are approximately in the range 7–11 dB. In practice, the modulator/demodulator circuits incur some implementation loss, some of which may be quantified – as discussed below. Other losses, such as those caused by interference from adjacent carriers and radiation losses caused by less-than-perfect circuit boards (e.g. imperfect grounding or shielding), are not easy to quantify. When designing satellite links, all implementation loss components are added to give an *implementation margin* – which is typically in the range 1–1.5 dB.

Error in carrier regeneration

We have seen that demodulation of PSK schemes necessitates generation of a carrier synchronized to the received carrier. In practice, invariably the synchronization is not perfect. It can be shown that for an error of $\Delta \phi$

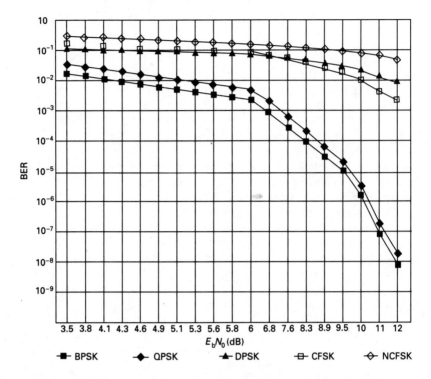

Figure 5.12 E_b/N_0 versus bit error rate (BER) for various modulation schemes. For QPSK, symbol error rate is given.

between the received carrier and the regenerated clock, the expression of bit error probability for BPSK due to thermal noise is modified to

$$P_e(\text{BPSK}) = \frac{1}{2} \sqrt{\frac{E_s}{N_0} \cos^2(\Delta\phi)} \tag{5.39}$$

Thus we see that as the phase error increases, bit error probability also increases.

Error in bit synchronization

Recall that for effective regeneration of baseband data, PSK receivers need a bit synchronizer. Referring to the demodulation implementation discussed above, it can be seen that if the integrate and dump circuit is not synchronized to the incoming bits, it may integrate part of the previous bit causing an erroneous decision. For an error of ΔT in bit synchronization, bit error probability caused by thermal noise and regenerated carrier phase error for BPSK is modified to

$$P_e(\text{BPSK}) = \frac{1}{2} \text{ erfc } \sqrt{\frac{E_s}{N_0} \left(1 - \frac{2\Delta T}{T_b} \right)^2} \tag{5.40}$$

where T_b = bit period.

RF bandwidth

It is necessary to assess the occupied RF bandwidth of transmission for a number of reasons. This information is used to estimate the transponder bandwidth, adjacent carrier spacing, and intra- and inter-system interference.

The bandwidth in each case can be determined by first obtaining the spectrum of the baseband signal and translating it into the carrier frequency. Consider the BPSK scheme and assume the baseband data to be a nonreturn-to-zero (NRZ) data stream varying between $+V$ and $-V$ with a bit time of T_b or $1/R$, where the bit rate is R bits/s (see chapter 7 for NRZ data description). Applying Fourier analysis, it can be shown that the power spectral density of the BPSK modulated signal is

$$P(f)_{\text{BPSK}} = \frac{P_s T_b}{2} \left\{ \left[\frac{\sin \pi(f - f_0)T_b}{\pi(f - f_0)T_b} \right]^2 \right.$$

$$\left. + \left[\frac{\sin \pi(f + f_0)T_b}{\pi(f + f_0)T_b} \right]^2 \right\} \tag{5.41}$$

where f_0 = centre frequency
P_s = power in each binary pulse
T_b = bit period.

The right-hand side of the equation is commonly encountered in communication theory and is known as the sinc^2 function. This function is shown plotted in figure 5.13. Note that equation (5.41) extends to infinity, although levels decrease as the offset with respect to centre frequency is increased. About 90% of the energy is contained in the main lobe of the signal. From the figure we note that the first side lobe is 14 dB below the main signal, and side lobes reduce gradually to 30 dB for the tenth side band. For closely spaced carriers this extent of interference into adjacent channels can be unacceptable. For this reason, the output of modulators is usually filtered. However, if signal bandwidth is reduced excessively, pulse shapes become distorted causing *inter-symbol interference* (see section 7.1). Inter-symbol interference may be reduced to a certain extent by the use of equalizer circuits which compensate the effects of distortion introduced by various filters in the satellite channels.

We can obtain the power spectral density of QPSK (or O-QPSK, which has the same spectral behaviour) by noting,

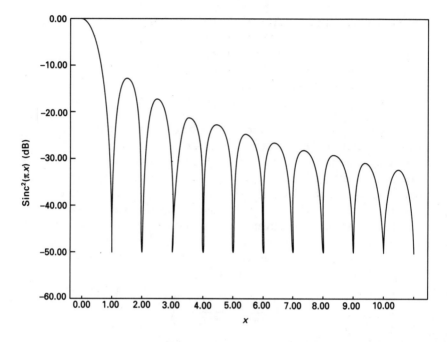

Figure 5.13   A plot of the sinc² function, which is commonly encountered
in practice.

1. the I and Q channels are identical;
2. each channel is identical to a BPSK channel operating at the symbol rate
   $(1/2T_b)$ where T_b is the bit period of the baseband signal.

Thus the power spectral density of the QPSK signal is the sum of the PSD
of each channel:

$$P(f)_{QPSK} = E_b \left\{ \left[\frac{\sin \pi(f - f_0)T_s}{\pi(f - f_0)T_s} \right]^2 \right.$$

$$\left. + \left[\frac{\sin \pi(f + f_0)T_s}{\pi(f + f_0)T_s} \right]^2 \right\} \qquad (5.42)$$

where $T_s =$ symbol period
          $= 2T_b$.

From equation (5.42) we note that the null-to-null bandwidth of a QPSK
(or O-QPSK, which gives the same result) is half the null-to-null bandwidth
of the BPSK scheme. Hence QPSK occupies half the bandwidth of a BPSK
scheme. This confirms the conclusions arrived at while comparing the theor-

etical spectral efficiencies of the various PSK schemes earlier in this section.

Equation (5.42) also shows that the signal levels extend to infinity, whereas it is necessary to maximize the use of the RF spectrum by spacing carriers as closely as possible to each other. As in BPSK, this is achieved by filtering the modulated signal, taking account of the fact that excessive filtering causes inter-symbol interference. Another consideration is that when signals, having side bands reduced significantly through filtering, are passed through a system with inherent non-linearity (e.g. high power amplifiers at the earth station and satellite), side bands are regenerated – defeating the purpose of the filtering.

The general form of expression for spectral occcupancy of an *M*-ary PSK system is derived following a similar approach and is given by

$$P(f)_{\text{MPSK}} = P_s T_s \left(\frac{\sin \pi f T_s}{\pi f T_s} \right)^2 \tag{5.43}$$

where T_s is the symbol period. Therefore the null-to-null bandwidth is given as $2(1/T_s)$ or $2f_b/N$. The reduction in bandwidth relative to a BPSK scheme is $1/N$ (note that $N = 1$ for BPSK).

A scheme where baseband data signals are filtered so as to avoid abrupt phase change is known as 'shaped O-QPSK', or more commonly as *Minimum Shift Keying* (MSK). MSK can be used to minimize the effect of system non-linearities. In this type of modulation the baseband signal is multiplied by a sinusoidal function to effect a gradual change in amplitude of such signals. The resulting sinusoidal-shaped data signals are then used as signal inputs to the *I* and *Q* channels, as in QPSK schemes. As a result, MSK modulated signals have about a 1.5 times wider main lobe than do QPSK, but the side lobes have a lower amplitude. Comparing them quantitatively shows that 99% energy in a MSK scheme lies within $1.2f_s$, whereas a QPSK scheme requires $8f_s$ (see Taub and Schilling, 1986 for details). The equation used for QPSK can also be used to represent MFSK, but here the baseband signal is multiplied by a sinusoidal term. The resulting equation can be re-arranged to resemble a FSK modulation scheme having minimum frequency difference satisfying frequency 'orthogonality' criteria. For this reason the scheme is commonly known as *minimum* frequency shift keying. Variants of the MSK system have been introduced into terrestrial systems. However, MSK has yet to be commonly used in satellite communications.

(b) Frequency shift keying (FSK)

In frequency shift keying systems, the frequency of a carrier is switched in accordance with message state changes. Binary Frequency Shift Keying (BFSK) modulation, the simplest FSK scheme, permits two frequency states, f_1 and

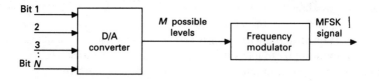

Figure 5.14   A block diagram of an *M*-ary FSK transmitter.

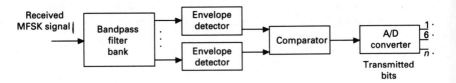

Figure 5.15   A block diagram of an *M*-ary FSK receiver.

f_h, corresponding to binary states '0' and '1' respectively. BFSK can be mathematically represented by

$$v_F(t) = A \cos [\omega_0(t) + a(t)\omega_f t] \tag{5.44}$$

where $a(t)$ is the message stream.

The BFSK concept can be extended to *M* level frequency shift keying called *M*-ary FSK. In a *M*-ary FSK scheme, *N* bits are combined to form a symbol. Therefore $M = 2^N$ possible symbols may be transmitted, each symbol being assigned a unique discrete frequency. For example, an 8-ary FSK ($N = 3$) will transmit any one of eight discrete frequencies, depending on the input data stream (e.g. $0,0,0 = f_0$; $0,0,1 = f_1$, etc.).

Figure 5.14 shows a block diagram for generating an *M*-ary FSK. *N* bits are combined in a digital-to-analog converter to form an analog equivalent of the *N*-bit input data. Therefore, for each symbol a unique analog voltage is developed. The output of the digital-to-analog converter is next fed into a frequency modulator, which then produces the corresponding symbol frequency at its output. One possible implementation of an *M*-ary FSK *receiver* is shown in figure 5.15. The received carrier is amplified, down-converted and fed into a bank of *M* filters where each filter is tuned to the centre frequency of one of the *M* discrete frequencies. Consequently, a signal appears only at the output of the filter tuned to the transmitted symbol. This output signal is then decoded to give the received bit stream.

Bandwith of BFSK

The spectrum of the BFSK signal can be obtained from simple mathematical manipulation of equation (5.44) – the rationale being to transform the equation into components of known power spectral densities. It can then be shown that the first two terms consist of impulses centred at f_h and f_l. The second two terms produce spectra of BPSK signals, these spectra being centred at f_h and f_l respectively. Assuming that most energy is contained in the main lobe of the spectrum, the required bandwidth of a BFSK is

$$\text{BFSK} = 4f_b \qquad (5.45)$$

where f_b is the input bit rate.

Bandwidth of M-ary FSK

The bandwidth of an M-ary signal can be obtained similarly. Here the power spectral density at each of the M discrete frequencies consists of an impulse function together with a $\sin(x)/x$ component. For determining the bandwidth we must first establish the optimum frequency separation between adjacent frequency components. It can be shown that the least probability of error in the received bit stream occurs when the centre frequencies $f_0, f_1, f_2, \ldots, f_m$ are *orthogonal* to each other. In general, orthogonality provides the largest discrimination between each signal, and therefore they are least susceptible to noise. The concept of orthogonal signals can be visualized by considering an X–Y plane. Here two signals are orthogonal to each other if one signal lies on the X-axis and the other on the Y-axis. Similarly, a set of m orthogonal frequencies, $f_0, f_1, f_2, \ldots, f_m$ may be mapped on an M-dimensional space, although such a space is difficult to visualize. Mathematically, two signals are said to be orthogonal in an interval when the integration of the product of these two signals is equal to zero within the period. For an M-ary FSK, a simple scheme for obtaining orthogonality is to use frequency separation equal to even harmonics of symbol frequency ($f_s = 1/T_s$), i.e. f_0, $(f_0 + 2f_s)$, $(f_0 + 4f_s), \ldots, (f_0 + 2mf_s)$.
The bandwidth B_{mfsk} of this signal is therefore

$$B_{\text{mfsk}} = 2Mf_s \qquad (5.46)$$
$$= 2^{N+1}(f_b/N)$$

Note that the bandwidth of an M-ary FSK *increases* with an increase in M, whereas in an M-ary PSK the bandwidth *reduces* with an increase in M.

Spread spectrum modulation

Spread spectrum modulation schemes are closely associated with code division multiple access and are therefore included as part of chapter 8.

5.6 Selection of modulation

Having developed a basic understanding of the various modulation schemes, the issues involved in the selection of suitable modulation are now addressed. The factors which influence the choice are summarized below. In addition it must be mentioned that one of the fundamental characteristics of a communication link is the possibility of trade-off between bandwidth and power to obtain a specified receive signal quality. In bandwidth limited links, spectrally efficient modulation is preferred at the expense of additional carrier power whereas, in power limited links, bandwidth is traded off with power (e.g. direct broadcast television systems). In a real system such as MSS, the situation is often not as distinctly defined. Links may be constrained both in power and spectrum. In such cases the optimization process needs careful practical considerations. In the final reckoning the choice in a commercial system is often influenced by economics.

(i) *Channel type*

Channel refers to the RF path between communicating ground stations including the satellite path (see chapter 4). The main sources of signal disturbance are propagation effects, thermal noise associated with satellite and earth station receivers, inter-modulation noise caused by system non-linearities, and intra/inter-system interference. The severity of such impairments depends on the type of link (and hence application), and influences the choice of modulation. Thermal and inter-modulation noise are important considerations in power limited links such as links to small earth stations. In addition to thermal noise, propagation effects have an important influence in the radio link design of mobile communications. Finally, with an increasing demand on a limited spectrum (this is true in particular for mobile communication systems), satellite links could eventually become severely interference limited and the choice would then favour modulation schemes, such as spread spectrum which are naturally resistant to interference.

When the choice is not obvious, various candidate schemes are studied, resorting either to theoretical analysis or more often computer simulation using the applicable channel model.

As an example, consider some of the channel related issues for selection of modulation of the mobile satellite service. The modulation scheme in the forward direction (i.e. fixed earth station to mobile) must be resistant to shadowing and multipath noise, and should be spectrally efficient. In the return direction, the uplink signals (mobile–satellite) undergo fading, but the satellite to fixed earth station path is stable. The choice of modulation scheme in this link is governed by the need to maximize the spectral efficiency with the constraint that the mobile earth station (MES) hardware should not be so complex as to make the MES too expensive or unreliable.

It has been observed that in general the simplest types of digital modulation scheme and frequency modulation are best suited for both mobile and fixed services on power limited links. When message signals are digital, coding is used to provide the desired link reliability.

(ii) Constraints imposed by earth stations

The modulation scheme must be able to operate within the constraints imposed by earth stations. The earth station characteristics depend on application. Some of the main features are summarised below.

(a) Fixed satellite service

Fixed satellite service uses large, medium and very small aperture terminals (VSAT). The permitted complexity in the modulator/demodulator implementation is highest for large earth stations and least for small earth stations. The power limited satellite–VSAT link is most susceptible to thermal noise because of the low G/T of the earth stations. Therefore in addition to selection of simple and robust modulation scheme, *coding* (see chapter 6) is commonly used. Coding provides extra protection to the user information bits.

(b) Direct broadcast service

Direct-to-home broadcast receivers must use simple and low-cost demodulators. Therefore the selected modulation schemes must be simple and provide a large signal-to-noise ratio with minimum distortion. At present, bandwidth efficiency is not a severe constraint on direct broadcast systems (DBS), as adequate bandwidth has been allocated to this service. Frequency modulation is well suited to provide this requirement and is therefore generally used by most direct broadcast systems at present. The ITU direct broadcast plan is based on the use of frequency modulation. With the introduction of digital television baseband it is likely that digital modulation schemes will be introduced. Spectral efficiency of modulation schemes together with efficient coding of television signals are expected to be important considerations. Coding here refers to analog-to-digital conversion of television signals.

Two other classes of television broadcasts are in use – medium-power direct-to-home broadcasts and broadcasts for television programme distribution. Both systems operate in the fixed satellite service band (whereas the DBS, mentioned above, operates in exclusive bands). Therefore, efficient use of bandwidth is an important criterion in both these applications.

(c) Mobile satellite service

The selection of modulation scheme for the mobile satellite service depends on the operational environment of the mobile – maritime, aeronautical and land. Generally, the constraint imposed by maritime earth stations is the least and that of land mobile earth stations the most. The criterion for the selection of modulation schemes both in the forward (satellite–mobile) and the return directions is to minimize complexity in the mobile earth station, together with the establishment of the highest possible spectral efficiency. However, choosing simple noise and fade resistant modulation schemes is often not sufficient, especially in a land mobile environment. Additional protection to the user data is provided by resorting to *coding* and *automatic repeat requests* (ARQ) schemes (discussed in chapter 6).

(iii) Hardware constraints

One of the main goals in system design is to use the most bandwidth-efficient modulation scheme. We have seen that higher-order digital modulation schemes provide spectrum economy, but this is achieved at the expense of the higher satellite EIRP/carrier. Satellite power is expensive and often limited either by technological constraints or by radio regulations. At present, QPSK is widely used and 8-ary PSK schemes are under investigation for fixed satellite service. The use of radio spectrum is growing steadily and the demand for spectrum in certain bands is much more than in others. In such bands, satellite systems are already becoming bandwidth limited and the use of more spectrally efficient schemes is likely to be necessary in the future. It has already been mentioned that hardware, and in particular demodulator, complexity increases, as a consequence of the increase in order of digital modulation schemes. To effect the power/bandwidth trade-off will require the use of high-power satellites, perhaps coupled with on-board regeneration and processing capabilities.

Finally, it should be mentioned that the satellite accessing technique (see chapter 8) has an important influence on the choice of modulation. For example, TDMA, a highly efficient accessing technique, is used with digital signals only. The digital modulation scheme for the purpose must be capable of providing the desired throughput with simple/realizable demodulators. Similarly for code division multiple access, spread spectrum modulation is an obvious choice. Demand-assigned or pre-assigned SCPC systems, on the

Table 5.1   Modulation schemes currently in common use.

Fixed satellite service	
Television, analog multiplexed	
telephony and SCPC:	FM
Digital data:	BPSK and QPSK
Direct broadcast service	
(a) Television programmes	
(i) Analog baseband	FM
(ii) Digital baseband	QPSK
(b) Direct sound broadcast	System not yet
	implemented
Mobile satellite service	
(a) Telephony (analog)	FM
(b) Telephony (digital)	O-QPSK
(c) Digital data	BPSK and O-QPSK

other hand, depend mainly on the type of baseband. Here, either digital or analog modulation schemes capable of optimal bandwidth/power trade-off may be used.

Table 5.1 shows the modulation schemes for various services, in general use currently. Note that BPSK and QPSK are the most widely used digital modulation schemes. Many of these applications also use coding to provide protection against noise.

Problems

1. Discuss the principle of SSB modulation and its suitability for satellite communications.
2. A direct broadcast satellite transmits television programmes for rural community reception and television transmission with the following characteristics:

Modulation – FM
Peak-to-peak frequency deviation = 20 MHz
Total improvement from weighting and pre-emphasis/de-emphasis = 16.3 dB
Specified signal-to-noise ratio (weighted) = 50 dB
Ratio of peak-to-peak luminance signal to composite television signal including syncpulse = 0.7
Maximum baseband frequency = 5 MHz

Determine the RF bandwidth of transmission and the desired carrier-to-noise ratio at the FM demodulator input.

3. With the help of block schematics illustrate the principles of the modulation and demodulation of BPSK and QPSK, and compare their spectral characteristics and performance in the presence of noise.
What is the limitation in using higher-order modulation schemes for satellite communication?

4. What are the main considerations in selection of modulation for a mobile satellite service? Suggest a possible modulation scheme for providing voice communication via a satellite to land mobiles, giving reasons for your choice.

References

Abramowitz, M. and Stegun, I.A. (eds) (1972). *Handbook of Mathematical Functions*, Dover Publications, New York.

Calvert, W.R. (1984a). 'Transmission systems: An introduction to frequency modulation', *International Journal of Satellite Communications*, Vol. 2, pp 209–214.

Calvert, W.R. (1984b). 'Transmission systems: An introduction to frequency modulation – part 2: Earth station technology, Energy Dispersal and Group Delay Tutorial', *International Journal of Satellite Communications*, Vol. 2, pp 305–310.

CCIR (1982). *Recommendations and Reports*, ITU, Geneva.

Lutz, E., Papke, W. and Plöchinger, E. (1986). 'Land mobile satellite communications – channel model, modulation and error control', *Proc ESA Workshop on Land Mobile Services by Satellite*, ESTEC, 3–4 June, pp 37–42.

Schwartz, M. (1987). *Information Transmission, and Noise*, McGraw–Hill, Singapore.

Taub, H. and Schilling, D.L. (1986). *Principles of Communication Systems*, McGraw–Hill, Singapore.

6 Coding

Chapter 5 covered various modulation schemes quantifying the relationship between carrier-to-noise ratio at the input of a demodulator and baseband signal quality. In many satellite communication applications the desired carrier-to-noise ratio is not achievable. For instance, a VSAT terminal may not have adequate RF sensitivity, or the signal received in a mobile terminal may fluctuate. In such applications a technique known as *coding* is often used. Coding is the technique of protecting message signals from signal impairments, by adding redundancy to the message signal.

In this chapter we discuss the subject of coding as applied to satellite communications. This subject has received extensive coverage in the literature. The treatment here is at the system level and briefly reviews only those aspects relevant to satellite communications. There are many interpretations of coding. As discussed here, coding refers to the process of providing protection to message signals against corruption of the RF signal by incorporating redundant bits within the baseband digital signals.

A brief introduction to information theory is included in the first part of the chapter to develop an appreciation of the fundamental communication capacity limitations of a channel. An understanding of the basics of information theory can broaden the outlook of a perceptive student. This is followed by a discussion of the basic principles of coding schemes commonly used in satellite communication systems.

6.1 Information theory basics

The subject of coding emerged following the fundamental concepts of information theory laid down by Shannon (Shannon, 1948). Information theory is a subject in its own right, the details of which are outside the scope of this book. Only the main results of interest will be reviewed briefly. Theorems laying down the fundamental limits on the amount of information flow through a channel are given. The significance of these for satellite system designers will become evident as the discussion progresses to the subject of coding. It is worthwhile noting that sometimes the coding efficiencies of practical systems are specified with respect to the theoretical bound.

We shall begin by defining the basic elements of information theory, followed by an explanation of the fundamental theorems.

The amount of *information* in a message is defined as the ratio

$$I = \log_2(1/p_k) \tag{6.1}$$

where p_k is the probability of occurrence of the kth message. Thus a less likely message carries more information. To illustrate the concept of information very subjectively in an everyday situation, consider a telephone conversation in which the person called recognizes the calling voice and knows the person well. If the caller gives his/her name and other personal details in an introductory message, the amount of information received by the listener is not significant. However if the caller is completely unknown to the called party, the information contained in the introductory message is significant.

When several different and independent messages (say M) are combined, the total information contained in the message set is the sum of all the individual messages:

$$I_t = \sum_{k=1}^{M} p_k M \log_2 \frac{1}{p_k} \tag{6.2}$$

The average information/message is called the *entropy*, given by

$$H = I_t/L \tag{6.3}$$

where I_t is the total information
        L is the total number of messages.

If r messages are generated per second, the average information rate of the source is

$$R = rH \text{ bits/s} \tag{6.4}$$

When we consider an M message set it can be shown that the maximum entropy is possible only when these messages are likely to occur equally, and its value is given by $\log_2 M$. When the messages are not equi-likely, the average information/bit reduces. Consider a possible application of the concept. Assume that the probability of occurrence of messages can be estimated. It should then be possible to *code* the signal in such a manner that the messages with low information content (i.e. higher probability of occurrence) are assigned lower redundancy than messages with lower probability. It can be shown that in a message coded in this manner the average information/bit is increased. This type of coding could be used in a K_a-band satellite communication link which, as known from chapter 3, suffers large rain-induced attenuation for a significant amount of time. To apply coding, it is assumed that the messages during fading intervals have a larger information content and therefore are assigned a larger number of redundant bits. Then, provided the channel conditions at the receiver are known, the redundancy in transmitted messages may be altered dynamically to match the channel fading. This is an example of an adaptive coding scheme wherein the bit rate is altered dynamically to match link attenuation.

Now we come to a fundamental theorem of communication theory known as *Shannon's theorem* (Shannon, 1948). Shannon's theorem states that when M equally likely messages ($M > 1$) are transmitted at an information rate of R bits/s through a channel of capacity C then provided $R < C$, it is possible to transmit information with any desired error by coding the message. This can be achieved by allowing the code length to increase so as to provide the desired bit error rate. The capacity of a channel limited by gaussian noise is given by the Shannon–Hartley theorem:

$$C = B \log_2(1 + S/N) \text{ bits/s} \tag{6.5}$$

where B = channel bandwidth

   S/N = signal-to-noise ratio at the receiver input.

Note that equation (6.5) is only valid for gaussian noise (i.e. radio links dominated by thermal noise), but the equation provides the lower bound of performance limit for non-gaussian channels.

Equation (6.5) provides the basis for a power/bandwidth trade-off. This trade-off is fundamental to the design of satellite communication links. For example, bandwidth and carrier-to-noise ratio trade-off in a FM system given by equation (5.15) is a consequence of Shannon's fundamental assessment. It can be seen that as $S/N \to \infty$, the channel capacity also increases to infinity. However when bandwidth is increased to infinity, equation (6.5) approaches

$$R = 1.44 \frac{S}{N_0} \tag{6.6}$$

where N_0 is the noise power density.

Increasing the bandwidth does not provide infinity capacity because noise is also increased simultaneously (see the logarithmic term in equation 6.5).

Shannon states that it is possible to design a system capable of providing the theoretical capacity (equation 6.5) by suitably *coding* the signal. Theoretically such a system can be realized by using M orthogonal signals, with M approaching infinity. Such a signal can be detected with the help of M correlators. (A correlator is a type of 'matched filter' receiver, which provides the maximum output signal-to-noise ratio when the received signal matches a locally generated replica of the signal.) As M is increased, the number of correlators increase correspondingly. It can be shown that for the limiting case when M approaches infinity, an error-free transmission is possible at a channel rate given by (6.5). However, the required bandwidth also increases to infinity. Also, as M is increased the symbol time increases directly and therefore approaches infinity as M tends to infinity. The receiver then has to wait a time approaching infinity, at the end of which all the messages are received simultaneously. An M-ary orthogonal FSK system is a good example of *M* orthogonal signals. From a *practical viewpoint* we note that as *M* increases the number of correlators required at the receiver increases, and very soon the complexity of the receiver becomes unmanageable.

With this basic background we now leave this topic and study more practical ways of implementing coding. Occasional references to this section will be made during the ensuing discussions.

6.2   Coding – background

While studying modulation, it became evident that the received signals suffer degradation in signal quality by noise contamination in the channel. In many satellite communication applications it becomes extremely difficult to achieve large carrier-to-noise ratios. Some measures to correct or at least detect bit errors are therefore desirable. Error detection and correction can be achieved with the help of *channel coding*. In most satellite communication applications coding is applied in such a way that error detection or correction is possible at the receiver without resorting to any feedback to the transmitter. This type of coding is called *forward error correction*. In a few applications error correction is achieved by providing a feedback to the transmitter for re-transmission of bits affected by noise. This type of technique is called *automatic-repeat-request* (or *ARQ*). Both techniques are discussed here.

Development of the subject of coding followed the publication of Shannon's work mentioned in the previous section. Here we will confine the treatment to a brief description of the principles of commonly used coding techniques. A section will cover the issues involved in selecting codes for specified satellite communications applications. As before, we will confine our discussions to a system level and issues of interest to satellite communications. The interested reader may refer to books and review-papers in the literature for more comprehensive coverage (e.g. Wu, 1971; Ristenbatt, 1973; Lin and Costello, 1983; Farrel and Clark, 1984; Taub and Schilling, 1986; Sweeney, 1991).

Coding of messages for avoiding detection, called encryption, is not considered. Encryption is commonly used in direct television broadcasts to avoid illegal reception or in military applications to minimize the probability of message interception. Coding of analog signals to digital waveforms is also not considered.

Coding is used in satellite communication links for the following reasons:

1. In power limited links, such as satellite–small earth terminal links, the desired fidelity in communication quality can only be achieved through coding. Small earth stations such as VSATs and mobile terminals oper-

ate in a power limited environment and consequently suffer excessive signal degradation. Additionally, mobile terminals have to contend with hostile propagation environments. Coding is now widely used to serve such terminals. In many types of mobile environment it is impractical to provide communication without coding.

2. Coding helps minimize the error rate in applications such as computer communications where a virtual error-free transmission is essential. In such applications the added complexity introduced by coding is compensated by the economic advantages offered.

3. Coding can be used to achieve better utilization of the channel capacity.

It is not surprising that coding has become an important part of satellite communication system design. It should be recalled here that coding is only applied to digital baseband signals.

Let us take a rather simple example to illustrate some basic elements of coding. Consider the transmission of a binary stream which has an equal probability of generating '0's and '1's, in a channel corrupted by gaussian noise. From previous discussions we already know that the probability of error can be reduced by introducing *redundancy*. One possibility is to repeat transmission of each digit several times, i.e. '0's are transmitted as 00 or 000 etc. At the receiver, the bits may be decoded by a suitable technique, say by 'majority voting' (i.e. take the detected bit as the one which is in the majority). It is well known that the tails of the gaussian curve extend to infinity. Therefore even after coding, some of the bits could be received in error. It should be possible to reduce the bit error probability to any arbitrary value by repeating the message. In the limit, when each digit is repeated an infinite number of times, we achieve an errorless transmission as predicted by Shannon. Even though this example is not of much practical significance, we can make some useful observations:

1. Error probability is reduced as the redundancy is increased.
2. Adding redundancy increases the number of transmission bits.
3. We can also deduce that it is possible to *detect* and *correct* errors if we have enough knowledge about the transmitted messages. In the above example, if transmissions are repeated say 3 times, we could conclude with a *reasonable* confidence that if we receive 001, the transmission should be a 0 – assuming of course that channel is not so noisy as to cause two errors. In this process, we have detected and corrected errors introduced in the channel. We also note that despite coding, errors are possible and there is a limit to error correction.

A general concept of coding is shown in figure 6.1: k information bits enter a coder which introduces r redundant bits producing a coded word of $(k + r)$ bits. If the message transmission time has to be retained, bit time

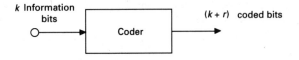

Figure 6.1   The general concept of coding.

must be reduced by a factor of $k/(k + r)$. As a result, the transmission bandwidth and the received noise increase. At the receiver it should be possible to obtain an improvement in message bit error rate, maintaining the same transmitted power as before, i.e. without coding. Unless proper considerations are taken, it may turn out that adding redundancy reduces the received carrier-to-noise ratio to an extent where no advantage is gained. A good system design ensures that the carrier-to-noise at the receiver is adequate to provide coding improvement. Typical advantages gained are of the order of 3–4 dB, i.e. to achieve a specified bit error rate a coded signal can be transmitted at 3–4 dB lower power than required for an uncoded signal. Viewing it from another aspect, bit error is improved by an equivalent of a 3–4 dB increase in power. For satellite communication systems, a coding overhead of the order of 10% can provide significant improvement in the bit error rate provided that the bit error rate probability is more than $\sim 10^{-2}$.

Shift registers

Shift registers and modulo-2 adders are widely used for implementing codes and therefore, prior to discussing the various types of codes, a brief description of this digital building block is included. Most of the codes used in practice are *linear* which are best generated by linear digital circuits. Shift registers, together with modulo-2 adders, form the main elements of linear code generators. A shift register consists of cascaded flip-flops so that every time a data bit is clocked into the register the existing bits in all the stages shift to the next stage. The number of shift register stages can be increased as necessary. In most applications each bit shift is triggered by an accurate and stable clock, in which case the data transfer is said to be *synchronous* because data transfer is synchronized to a clock. When the bit shifts between stages are independent of each other the shift register is *asynchronous*. The cyclic property of shift registers is useful in generating several codes. The modulo-2 adders mentioned above perform binary addition.

A code is generated by tapping off outputs from specific shift register stages and adding these outputs in one or more modulo-2 adders. The output bit stream from adders is represented as the coded version of the input stream. The number of stages in the shift registers, the number of modulo-2 adders and the position of the tapping stages are all functions of the code.

The decoder utilizes certain unique properties of the transformation at the coder to decode the received bit stream.

6.3 Classification of codes

There are two families of code used in satellite communications. These are called *block code* and *convolution code*.

(a) Block code

These types of code operate on groups of bits organized as blocks. A coded block comprising n bits consists of k information bits and r redundant bits:

$$n = k + r$$

Such a code is designated a (n,k) code. A block of 8 bits having 5 information bits and 3 redundant bits is therefore designated as an (8,5) block code. The *rate* of code r_c is specified as k/n. As mentioned in the introduction, when the bit rates of the coded and uncoded streams are the same, information flow rate reduces by a factor r_c. Also, if the coded word needs to be transmitted in the same time interval as the uncoded word, the bit rate of the coded stream must increase by n/k relative to the uncoded word.

In the next section we shall briefly explain certain basic concepts used in block coding.

Hamming distance

The concept of *distance* between two coded blocks is defined first. Consider two sequentially generated words of a (8,5) code. Let the first coded word be 10001001 and the second **01**001001. Observe that the two blocks differ from each other in the first and second bit positions (shown in bold). The distance between these two coded words is said to be 2. In a general sense, the distance between two code words is defined as the number of bits in which the words differ. *Hamming distance*, d_h, is the *minimum possible* distance between two coded blocks. In the detection process, two coded words separated by d_h are most likely to be mistaken for each other. The Hamming distance, therefore, sets a limit on the detection and correction capability of a block code as follows:

1. A code has a capability of *detecting* all coded words having e_d bits in error, given by

$$e_d < (d_h - 1) \tag{6.7a}$$

2. A code has a capability of *detecting* and *correcting* e_{dc} bits, given by

$$e_{dc} = (d_h - 1)/2 \tag{6.7b}$$

When equation (6.7b) is not an integer, e_{dc} is taken as the next lowest integer. For example, if for a code $d_h = 4$, the code can detect 3 errors and correct 1 error. From these code properties it can be deduced that for a code to be correctable the Hamming distance must be at least 3. Similarly, for a code to be at least detectable the Hamming distance must be 2.

3. A code can also be used to correct t and detect e errors. For this case the Hamming distance d_h is given by

$$d_h = t + e + 1 \tag{6.7c}$$

where $t \leqslant e$.

Code generation

There are two main alternatives for generating block codes – the use of *orthogonal signals* or *algebraic coding* (Ristenbatt, 1973). The decoding of orthogonal signals requires M correlators for a system with a total of M possible codes. As the implementation of a correlator is rather complex, the decoding complexity of block codes using orthogonal signals becomes increasingly complex as the number of code words increases. In practice, the number of code words must be quite large for any meaningful communication, making the use of orthogonal block codes difficult. Algebraic codes, on the other hand, use simpler decoding algorithms utilizing modulo-2 arithmetic and this technique is therefore preferred in practice.

There are a number of methods for generating algebraic codes. One implementation technique is the use of a look-up table containing a complete list of coded words – an identical table is used at the receiver for decoding. This type of coding scheme becomes unwieldy as the number of code words increases. A practical solution is to use *linear codes* which are generated by a linear mathematical transformation of the incoming information words. Here, r redundant bits are calculated from k information bits. This type of code leads to simple decoding implementation. All codes used in practice are linear codes – non-linear codes are only of theoretical interest.

Matrix algebra is useful for analysis of block codes because matrix transformations provide the simplest relationships. The elements of matrices are taken as *logical* variables. Therefore, all mathematical operations such as multiplication and addition follow the rules of Boolean algebra. Remember that a multiplication is equivalent to an AND operation and an addition, to an exclusive-OR operation.

A k-bit uncoded word can be represented as a vector consisting of a single row matrix $\bar{A}$:

$$\bar{A} = [a_1, a_2, a_3, \ldots, a_k] \tag{6.8}$$

The generation of a codeword $\bar{C}$ involves creation of a *generator matrix* G and multiplication of $\bar{A}$ by $\bar{G}$:

$$\bar{C} = \bar{A}\,\bar{G} \tag{6.9}$$

The elements of the code generator matrix $\bar{G}$ are constants and also known at the decoder. The elements in this matrix determine the effectiveness of a code. There are no specific rules for obtaining its element values beyond the obvious requirement that the coded word must be able to operate with a specified error probability in the presence of channel noise, using as few redundant bits as possible and be easy to decode. The effectiveness of the code depends on the ingenuity of the code inventor and therefore a code is often named after its inventor. Codes generated through the use of equation (6.9) preserve the information bits in their original form. Such codes are known as *systematic codes*. In some codes the information bits are not explicit in the code-word and such codes are called *non-systematic codes*. The main advantage of a systematic code is that the information word can be estimated by inspection.

The code generator matrix consists of two sub-matrices, a ($k \times k$) identity matrix (all elements of an identity matrix, except diagonals, are zeros) and a ($k \times r$) matrix. The error pattern of such codes contains r bits. Thus the decoding table need contain only 2^r elements rather than 2^k, as required in the look-up table method.

Example 6.1

The generator matrix G of a (7,4) Hamming code is given by

1000 111
0100 110
0010 101
0001 011

Identity
Matrix

Determine the coded word for the sequence

$$A = [a_1 \quad a_2 \quad a_3 \quad a_4]$$

Solution
From equation (6.9) a coded word is obtained as

$$\bar{C} = \bar{A}\,\bar{G}$$

Notice that the generator matrix consists of two sub-matrices, a 4×4 identity matrix whose function is to retain the message bits; followed by a 4×3 matrix which generates the redundancy bits.

Substituting the elements of the matrix:

$$\bar{C} = [a_1 \quad a_2 \quad a_3 \quad a_4] \begin{array}{l} 1000\ 111 \\ 0100\ 110 \\ 0010\ 101 \\ 0001\ 011 \end{array}$$

The coded bit stream is

$$c = a_1, a_2, a_3, a_4, (a_1 \oplus a_2 \oplus a_3), (a_1 \oplus a_2 \oplus a_4), (a_1 \oplus a_3 \oplus a_4)$$

where $\oplus$ is an exclusive-OR operation. Note that the first four bits represent the message bits, followed by three bits which add redundancy. Therefore this is a systematic code.

Decoding

Decoding can be performed by using a look-up table containing a complete map of the coded words and the corresponding decoded messages. As already mentioned, such a decoding process becomes cumbersome as the number of information bits is increased. Linear algebraic codes simplify decoding. The basis of decoding algebraic codes is the code property given by

$$\bar{H}\,\bar{C}^{\mathrm{T}} = \mathrm{O} \qquad (6.10)$$

where $\bar{C}^{\mathrm{T}}$ is the transpose of the coded word (transpose of a matrix is obtained by interchanging its rows and column)

$\bar{H}$ is a known matrix.

Assume that a corrupted version of $\bar{C}$, $\bar{C}_n$, is received at the decoder. The decoding process begins by obtaining a *syndrome* $\bar{S}$ of the received word, given by

$$\bar{S} = \bar{H} \, \bar{C}_n^T \tag{6.11}$$

The decoder checks if the syndrome $\bar{S}$ is zero. (Note that $\bar{H}$ is known at the decoder.) If $\bar{C}_n$ is corrupted, the check is likely to provide a non-zero result. In some unlikely instances, the values may be zero if the corrupted word matches another coded word. In such cases the received message is bound to be in error. When the syndrome is non-zero, it can be shown that

$$\bar{S} = \bar{H} \, \bar{E}^T \tag{6.12}$$

where $\bar{E}$ is known as an error matrix. Any non-zero element in this matrix shows that the corresponding received bit is in error. For example, if only the third element of the matrix is '1' this would imply that the third bit of the received block is in error and requiring correction. Thus, we obtain the decoded word as

$$\bar{C} = \bar{C}_n + \bar{E} \tag{6.13}$$

From equation (6.13) it can be observed that to recover the transmitted word, $\bar{E}$ must be estimated. Examining equation (6.12) it is noted that $\bar{S}$ and $\bar{H}$ are known and it may appear that equation (6.12) can be easily solved to obtain $\bar{E}$. However, it can be shown that the number of equations obtained after simplification of (6.12) is less than the number of unknown error elements. The solution is simplified by utilizing the knowledge that the error detection/correction capability of a code is bounded according to equation set (6.7) (Hamming distance equations) and, in practice, error probabilities are adjusted to acceptable levels by proper link design. The vector $\bar{E}$ can be estimated by solving the equation starting with an error of 1 and increasing to a maximum error limited by (6.7). This simplifying assumption reduces the computational requirements enormously.

From information theory it is known that the error rate can be reduced as the message size in each block is increased, but such a requirement imposes increasingly more decoding complexity. In real systems, coding advantage is limited by the availability of hardware at affordable costs. The advent of low-cost very large scale integration (VLSI) has greatly improved the implementation of coding.

Cyclic codes

Even though the syndrome decoding technique described above reduces the decoding complexity, additional improvements to the alegebraic code structure are desirable for more efficient practical implementations. Cyclic codes provide additional structure which permits further simplicity in decoding algorithms. Here a code-word is generated for each clockwise or counter-clockwise shift of bits in a circular representation of codes. Such a configuration can be implemented by connecting the start and end stages of a shift register. A large number of useful codes belong to this class. Some examples of cyclic codes are Hamming; Bose, Chaudhari and Hocquenghem (BCH); Reed–Solomon; and Golay. The coding/decoding process of cyclic codes is briefly described here because of the importance of this class of code in satellite communications.

In a cyclic code the message $A(x)$ is expressed as a polynomial

$$A(x) = A_0 \oplus Ax \oplus A_2 x^2 \oplus \ldots \oplus A_{n-1} x^{k-1} \tag{6.14}$$

where $\oplus$ shows a modulo-2 addition.

A n-bit coded word is expressed as the polynomial

$$C(n) = C_0 \oplus K_1 x \oplus K_2 x^2 \oplus \ldots \oplus K_{n-1} x^{n-1} \tag{6.15}$$

The generating polynomial is represented by

$$g(x) = 1 \oplus g_1 x \oplus g_2 x^2 \oplus \ldots \oplus g_{r-1} x^{r-1} \oplus x^r \tag{6.16}$$

The coded word is obtained by using the relationship

$$C(n) = g(x) \, A(x) \tag{6.17}$$

The coefficients of $g(x)$ are obtained by factorizing the equation

$$f(x) = x^n \oplus 1 \tag{6.18}$$

where n = number of bits in the code word.

Since $C(n)$ has a degree of freedom $(n - 1)$ and $A(x)$ a degree of freedom $(k - 1)$, $g(x)$ must have a degree of freedom $(n - k)$. Therefore, the factor of (6.18) which has a $(n - k)$ degree of freedom is taken as the code-generating polynomial $g(x)$.

Examples of algebraic code

Single parity check
Single parity check code consists of one parity bit added to a block such that the number of '1's in the block equal an even number. Such a code can detect one error. Parity check codes are applicable when the expected number of errors is expected to be very small – for example, in a computer local area network. This type of code is of very limited use in satellite communications.

Hamming code
In a *Hamming code*, the Hamming distance is 3. Therefore, from equations (6.7a) and (6.7b) it is noted that Hamming codes can detect 2 errors and correct 1 error. The number of bits in the uncoded word, the redundant parity words and the coded word are related as follows:

$$n = 2^r - 1 \tag{6.19}$$

$$k = 2^r - 1 - r \tag{6.20}$$

For $r = 3$: $n = 7$ and $k = 4$. This is the (7,4) Hamming code which was illustrated in the example above. Hamming code is also a type of cyclic code.

BCH codes
Bose, Chaudhari and Hocquenghem (BCH) codes are a class of cyclic code often used in satellite communication. (Reed–Solomon code is another example of this class of code.) Some of the characteristics of this code are

Number of codes which may be corrected, $e_{dc} = r/m$

where m is related to the number of coded bits n by

$$n = 2^m - 1 \tag{6.21}$$

The Hamming distance is related to e_{dc} according to equation (6.7b).

Codes for a channel with error bursts

The codes discussed in the preceding section are suitable for use when errors are randomly spaced, i.e. those caused by thermal noise. When noise occurs in bursts, several words adjacent to each other are severely contaminated with long intervening sequences which remain relatively noise free. The codes discussed above are unsuitable for channels contaminated by bursty (or impulsive) noise. There are many types of satellite communication link in which impulsive noise occurs. Examples of such links follow.

(i) *Links above 10 GHz*
Satellite links above 10 GHz suffer fading caused by precipitation. When fade margins are inadequate or marginal, signal fades occasionally cause carrier level to go in and out of receiver threshold, producing error bursts at the output of the demodulator.

(ii) *Mobile satellite links*
In such links, fading is caused by shadowing and multipath. Fade margins are often inadequate and therefore, in particular, land mobile satellite links

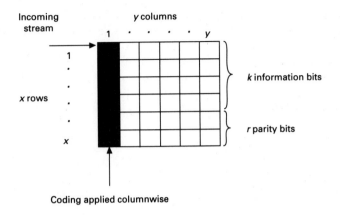

Coding applied columnwise

Figure 6.2　One possible implementation scheme of interleaving.

are often contaminated by impulsive noise, as the mobile moves in and out of obstructions or multipath causes large signal fluctuations.

(iii) *Links affected by scintillation*
Scintillation fades affect links below 2 GHz. This type of fading is quite rapid, having spectral components up to 10 Hz. In channels affected by scintillation, error bursts could be caused by the reason given in (i).

An effective solution for minimizing the effect of noise bursts is to 'spread' each message in time. The net effect is a reduction in the average error/message. This is achieved by a technique known as *interleaving*.

A possible implementation of interleaving is shown in figure 6.2. Input data stream is fed into shift registers arranged as a matrix of x rows and y columns. The input data stream fills this matrix row-wise, data folding into the succeeding row as each row is filled. The separation between adjacent elements in any column is therefore y bits. This separation is called the *interleaving depth*. Next, coding is applied to each column and finally, the coded word is transmitted row-wise. At the decoder the received bits are assembled in an identical shift register matrix. The message word can be retrieved by decoding the assembled word columnwise and reading out the bits from the shift register matrix row-wise.

Let us investigate the mechanism which provides resistance to noise bursts. Consider a noise burst causing error to all bits in a single row of the interleaved word. The duration of noise impulse is yc seconds (where c = bit period). Such an error burst corrupts only one bit of the coded word, noting that coding is applied columnwise. From our previous discussions we know that a single bit error can be easily corrected. Thus all y bits of the affected row can be corrected individually thereby completely re-constructing the corrupt row which would have otherwise been irretrievably lost. In general, if the code is able to correct t errors then the maximum length of noise burst we can correct is $t_m < yt$ bits. Here we notice that a sound knowledge of noise characteristics is essential for developing an optimum interleaving depth and coding.

Reed–Solomon code (or *RS code*) is a useful error correction code for channels affected by impulsive noise. The codes discussed in the preceding sections operate on groups of bits. A *Reed–Solomon code* instead operates on groups of n bits called *symbols*. A block in RS code consists of k information symbols and r parity symbols, and a code word consists of $(k + r)$ symbols. The code rate is k/n. A single bit error occurring in a symbol implies that the symbol is in error. The number of symbols/code words are given by

$$n = 2^s - 1 \qquad (6.22)$$

where s = number of bits/symbol.

An RS code can correct errors in $r/2$ symbols. For example, consider a code in which 6 bits are organized as a symbol, then there are 63 symbols in the code word. Now, if it is desired that the code be able to correct 2 symbols, then $r = 4$. The total number of information symbols is therefore $(63 - 2^4)$ or 47. The total number of consecutive bit errors which can be corrected by this code is 2 symbols or 12 bits. A code with similar error-correcting capability operating on single blocks could have corrected 2 consecutive errors at most. The power of the code can be increased according to requirements by enhancing the error-correcting capability of the code.

The burst error capability of RS code can be greatly enhanced when RS codes are used in conjunction with interleaving. The number of consecutive bits correctable is then increased by a factor s. The total number of correctable burst error is

$$E_c = sty \qquad (6.23)$$

Another advantage of RS codes is their capability to be coded into a multilevel modulation scheme.

RS codes can be decoded efficiently. The decoding algorithm permits the use of long codes and therefore RS codes are quite useful in practice.

A limitation of RS codes is their inefficiency in correcting random errors. Also, noise bursts must be separated by relatively large periods of noise-free region in order that the RS code can work efficiently. In many types of channel, random and impulse noise may coexist. In such cases, interleaving with RS codes does not give satisfactory results. Error correction in this type of channel should be performed by *cascading* two types of code – one for correcting random errors and the other for correcting bursty errors. This type of code organization is called *concatenation*. For example,

an incoming bit stream is first interleaved as described above. Then coding
is applied row-wise to correct random errors and columnwise to correct burst
errors.

(b) Convolution code

The second family of commonly used code is known as convolution code.
Unlike block codes which operate on each block independently, convolution
codes retain several previous bits in memory which are all used in the coding process. Convolution codes are formed by convolving the information
bits with the impulse response of a shift register encoder. Impulse response
of the encoder is defined as the encoder response when a single 1 followed
by all '0's are entered into the encoder.

Figure 6.3 illustrates the main functional blocks of a convolution coder.
An input bit b_0 progressively moves to the left in a shift register as the
subsequent information bit b_1, b_2, ... etc. arrive, until b_0 reaches the last
stage at which point the bit is 'dropped' out of the encoder. The outputs
from certain stages of the shift registers are combined in v exclusive-OR
adders. For each information bit entered in the shift register, v bits appear
at the output, which are then transmitted sequentially. In figure 6.3 the switch
S samples the output of gates v_1 and v_2 during each bit interval. Thus each
message bit produces two coded bits, C_1 and C_2. This is therefore a 1/2 rate
coder. The coded bits C_1 and C_2 of figure 6.3 can be defined by the equations

$$C_1 = b_1 \oplus b_2 \oplus b_3 \qquad\qquad (6.24a)$$

$$C_2 = b_1 \oplus b_3 \qquad\qquad (6.24b)$$

Exclusive-OR is a linear operation and therefore convolution codes, like
block codes, are also linear. In general, a convolution coder consists of K
stage shift registers and v exclusive-OR adders. k information bits are entered
simultaneously for which v output bits are produced. The rate of the coder is
then given as k/v. In the example encoder of figure 6.3, $k = 1$ and $v = 2$;
the rate of the coder is 1/2, as already mentioned. Code rates 1/2, 1/3 and
3/4 are in general use. It is necessary to clear the shift register at the end of
each message by feeding in a string of '0's. This incurs overheads. Therefore
a convolution coder is not well suited for very short messages. A coder with
K shift registers and v exclusive-OR gates fed with M message bits produces
$(K + Mv)$ output bits.

An important parameter in estimating the error-correcting property as well
as decoding complexity of a convolution code is the *constraint length* of the
code. The *constraint length* (or *span*) of a convolution encoder is defined as
the number of information bits which influence the encoder output. The

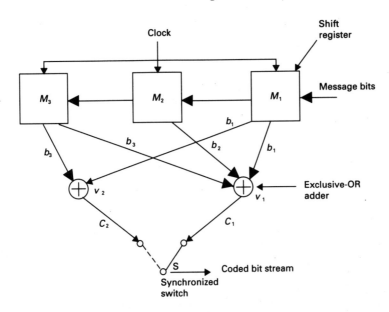

Figure 6.3   The main functional block of a convolution encoder.

constraint length is therefore determined by the number of shift register stages K. In figure 6.3 there are three shift registers and therefore each information bit influences three coded bits in the encoding process and hence the constraint length is 3. A convolution coder is a feed-forward type of coder which produces output *continuously*, whereas in a block code scheme information bits are assembled (i.e. stored) into blocks before applying the coding.

The concept of Hamming distance can be extended to convolution code, although some modification is necessary because convolution codes cannot be clearly defined in terms of blocks. The *minimum distance* d_{min} is defined as the minimum number of bits in a code that have to be altered to obtain another valid code word, taking the constraint length of the code as the block size. When a semi-infinite sequence is considered the minimum number of bits to be changed in order to form a valid code word is called the *free distance*. The applicable definition depends on the decoder used. For example, when considering a Viterbi decoding algorithm, often used in satellite communications, the definition of free distance is used.

The selection of the optimal number of shift registers and feedback connections to exclusive-OR is generally performed through computer simulation. The error correction of the code increases with increase in the constraint length of the code. However, increasing the constraint length increases the decoding complexity of the code. Therefore a compromise has to be exercised between the error-correction capability and the decoding complexity.

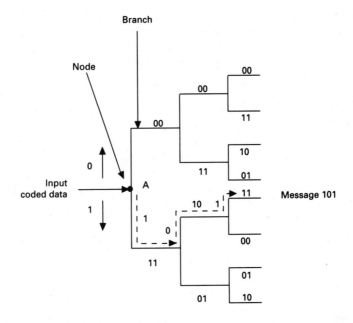

Figure 6.4   A code tree for a convolution coder using three-stage shift
registers and two exclusive-OR gates ($K = 3$, rate = 1/2).

The convolution coding process is represented as a *code tree*. Each code
has its unique code tree obtained by use of a code definition such as that
given by equation set (6.24). The code tree of a convolution coder of span
3 and rate 1/2 is illustrated in figure 6.4. The vertical lines called *nodes*
represent the instants when information bits are entered into a coder. Branches
emerging from each node denote coded bit output corresponding to the information bit entered at the node. By convention, if the information bit is 1
the corresponding coded bit is represented in the lower branch, and when
the bit is 0 the coded bit is shown in the upper branch.

To decode a message, the decoder moves along the tree, applying the
decoding rule given above as each coded word is received. Consider the
decoding of the coded bit stream using the code tree given in figure 6.4:

    11   10   11

The entry point in the code tree is the first node marked A. The first coded
word is 11 so the decoder takes a path in the downward direction towards
the 11 branch; the message bit is therefore 1. The second received coded
word is 10 indicating an upward move in the code tree which implies that
the message bit is 0. The decoding process is continued similarly until the

message is recovered. In figure 6.4, the decoder path through the tree is shown as a dashed line and the given stream is decoded as 101.

When the received bits contain errors this simple decoding scheme is not adequate. It has been mentioned above that each message bit influences Kv coded bit implying that the knowledge of the state of each message bit gets distributed to Kv bits. This distribution of knowledge makes convolution codes robust to noise. Intuitively it would therefore appear that to exploit the full capability of a convolution code we need to examine all Kv bits before making a decision regarding each message bit. This is indeed true and is effected in practice by the use of digital correlation as follows.

Each convolution code permits only a limited number of coded words. In performing a digital correlation, the decoder compares each permitted path in the code tree with the received coded message stream of Kv bits. The path which is estimated as the closest to the received coded word is taken as the valid path. The message bit is then retrieved, following the convention given above. The process is continued for decoding the next message bit and so on. When the channel noise is white gaussian, such a receiver (receiver is a general term for this technique) gives the largest possible signal-to-noise ratio and is known as an *optimum receiver*. When the probability of occurrence of channel symbols is equally likely, the receiver is a *maximum likelihood* receiver. Implementation of such a receiver becomes impractical as message length increases. An algorithm proposed by Viterbi and discussed later in the section can provide a near-optimum decoding at an acceptable implementation complexity. This is achieved by removing several paths which are unlikely to represent the message.

The probability of error of convolution code reduces with an increase in the constraint length of the code. However this is accompanied by an increase in the complexity of the decoding process, because the number of code branches to be examined is given as 2^z where z is the constraint length. For example, if the constraint length is increased from 3 to 5, the number of branches to be examined increases from 8 to 32.

Sequential decoding eliminates the need to examine every branch of the tree by beginning error checks at the starting point and sequentially following the branches which are *closer* to the received word. The numbers of errors at each node are stored. As the decoder traverses the tree a cumulative count of error is maintained. The assumption is that if a wrong path is taken at any node, the number of accumulated errors will rise very rapidly. Hence whenever the cumulative error count exceeds a pre-set threshold, the decoder jumps back to the previous node and takes the alternative path. If the error count is again exceeded, the decoder jumps back two steps and so on. Thus after some trial and error the complete path is traversed. The decision on the first message bit is taken on the basis of this estimated path. The decoding of the subsequent message bit follows the same algorithm.

The sequential decoding algorithm results in a greatly reduced computation complexity over direct coding because the technique avoids the need to perform checks on each branch of the code tree. It also utilizes the knowledge gained when it makes an error by avoiding certain sections of the code tree. A sequential decoding algorithm makes the scheme rather insensitive to the constraint length. Thus very low error probability may be achieved. It may be noted that decoding time increases with an increase in channel noise as a greater number of searches need to be performed. However the major issue with sequential decoding is the need to store bits during the trial and error process. Whenever the storage requirement exceeds the storage capacity of the buffer the decoding process stalls. This buffer overflow problem occurs whenever the number of trials becomes too large. The problem can be reduced by using larger computational speed. Sequential decoding techniques usually employ *hard decision decoding* because of the added complexity associated with *soft decision decoding*. This in turn increases the size of the buffer and the computation complexity.

A brief explanation of *hard* and *soft decision* coding is included here to develop an appreciation of the underlying mechanism. Reference to these techniques will be made at several places. Hard decision decoding refers to the decoding technique in which a received code word is assembled through a decision taken on a bit-by-bit basis. A demodulator provides an analog equivalent of the binary digit at its output (this is the unshaped output of the demodulator) which is then fed into an analog-to-digital converter (a comparator, for example) in which the decision regarding the state of each received bit is taken on a bit-by-bit basis. The resulting binary sequence is applied to the decoder. In soft decision decoding the binary stream is estimated by correlating the demodulated analog signal (or its quantized version) with locally generated binary sequences and selecting as the detected binary stream the sequence which provides the highest correlation. We shall see later that soft decision decoding has about a 2 dB performance advantage over hard decision decoding. However it is evident that implementation of soft decision is more complex.

The *Viterbi algorithm* (Viterbi, 1971; Heller and Jacobs, 1971) is a cost-effective decoding algorithm widely used in satellite communication systems. This decoding technique uses a maximum likelihood technique and provides medium bit error rates. Viterbi algorithms estimate the path of the coded message by truncating the depth of search to typically 4–5 times the constraint length. Only those paths which give the least error counts are retained and others discarded so that the complexity is thereby reduced. The process of discarding paths and truncation makes the system sub-optimal. However for all practical purposes the performance degradation is negligible compared with an optimal scheme.

Both sequential decoding and Viterbi algorithms can operate at high bit rates but Viterbi algorithms are preferred in many applications on balance, because of the cost-effectiveness of such algorithms. However, it should be noted that the sequential decoding technique has a potential of lower error bit rate probability.

6.4 Coding gain

A system designer is interested in quantifying the advantage in using a code in terms of permitted reduction in the carrier-to-noise ratio. For quantifying the advantage a parameter called *coding gain* is defined as follows:

$$G_c = (E_b/N_0)_u/(E_b/N_0)_c \qquad (6.25)$$

where $(E_b/N_0)_u$ and $(E_b/N_0)_c$ are carrier energy/bit-to-noise-power-density ratio of the uncoded and coded word, respectively, to give the same bit error rate at a receiver. The assumption is that the rate of information transmission remains the same in both cases. This implies that the bit rate of the coded signal is higher than that for the uncoded case.

For a gaussian channel the coding gain can be approximated by (Farrel and Clark, 1984);

$$G_c = 10 \log_{10} (Rd) \qquad (6.26)$$

where R is the code rate (k/n)
$\quad\quad\quad d$ is the minimum code distance for the block or convolution code
$\quad\quad\quad Rd$ is the code quality factor.

The coding gain increases as the code distance d is increased. The magnitude of d can be increased by increasing the block size (or constraint length) of a code. This is in agreement with Shannon's theorem which states that the error rate can be made arbitrarily small by increasing the code length. In practice, increasing code length increases the decoding complexity. Practically, coding gains in the range 4–7 dB are feasible (Farrel and Clark, 1984).

6.5 Automatic-repeat-request

Concept

The forward error correction schemes discussed above have a capability to detect and correct errors. The complexity of FEC codes increases and the efficiency reduces as the carrier-to-noise ratio reduces, or there is a need for very few errors. *Automatic-repeat-request (ARQ)* schemes are suitable in such cases. In ARQ schemes when an error is detected the receiver requests re-transmission of the corrupted message via a feedback channel. In a satellite communication environment, two-way propagation delays are of the order of 0.5 s and therefore use of ARQ schemes leads to a large delay in message transfer. Therefore ARQ schemes are best suited for data communication applications which may tolerate delays of more than 1/2 second. It is also possible to add some FEC coding in order to reduce the number of repeat requests.

Performance evaluation

The parameters used for assessing the performance of ARQ schemes are *error probability* and *throughput* (or *transmission efficiency*).

(a) Error probability

Block codes are better suited for short transmissions and so are commonly used for error correction in ARQ schemes. For this discussion we shall consider an ARQ system transmitting an (n,k) block coded data stream. In such an ARQ scheme an error can only occur when a transmitted word is so corrupted that it is received as another coded word. All other errors in the received word can be detected by the block code and flagged. The upper bound on the error probability of an ARQ scheme is the probability of receiving a different coded word than the transmitted word. In an (n,k) block, there are 2^k possible message words and 2^n possible received words. Out of 2^k coded message words, only one word is valid and the reception of the remaining $(2^k - 1)$ code words causes an error. (In practice, the probability of reception of another valid code word is very low but not absent.) The probability bound is given as the ratio of the maximal possible numbers of errors to the maximal possible number of received words:

$$P_e \leqslant (2^k - 1)/2^n \qquad (6.27)$$

As an example, consider an ARQ scheme using a (64,32) block code. The probability of error in this scheme is bounded by

$$P_e \leqslant (2^{32} - 1) / 2^{64}$$
$$\leqslant 2.33 \times 10^{-10}$$

There are various techniques used for ARQ implementation but P_e, given by equation (6.27), does not depend on the implementation technique used. However, the throughput of the system does depend on the type of ARQ scheme used.

(b) Throughput

The throughput T_{arq}, is defined as

$$T_{arq} = n_i/n_{arq} \qquad (6.28)$$

where n_i and n_{arq} are respectively the information bits received per unit time, with and without the ARQ scheme.

ARQ schemes

There are a number of ways in which ARQ systems may be implemented. The basic techniques follow.

(i) *Stop and wait ARQ*
In this scheme a transmitter sends a message only when the correct receipt of the previous message is acknowledged by the receiver. The delay in this scheme is therefore very large.

(ii) *Go-back N ARQ*
In this scheme, the message transmissions are continuous until a repeat request for a message (say the nth) arrives at the transmitter. Then all messages beginning from the nth message are re-transmitted. In this scheme we note that several message transmissions are unnecessarily duplicated.

(iii) *Selective-request ARQ*
This scheme is similar to go-back N ARQ but only the corrupted message is re-transmitted on receipt of a repeat request. This is the most efficient of the basic ARQ schemes and is best suited for application in satellite communication.

The mean time T_m for transmission of a word for selective request ARQ is given by

$$T_m = T_w/P_A \qquad (6.29)$$

where T_w = time for transmission of a block
P_A = probability that a message is accepted by the receiver.

The throughput efficiency of a selective repeat request scheme η_t is given by

$$\eta_t = \frac{n}{k} P_A \qquad (6.30)$$

where k = number of message bits in a block

n = total number of a bits in a block.

For satellite communication selective-request ARQ should provide an effective solution when a store-and-forward system is acceptable. An example of the application of an ARQ scheme in satellite communication is the INMARSAT-C system which provides store-and-forward data communication to mobile terminals. But ARQ schemes have not been in general use for satellite communications to date. Several store-and-forward messaging systems are being considered for mobile data messaging systems. ARQ schemes could be used effectively in such systems.

6.6   Selection of coding

From the above discussions it is clear that coding hardware can be complex. In particular, decoding can be very complex, especially when maximal benefits must be derived by the use of FEC. A trade-off study between the complexity introduced and the advantages offered by the coding is essential.

A system designer begins by developing a link budget with constraints on the spacecraft/earth station EIRP and G/T, and the necessary link margin, included. This evaluation provides knowledge about whether the coding is necessary. A preliminary estimate of the desired coding gain can also be obtained. If the use of coding is justified, the characteristics of the channel need careful evaluation to obtain the best suited code. In particular, the channel characteristics of a mobile satellite service are the most challenging (see chapter 3). If a mathematical description of channel behaviours is not easy to make, stored measurement data can be used together with a computer simulation.

Some of the important channel characteristics used for evaluating coding (and modulation) schemes are:

1. Carrier-to-thermal noise.
2. Amplitude fading characteristics, e.g. Ricean, Rayleigh.
3. Time characterization
   • How rapidly do signal fades occur at various fade levels?
   • For what duration does the signal level remain below the specified fade levels?
4. Characteristics of the noise (whether thermal only or impulsive, etc.).

Figure 6.5 (Ristenbatt, 1973) shows a comparison of the bit error performance in gaussian noise of the following coding and modulation schemes:

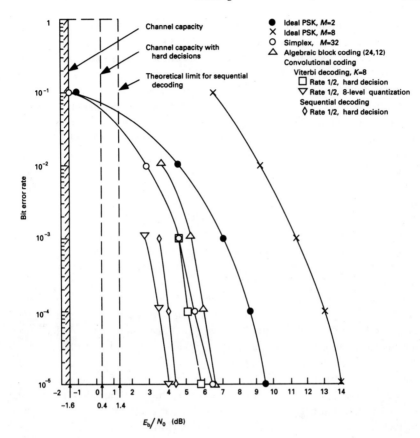

Figure 6.5   A comparison of bit error rate versus E_b/N_0 of various coding schemes (Ristenbatt, 1973).

Ideal BPSK;

Ideal 8-phase-PSK;

32-level, 'simplex' signal (Simplex signals have the maximal available negative correlation with each other. In other words, for a given energy, simplex signals take up the maximal distance between each other (Ristenbatt, 1973). Pseudo-random sequences generated by using shift registers are an example of digital simplex signals. For a specified M, a simplex signal set gives the least bit error rate in gaussian noise.);

Rate 1/2, algebraic block coding (24,12);

  Convolution coding, using two types of decoding algorithm –

  (a) rate 1/2, Viterbi decoding with constraint length 8 using both hard decision and soft decision (8-level quantization refers to soft decision); and

  (b) rate 1/2, sequential decoding.

The main interest here is the performance comparison of block and convolution code with each other and the uncoded BPSK signal. Included in the figure is the theoretical channel limit estimated by using Shannon's theorem both for hard and soft decision and the estimated theoretical limit for sequential decoding.

From the figure it can be observed that according to Shannon's theorem the minimum values of E_b/N_0 required for error-free transmission using soft and hard decisions respectively are -1.6 dB and 0.4 dB. Shannon's bit error rate curve steps from a bit error of 1 to 0 at these threshold E_b/N_0 values. Note that the loss in performance due to hard decision is approximately 2 dB relative to soft decision decoding. The same order of degradation is observed when comparing hard and soft decision Viterbi algorithms shown in the figure. Therefore soft decision decoding is preferred when it can be implemented and the associated additional complexity is acceptable. Not surprisingly, codes which can utilize soft decision decoding are preferred. In this respect convolution codes have an advantage over block codes. It is noted that for the code considered, convolution codes provide approximately 1–2.5 dB advantage over algebraic block codes at a bit error probability of 10^{-5}. However, it has been shown that the performance of larger block codes is similar to convolution codes (Farrel and Clarke, 1984). Comparing the coding advantage of various schemes relative to BPSK it can be seen that at the bit error rate probability of 10^{-5} coding provides an advantage of approximately 3 dB (algebraic block code) to 5.5 dB (convolution code with rate 1/2, soft decision Viterbi decoding). It can also be seen that the advantage gradually decreases as the bit error rate probability is reduced. In fact it has been observed that bit error performance with coding is worse at high bit error probability (10^{-1} to 10^{-2}).

A comparison of coding gain at bit error rate probabilities of 10^{-5} and 10^{-8} using BPSK or QPSK in gaussian noise for several coding techniques applicable to satellite communications obtained from a different source is given in table 6.1 (Taub and Schilling, 1986). In general, the results in the table compare well with the results quoted above (figure 6.5).

Some other general conclusions of interest are (Sweeney, 1991):

1. Coding gain offered may not always be realizable in practice because demodulators cannot operate satisfactorily below a certain carrier-to-noise ratio. (Carrier recovery and bit synchronization circuits require a specific minimum carrier-to-noise ratio for reliable operation.)
2. Low rate code (i.e. codes with larger redundancy) provide better performance than high rate code when bit error rate is high, whereas at medium/low bit error rate high rate coders perform better.
3. In general, decoding is a more complex process than coding and consequently future major developments are expected to be in improving decoding techniques rather than in developing new codes.

Table 6.1   Theoretical values of coding gain of various coding technique in a gaussian channel.

Coding type	Coding gain (dB)	
	$BER = 10^{-5}$	$BER = 10^{-8}$
Block code (hard decision)	3–4	4.5–5.5
Convolution coding with sequential decoding (hard decision)	4–5	6–7
Convolution coding with sequential decoding (soft decision)	6–7	8–9
Convolution coding with Viterbi decoding (hard decision)	4–5.5	5–6.5
Convolution coding with Viterbi decoding (soft decision)	6–7.5	7–8.5
Concatenated codes (RS and convolution coding with Viterbi decoding) (hard decision)	6.5–7.5	8.5–9.5

Summarizing, we note that the choice of codes is governed by the type of error in the channel, the permitted implementation complexity and the type of message. For high rate code with a large correction capability, block codes can provide better speed with a simpler implementation. Block codes also offer advantages for short messages in overheads (in terms of extra bits) and speed. Convolution codes are not suited for short messages because of the extra overhead arising from the need to clear the shift register at the end of each message.

On the other hand, it has been noted above that a major advantage of convolution codes is their ability to obtain a performance advantage by the use of soft decision decoding. Convolution codes have therefore been favoured for satellite communication systems where possible.

6.7 Summary of coding

This section summarizes the salient features of coding:

1. Coding adds overheads in terms of redundant bits and increases implementation complexity. Therefore a trade-off analysis is necessary to study if a net advantage is available by the use of coding.
2. The following considerations apply while selecting a code for a given application –

- a code designed to minimize error in a channel limited by thermal noise may not be suitable for a channel which suffers fading additionally;
- for an application demanding a virtually error-free transmission it may be necessary to introduce more than one type of coding;
- in a channel where noise occurs in bursts it may be sensible to spread the information word over a long period (i.e. to apply interleaving) such that an impulse noise corrupts only a small portion of a word which could be corrected – the 'depth' of coding depends on the channel type.

3. Either block or convolution code may be used when a channel is dominated by thermal noise.

4. Concatenated codes are useful when a channel suffers both random and impulse noise.

5. Soft decision provides better performance than hard decision but is more complex.

6. When an extremely low bit error rate is required or for links in which the RF signal could be completely lost, an automatic-repeat-request (ARQ) scheme is useful. However an ARQ scheme requires an additional feedback channel and adds delay. Therefore ARQ schemes are best suited for data transmissions which can tolerate delays >0.5 second.

Problems

1. What are the advantages of coding in satellite communications applications? Give examples of applications of coding.

2. What do you understand by:
   (a) Hamming distance;
   (b) constraint length of a code;
   (c) hard and soft decision decoding;
   (d) a coding gain of 3 dB?

3. Describe a method of generating and decoding:
   (a) linear algebraic code;
   (b) convolution code.
   Compare their performance and give possible applications of each.

4. An L-band satellite downlink used for store-and-forward communication is known to be affected by fading. An analysis of propagation data shows that fades causing the carrier-to-noise ratio to go below threshhold occur for up to 5 seconds. Suggest a technique to overcome the effect of fades. To illustrate the principle, a transmission bit rate of 600 bps may be assumed.

References

Farrel, P.G. and Clark, A.P. (1984).'Modulation and coding', *International Journal of Satellite Communications*, Vol. 2, pp 287–304.

Heller, J.A. and Jacobs, I.M. (1971). 'Viterbi decoding for satellite and space communication', *IEEE Trans. Commun. Technology*, COM-19, October, pp 835–848.

Lin, S. and Costello, D. (1983). *Error Control Coding: Fundamentals and Applications*, Prentice–Hall, Englewood Cliffs, New Jersey.

Ristenbatt, M.P. (1973). 'Alternatives in digital communications', *Proc. IEEE*, Vol. 61, June, pp 703–721.

Shannon, C.E. (1948). 'A mathematical theory of communications', *BSTJ*, Vol. 27, pp 379–623.

Sweeney, P. (1991). *Error Control Coding: An Introduction*, Prentice-Hall, Hemel Hempstead, UK.

Taub, H. and Schilling, D.L. (1986). *Principles of Communication Systems*, McGraw–Hill, Singapore.

Viterbi, A.J. (1971). 'Convolution codes and their performance in communication systems', *IEEE Trans. Commun. Technology*, COM-19, October, pp 751–772.

Wu, W.W. (1971). 'Applications of error-coding techniques to satellite communications', *COMSAT Technical Review*, Vol. 1, No. 1, Fall, pp 183–219.

7 Baseband Signals

In previous chapters we have discussed the basics of radio link design, modulation and coding without considering the specific characteristics of the actual message (or information) signals. This chapter outlines the basic electrical characteristics of the main types of message signals which are carried by satellite communication networks.

Signals carrying actual information are called *baseband signals*. Examples of baseband signals are signals from a telephone set, a computer terminal, a television camera and even several such signals grouped together. It is evident that the radio path is merely a medium to carry baseband signals and hence it is important that a satellite communication system designer be familiar with baseband signal characteristics.

It will be seen that the baseband signal (or simply, baseband) characteristics influence the choice of modulation; multiple access scheme; RF characteristics such as the occupied bandwidth; and the desired immunity from noise.

The type of baseband is determined by the communication requirements of the users. The main types of baseband signals used in satellite communications are *telephony, television* and *data*. In a fixed satellite service, baseband signals may be multiplexed telephonic signals in analog or digital form, or data signals from computers. In a broadcast satellite service, baseband signals consist of television or sound signals.

This chapter begins with a brief introduction to digital signals. We have seen in chapter 5 that the RF spectrum of modulated digital signals is directly proportional to the baseband spectral characteristics. The discussion therefore includes the spectral characteristics of typical digital signals and the effect of filtering baseband signals.

Analog-to-digital conversion of telephone signals has traditionally been performed at telephone exchanges using pulse code modulation and the derivatives of this scheme. The basics of several commonly used analog-to-digital conversion techniques are discussed next.

In recent years, increasingly, analog-to-digital conversion is being done at the telephone set. The trend is being set by the mobile communication communities – both terrestrial and satellite – in order to profit from the advantages offered by digital transmissions. The principles of the voice coding technique used in such applications are outlined.

Television transmissions constitute a significant amount of satellite communications traffic. A section covering the basic characteristics of television signals is included.

When telephony or data traffic between two points is large, it is advantageous to combine signals into a composite baseband. This process is known

188

as *multiplexing*. The principles of digital and analog multiplexers are also discussed.

Finally, traffic considerations are reviewed briefly. This topic is not directly related to the characteristics of baseband signals but provides a basis for predicting the number of satellite channels required for communications between two points, given their traffic needs and the acceptable level of call congestion. Provision of satellite channels is expensive and all telecommunication authorities attempt to lease only as many channels as necessary. Therefore estimation of the number of channels to satisfy the traffic demand is essential. The information is also of paramount importance at the planning stage of earth stations to size the earth station properly and also to incorporate provision for future traffic growth.

7.1 Data

General

Digital signals consist of a series of bi-state pulses or 'bits' – the *low* state is called '0' and the *high* state '1'. Information is superimposed on a digital stream by arranging groups of bits called *words* to represent a quantity. For example, analog signals from a telephone or alphanumeric characters from a computer keypad may be coded into groups of n bit words for transmission.

There are several advantages in using digital systems. These include: high reliability, low cost, less susceptibility to noise, the possibility of using coding to increase noise resistance and the capability of utilizing the advantages of VLSI technology. The main impetus has been the availability of high-reliability, low-cost, light-weight digital integrated circuits operating at low voltages with low-power consumption. It is not surprising that the use of digital signals in telecommunication networks is increasing rapidly. Facsimile, electronic mail and computer interworking are some examples of data communication applications that are expanding rapidly.

An added advantage of digital signals is their ability to be combined, irrespective of the originating source. For example, digitized telephones, television signals, computer data, and facsimiles could all be combined into a single time division multiplexed (TDM) digital stream and transmitted on a single RF carrier. Eventually, most telecommunication networks are expected to become parts of an Integrated Switched Digital Network (ISDN).

Digital multiple access schemes offer distinct advantages. For example, Time Division Multiple Access (TDMA) is an efficient form of accessing a satellite and is increasingly being used in satellite communication networks. The baseband used in this scheme is digital and when input signals are available in analog form, their conversion into digital form becomes essential. Digital signals are now increasingly replacing analog signals in SCPC channels (see chapter 8) both in fixed and mobile communication networks. It may be recalled here that modulation schemes used for digital signals are generally categorized separately to analog signals as optimization of digital data requires a different approach to that for analog signals (see chapter 5).

The type of digital stream, transmission bit rate and occupied bandwidth all have an important bearing on the RF signal. These aspects will be covered in the following sections.

Characteristics of digital signals

A binary digital stream may be transmitted in a number of different ways. When '1' is transmitted as $+V$ volts and '0' as $-V$ volts the signal is said to be a *Non-Return to Zero* (NRZ) digital stream. An NRZ digital stream is shown in figure 7.1(b). Also shown in the figure is a *clock* signal (figure 7.1(a)). The function of the clock is to mark the start and end of each bit in a digital stream. Note that the clock period is equal to the bit time. An identical clock, *synchronized* to the transmitter clock, is used at the receiver to retrieve the digital data. (The concept of bit synchronization can be extended to *network* synchronization. In a synchronized digital network all the earth stations use the clock of a designated reference earth station. An example is the TDMA mode of accessing a satellite by earth stations – all the earth stations accessing a satellite in a time division multiplexed scheme are synchronized to the clock of a reference station.)

In another type of NRZ transmission, a '0' corresponds to 0 volt instead of $-V$ volts. A system which consists of transformers/capacitors in its path is unable to pass DC (i.e., '0's in this form of NRZs). This is the case for many satellite communication sub-systems, such as modulator/demodulators, and for this reason an NRZ with $+V$ to $-V$ excursion is preferred for satellite communications. Figure 7.1(b) shows a bit stream 01010. Note that a continuous stream of '0's and '1's appear as a continuous DC voltage.

Long strings of '0's or '1's cause difficulty in bit synchronization at the receiver, because circuits used for bit synchronization derive the timing information from bit transitions. Also, long continuous strings of '0's and '1's when transmitted produce RF carrier spikes. Under such circumstances, all the RF energy is concentrated within a very narrow bandwidth and therefore the potential of interference to other systems sharing the same frequency is significantly increased. Such a condition is avoided by multiplexing the information digital stream with a 'pseudo-random' sequence which has uniformly distributed '0's and '1's. Thus in the presence of long strings of '0's (or '1's) the pseudo-random sequence continues to spread the carrier power thereby avoiding concentration of RF power at specific frequencies.

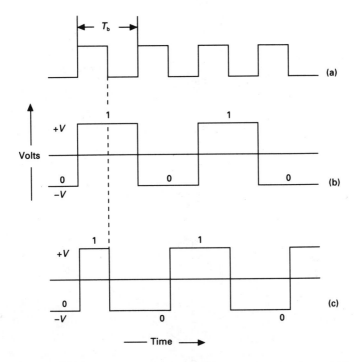

Figure 7.1 Digital stream: (a) clock used for synchronization of bits; (b) NRZ bit stream; (c) bi-phase data stream.

Digital data can also be transmitted as a *bi-phase digital* stream. In a bi-phase digital stream, the data signal returns to '0' in the middle of *each* bit period, as shown in figure 7.1(c). The problem associated with continuous transmission of '0's or '1's in NRZ is avoided in bi-phase transmission. However, because of the higher bit transition rate, bi-phase transmissions require larger bandwidths than do NRZ transmissions. As the radio spectrum is an important resource NRZ signal format is generally favoured in satellite communication systems.

The power spectral density of an NRZ signal, assuming that '0's and '1's are equally distributed, is given by

$$P_{NRZ}(f) = V^2 T_b \left(\frac{\sin \pi f T_b}{\pi f T_b} \right)^2 \tag{7.1}$$

Figure 7.2(a) shows the plot of the power spectral density. For comparison, the power spectral density of a bi-phase data stream is shown in figure 7.2(b). Theoretically, the spectrum extends up to infinity. In practice, the bandwidth is limited intentionally because spectrum is a limited resource

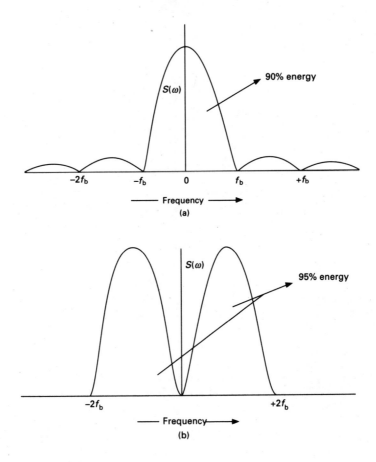

Figure 7.2   Power spectral density $S(\omega)$ of digital signals: (a) NRZ data;
(b) bi-phase data.

and physical devices have limited bandwidth. The consequence of band-limiting a pulse stream is shown in figure 7.3. The band-limited pulses are broadened, resulting in spill-over of energy into adjacent pulses. This type of noise is called *inter-symbol interference* (or ISI). ISI can be minimized by using a filter possessing a rectangular-shaped amplitude response, as shown in figure 7.4(a) (adapted from Schwartz, 1987). Its impulse response is shown in figure 7.4(b). The response is seen to cross zero every $1/2B$ seconds, where B is the filter bandwidth. If pulses are transmitted at a rate of $2B$ Hz and sampled every $1/2B$ seconds, the zero cross-over point, at the receiver, the samples should exhibit zero ISI. Figure 7.4(b) shows the response of two adjacent pulses and the sampling instants giving zero ISI (see Schwartz, 1987 for more details). Thus the theoretical maximum bit rate of a channel with bandwidth B is $2B$. This is known as the *Nyquist rate*. The rectangular

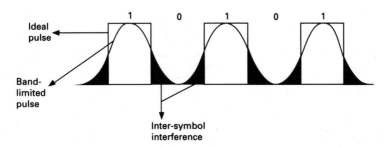

Figure 7.3   Inter-symbol noise in a digital stream as a consequence of band-limiting a pulse stream.

filter shown in figure 7.4(a) cannot be realized in practice. Moreover, the required timing accuracy for sampling is extremely difficult to achieve in real circuits.

Nyquist has proposed one possible class of practically realizable filters capable of achieving zero inter-symbol interference. The impulse response of such filters is given by,

$$h(t) = \left(\frac{\omega_c}{\pi} \right) \frac{\sin \omega_c(t)}{\omega_c(t)} \cdot \alpha(t) \qquad (7.2)$$

where ω_c = angular cut-off frequency of the rectangular filter (see figure 7.4(a))

$\alpha(t)$ = damping factor.

The impulse response consists of a $\sin(x)/x$ term which ensures zero crossings every $1/2T$ seconds multiplied by a factor $\alpha(t)$ which dampens the response as a function of time, thereby reducing the amplitude of the tails of the $\sin(x)/x$ terms. The effect of clock jitter is therefore reduced, because the inter-symbol noise caused by imprecise sampling times affects the tails of the $\sin(x)/x$ response.

One example of this class of filter is the *raised cosine filter*. This type of filter is commonly used in satellite communication systems. A raised cosine filter is shown in figure 7.4(a). Its frequency response is given by

$$H(\omega) = \frac{1}{2} \left[1 + \cos \left(\frac{\pi\omega}{2\omega_c} \right) \right] \quad |\omega| \leqslant 2\omega_c \qquad (7.3)$$

$$= 0 \text{ elsewhere}$$

A compromise can be exercised between the bit rate b, the transmission bandwidth B and the filter implementation complexity, defined in terms of roll-off factor, r. The roll-off factor determines the slope of the filter

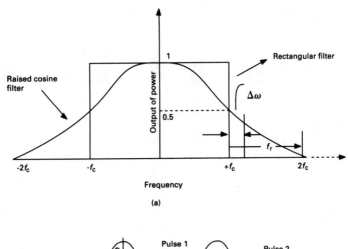

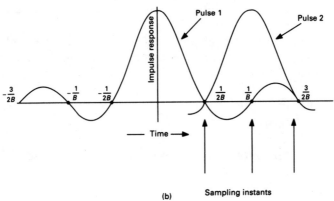

Figure 7.4 (a) Frequency response of a rectangular filter and a raised cosine filter. The roll-off factor is given as f_r/f_c (output power $= 0$, for $\Delta\omega > f_r$).

(b) Impulse response of a rectangular filter showing sampling instants when inter-symbol interference is zero.

response. The following relationship applies:

$$B = (1/2T)\,(1 + r) \tag{7.4}$$

where B = bandwidth
         T = bit time
         r = roll-off factor.

Defining bit rate b as $1/T$, (7.4) can be re-written as

$$B = b\,(1 + r)/2 \qquad\qquad (7.5)$$

The roll-off factor, r, is given as the ratio f_r/f_c where f_r is the offset from the cut-off frequency ω_c of the ideal rectangular filter where power reduces to zero – as shown in figure 7.4(a). The roll-off factor, which can assume any value between 0 and 1, determines how rapidly the filter cuts off. The value of 0 corresponds to a filter with a rectangular passband. Thus the implementation complexity of the filter reduces as its roll-off factor is increased, but from equation (7.5) we note that the required transmission bandwidth also increases for a given bit rate. Clearly a compromise may be exercised.

The treatment given here for minimizing ISI is also applicable to an RF signal modulated by a digital stream. This happens to be so because digital modulation is achieved by multiplication of the baseband signal with a high-frequency carrier. The spectrum of the product signal consists of the spectrum of the baseband signal transferred around $+f_c$ and $-f_c$, where f_c is the RF frequency. To obtain an optimal carrier-to-noise ratio at the receiver, the raised cosine filter can be split between the transmitter and the receiver, each end using a *square root raised* cosine filter.

7.2 Telephony

In spite of a large growth in data traffic in recent years, voice traffic exceeds data traffic and forecasts indicate that voice traffic will continue to remain dominant in the near future. At present, signals from most telephone sets are analog. Telephone signals are sent to a local exchange in an analog form where the signals are digitized for transmission on trunk routes. But in many applications, telephone sets capable of directly providing digitized voice signals are being introduced. In this section we shall outline the main characteristics of both analog and digital telephone signals.

Analog telephony

Based on extensive studies of voice signals, it is now well known that most of the speech energy during normal conversation lies between 0 and 3400 Hz. Therefore a bandwidth of 3–4 kHz is allocated to each voice channel. There are minor variations in the precise start and end frequencies used between systems of various telecommunication organizations. The CCITT, the international committee responsible for maintaining standards for telephone and telegraph systems, recommends the range of telephony signals as 300–3400 Hz.

It has been seen that baseband signal levels affect the characteristics of the modulated carrier when frequency modulation is used. It is therefore also necessary to characterize the amplitude of the speech signals. Unfortunately speech signal levels are random in nature and therefore the mean and the maximum levels are often used to quantify speech signal power. Studies show that mean speech levels vary in the region of -15 to -18 dBm, with maximum excursions up to 0 dBm. The CCIR recommends the mean speech level as -18 dBm0. Other organizations recommend different values. Modulators are usually aligned with a 800 Hz, 0 dBm0 sinusoidal test-tone corresponding to the peak speech signal level. A sinewave is used for alignment because of the difficulty in making adjustments with randomly varying speech signals (see also section 5.4 for speech levels).

As there are several points where speech signals may be accessed before reaching an earth station modulator, it becomes necessary to define a reference point, with respect to which other points in the link are specified. A 0 dBm signal is given the unit of 0 *dBm0* at the reference point.

Digital telephony

As mentioned above, analog telephone channels are digitized and multiplexed at a telephone exchange for transmission on trunk routes. In recent years, considerable effort has been spent in digitizing voice at the telephone set. The main need has arisen from the advent of the mobile satellite service (MSS) and the terrestrial cellular mobile communication system, both of which operate in a radio frequency band which is in great demand. Consequently, considerable effort is being spent on improving the spectral efficiency of transmission, as the spectrum becomes scarcer. The use of spectrally efficient voice coding coupled with efficient modulation/multiple access techniques provides a viable solution. For example, INMARSAT's standard-M mobile system provides voice communication to small mobile terminals using voice coders operating at about 6 kbps.

The following sections briefly review various analog-to-digital conversion techniques used for digitizing voice signals. Some of these techniques are applicable for digitizing any type of analog signal.

Voice coding (*Bayless* et al., *1973; Flanagan* et al., *1979; Holmes, 1982*)

A fundamental requirement of a voice coder is to provide the highest voice quality at the lowest possible bit rate and least cost. However, the voice quality generally reduces and implementation complexity increases as the bit rate of a voice coder is reduced. In practice, therefore, a compromise is

Table 7.1   Main characteristics of broadcast and toll quality signals.

Application	Bandwidth (kHz)	Signal-to-noise ratio (dB)	Harmonic distortion (%)
Brodcast	> ~7	> 30	2–3
Toll	~3.1	> 30	2–3

exercised between bit rate reduction and cost. All voice bit rate reduction techniques attempt to minimize the inherent (significant) redundancy in normal speech. It may be worth noting that human cognitive process cannot follow information rates of more than a few tens of bits/s (Scharf, 1970). However, reducing the bit rate to such low levels makes cognitive processing difficult, as humans have extremely individualistic attitudes towards speech reception (Holmes, 1982).

There are two broad categories of speech coders – *waveform* coders and *source* coders:

1. *Waveform coders* sample the input analog waveform and convert each sample into a binary code. At the receiver, a digital-to-analog converter is used to reconstruct the analog speech. In this type of coding the shape of the waveform is preserved and consequently the technique can be used for any analog signal. Bit rates for speech coding typically range between 16 and 64 kbps.
2. *Source coders*, also called *Vocoders*, do not attempt to preserve the shape of the speech signal but extract specific characteristics of speech which are each coded and transmitted. The bandwidths of such parameters are relatively small and therefore source coders can code speech at very low bit rates. Bit rates of the order of 0.05–4 kbps are possible.

Some types of coders are based on combining the best features of waveform and source coders. These are called *hybrid coders*.

The choice of a voice coding technique depends on the application. The main categories of application are broadcast and telephony. Table 7.1 provides a basic set of specifications for these applications. The quality of signal used in public telephone systems is known as toll quality. Reduction in specification below the toll quality can continue to produce intelligible speech adequate for communication – sometimes such a quality is referred as 'communication quality'. Communication quality is acceptable in applications where constraints such as limited bandwidth, cost or interference

Table 7.2 Main characteristics of voice coders.

Speech coder	Main features
Waveform	• Reproduces waveforms • Independent of signal type (speech, music, video, etc.) • Robust to talker characteristics • Bit rates range from 16 to >64 kbps: moderate bit rate economy possible • Toll to broadcast quality possible • Complexity increases with reduction in bit rate
Vocoders	• Parameters of speech model generated (a reasonably accurate description of speech possible with data rate ~1–3 kbps) • Talker-dependent characteristics • Bit rate ~0.05–4.8 kbps • Synthetic quality • High economy in transmission bandwidth
Hybrid	• Combines features of source waveform • Some advantages of both • Bit rate ~4.8–16 kbps • Communication/toll quality

override the need for a high voice quality. Examples of such applications are closed user environments (for example, internal communication in airports or taxi fleets) and military communication. When coding bit rates are reduced considerably (currently in the range ~2–4 kbps) the voice begins to lose recognizability and appears synthetic, i.e. unnatural and 'robotic'. Synthetic quality speech is used in robotics, toys and machine-generated speech.

Table 7.2 (Flanagan *et al.*, 1979) summarises the main features of various types of speech coders.

Waveform coders

Waveform coders can operate in the time or the frequency domain. The main types of *time domain coders* are

(i) Pulse code modulation (PCM)
(ii) Differential PCM (DPCM)

(iii) Adaptive DPCM
(iv) Delta modulation.

The main type of *frequency domain coder* is known as a sub-band coder.

(i) *Pulse code modulation*
A basic PCM coder consists of a sampler, a quantizer and a coder. The first stage consists of a sampler. The process of sampling involves reading off input signal levels at discrete points in time. Hence, the sampled signals consist of electrical pulses which vary in accordance with the amplitude of the input signal. An analog signal of bandwidth B Hz must be sampled at a rate of at least $1/2B$ s to preserve its wave shape when reconstructed. This is known as the *Nyquist sampling rate*, based on Nyquist's sampling theorem. (The Nyquist theorem states that when a signal is band-limited to B Hz, the information content of the signal is not lost, provided that the signal is sampled at least at $2B$ Hz.)

The amplitude of each pulse is then quantized into one of 2^N levels where N is the number of bits used to code each sampled pulse. In the quantization process the sampled signal is approximated to the nearest quantization level. For example, an 8-bit PCM uses 2^8 (256) quantization levels. The granularity of the quantized signal reduces as the number of bits (N) used for coding is increased. The quantized signal is then coded into its equivalent N-bit PCM digital *word*. In all, for a signal of bandwidth B the PCM bit rate is $2BN$.

A PCM bit stream is decoded by mapping each received word into the corresponding quantized level. Finally, the analog message signal is recovered by low-pass-filtering the decoded quantized signal.

The process of quantization introduces distortion into the signal, making the received voice signals raspy and hoarse. This type of distortion is known as 'quantization noise'. Note that quantization noise is only present during speech. When a large number of quantization steps, each of ΔS volts, are used to quantize a signal having an rms signal level S_{rms}, the signal-to-quantization noise ratio S_{qn} is given by

$$S_{qn} = \frac{(S_{rms})^2}{(\Delta S)^2/12} \qquad (7.6)$$

A large number of bits are necessary to provide an acceptable signal-to-quantization noise ratio throughout the amplitude dynamic range. Analysis of speech signal shows that smaller-amplitude levels have a much higher probability of occurrence than high-amplitude levels. Based on this premise, it would appear that a scheme which uses progressively finer quantization step size as the amplitude of the signal is reduced would give an overall reduction in the number of quantization levels and hence coding bit rate. This is in fact true and is achieved in practice by a process called *companding*. A *compandor* consists of a *comp*ressor at the transmitter and an ex*pandor* at the receiver. A compressor possesses a non-linear level response such that higher-level signals are compressed relative to the lower-level signals, whereas the expandor has a complementary response so that the signal amplitudes are restored at the output of the expandor. To ensure that systems can interwork, compandor non-linearity curves have been standardized (for example, see Freeman, 1981). At this stage we digress briefly from PCM in order to distinguish between the '*instantaneous*' compandors mentioned here and *syllabic compandors*. Whereas the compandors mentioned here operate on instantaneous signal amplitudes, syllabic compandors operate on the same principle but compression/expansion are performed on *syllabic* variations which are basically enveloped variations of speech. A subsequent section discusses syllabic companding in more detail.

PCM is widely used in FSS when digital baseband signals are necessary; for example, when using a time division multiple accessing scheme baseband signals are often sent to an earth station in an analog form from a switching centre. These signals are converted into digital form using PCM, multiplexed and suitably formatted in the earth station *interface unit* before transmission.

(ii) *Differential PCM*

Differential PCM schemes take advantage of the high degree of correlation between adjacent speech samples. It can be shown that if

$$D = x(r) - x(r - 1) \tag{7.7}$$

where　$x(r)$　　$= r$th speech sample
　　　　$x(r - 1) = (r - 1)$th speech sample
then

　　　　Variance of (D) < Variance $x(r)$

In other words, the dynamic range of difference signal is less than the actual speech sample. The difference signal can therefore be digitized by using a lower number of bits. In differential PCM this difference signal is transmitted instead of the actual speech signal, giving differential PCM a coding bit rate advantage over PCM. On the other hand, if the same number of bits is retained in both schemes, the signal-to-quantization noise of the differential PCM scheme increases over the PCM by a factor G:

$$G = 1/2(1 - C_1) \tag{7.8}$$

where C_1 is the correlation between adjacent samples.

Further improvement can be obtained if the difference D is taken between the present value $x(r)$ and its predicted value $ax(r)$. Prediction is based on the basis of assigning weights, obtained by statistical analysis of the speech signals of previous several samples. Once selected, these weights remain fixed for a system. The differential PCM technique provides a 2-bit coding advantage over the PCM scheme.

(iii) *Adaptive differential PCM (ADPCM)*

In *adaptive DPCM* the prediction coefficients are updated regularly (every 10–30 ms) in accordance with speech statistics. ADPCM can provide coding advantages of the order of 3 bits over PCM.

(iv) *Delta modulation*

The *delta modulation digital* conversion technique is a very simple scheme, using a one-bit quantizer. To understand the digital conversion technique, we define a difference signal $\Delta m(t)$:

$$\Delta m(t) = m_c(t) - m_c'(t) \tag{7.9}$$

where $m_c(t)$ is the message signal;

$m_c'(t)$ is a reconstructed version of $m_c(t)$ obtained by integrating the output of the encoder.

The difference signal is fed into a one-bit quantizer which produces a positive pulse (1) when $\Delta m(t)$ is positive and a negative pulse (0) when $\Delta m(t)$ is negative. The output stream then constitutes the coded bit stream $m_c(t)$. The signal is recovered at the receiver, simply by integrating the received bit stream. The delta modulation scheme has two inherent problems – *slope overload* and *idle channel noise*. Whenever the input signal amplitude changes more rapidly than the time constant of the integrator in the coder's feedback loop, the reconstructed signal becomes distorted causing a signal distortion termed *slope overload*. Slope overload occurs whenever the slope of the message signal exceeds G_d/T_s (where G_d = coder's feedback loop gain, and T_s = sampling time of the coder). This problem can be solved by reducing the sampling time interval or increasing the gain of the feedback loop. At the other extreme, when the level change in the input signal is too low, a delta modulator suffers from *idle noise*. Again, the problem is solved by adjusting the gain of the feedback loop.

Adaptive delta modulation techniques change the gain of the modulator in real time according to the requirements. Using an adaptive modulation technique, it is possible to code speech with acceptable quality at bit rates around 20 kbps. While this order of bit rate can be readily achieved by other waveform coding techniques, the main advantage of delta modulation is its extremely simple hardware.

Frequency domain coders

In frequency domain coders, speech signal is segmented into small-frequency bands, and each band is then coded separately using a waveform coder. It is therefore possible to use finer quantization steps in specific frequency bands to which listeners are more perceptive. Additionally the encoding accuracy may be altered dynamically between bands to suit specific requirements of the speech waveform. For example, encoding accuracy may be reduced considerably in bands containing very little energy. There are a number of techniques for implementing frequency domain coders, ranging from simple to complex. For example, in one sub-band coding scheme the speech is filtered into 4–8 sub-bands, each sub-band then being down-converted so that its lower band edge corresponds to 0 Hz. Next, each channel is encoded individually using an ADPCM scheme. Finally all the channels are multiplexed and transmitted, A reverse process is employed at the decoder. More complex techniques use varying degrees of signal processing and prediction (Flanagan *et al.*, 1979).

(v) Source coders (or Vocoders)

Source coders (also called Vocoders) do not code the speech waveform, but instead the parameters of a speech-generating model. Most Vocoders use the model shown in figure 7.5 (Flanagan *et al.*, 1979; Holmes, 1982) as the basis for speech generation. The main difference between Vocoders lies in the technique used for extracting the model parameters at the coder and synthesizing the received parameters to recreate the speech at the decoder.

The model is based on the following assumptions:

1. There are two independent mechanisms involved – (a) sound generation, called the *source*, and (b) intelligence modulation, called the *system*. (Note the analogy with amplitude modulation, discussed in chapter 5.)
2. These mechanisms can be linearly separable from each other. Thus each parameter can be extracted, and transmitted separately and reassembled at the receiver to give the synthesized speech.
3. Speech sounds are either 'voiced' or 'unvoiced'. Voiced sounds are generated by quasi-periodic vibrations of the vocal cord – the fundamental frequency of vibrations is called *pitch* (e.g. the letter 'a' – pronounce this letter and the vibrations in the vocal chord can be sensed). Unvoiced sounds are produced by air turbulence caused through sudden release of air by the lips or the tongue etc. (e.g. pronounce the letter 's' as in 'miss'). These sounds have different spectral characteristics. The short-term (a few tens of milliseconds) spectrum of a voiced sound is periodic whereas that of an unvoiced sound is similar to the spectrum of random noise.

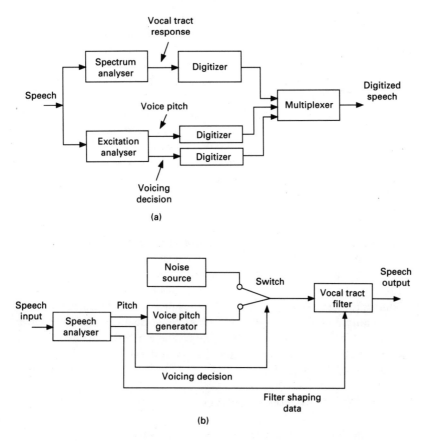

Figure 7.5   The main blocks of a Vocoder: (a) coder; (b) decoder.

Summarizing, it is noted that speech consists of a sound-carrying component which is modulated by the intelligence-carrying component. These components can be separately transmitted and synthesized.

The model used by Vocoders is illustrated in figure 7.5. The coder (figure 7.5a) consists of an excitation analyser which extracts information on the
type of sound, i.e. voiced or unvoiced, and the pitch frequency if the sound
is voiced. The spectrum analyser extracts the envelope of the sound signal
giving the intelligence component of speech. These parameters are digitized
and multiplexed, giving the digitized speech. The decoder (figure 7.5b) consists of a speech analyser which gives an estimate of the speech parameters.
The voicing decision drives a switch to select the desired sound-generating
source. When the speech is voiced the switch is connected to the voice pitch
generator which also receives the pitch information. The unvoiced sound is
generated by a noise source. The sound is then passed through a vocal response filter in which the intelligence component is combined with the sound

Table 7.3   Comparison of complexities of various types of coder in terms
of approximate relative counts of logic gates (Flanagan *et al.*,
1979).

Relative complexity	Coder
1	Pulse code modulation
1	Adaptive pulse code modulation
5	Sub-band coders
50	Adaptive predictive coders
100	Hybrid coders
100–500	Vocoders

through the filter-shaping data obtained from the speech analyser. The output of the system provides a synthesized version of the transmitted speech
signal (see the literature for details, for example, Flanagan *et al.*, 1979).

In recent years, significant research effort has been devoted to improving
Vocoder implementation techniques. The result is a steady improvement in
speech quality and a reduction in bit rates. Good quality voice at 2.4 kbps
seems a possibility in the next few years. There is significant interest in
using such coders in satellite communication systems.

Comparison of voice coding techniques

A system designer is interested in comparing various coding techniques
with a view to selecting the best on offer for the specified application.
Table 7.2 has summarized the salient features of each type of voice coding
technique and table 7.3 (Flanagan *et al.*, 1979) gives a comparison of the
various types of coder in terms of relative complexity, stated in terms of
an approximate number of logic gates. However when choosing a coder it
must be realized that voice coding is one of the most rapidly growing
areas. With the availability of low-cost VLSI having powerful signal processing
capabilities, the cost of voice coders is reducing very rapidly and this cost
reduction may not be directly proportional to the logic gate counts given
in table 7.3. It seems likely that voice coders at 2.4 kbps will be introduced
in satellite mobile terminals during the 1990s. It is worth noting here that
reduction in coding bit rate in itself may not be adequate for accepting a
coder for a mobile terminal. Important criteria in the selection of a voice
coder for mobile communication are the resistances of the coder to noise
introduced by (a) the local acoustic environment and (b) an RF fading environment.

Generally, the voice quality of waveform coders begins to reduce below about 16 kbps with rapid reduction occurring below about 8 kbps. Vocoders achieve the highest quality at about 4.8 kbps without any further improvement in quality above this bit rate. In the intermediate range, between 4 to 16 kbps, hybrid Vocoder methods provide a good compromise, benefiting from the advantages of each.

Satellite communication systems benefit greatly by the use of reduction in voice coding bit rate because transmission bandwidth and hence transmission cost directly depend on voice coding bit rate.

Syllabic compandors

Syllabic compandors have been mentioned in the preceding section dealing with PCM. A distinction was made there between *instantaneous* companding used in waveform coders and *syllabic* companding. A syllabic compandor consists of a compressor at the transmitter, which reduces the dynamic range of speech signals by imparting more gain to weaker syllables; and an expandor at the receiver, which restores the speech signal to its original levels. It has been seen in chapter 5 that when frequency modulation (FM) is used, the received signal quality depends on the baseband amplitude of a signal. Therefore, in FM lower-amplitude signals will operate at a lower signal-to-noise ratio during a conversation. Because the probability of occupancy of lower signal levels in normal speech is higher, companding provides an improvement in the average signal-to-noise ratio by boosting the signal-to-noise ratio of signals at lower amplitude, when using FM. In a multicarrier environment, companding avoids carrier overload and cross-talk by reducing the dynamic range of speech.

Compression (or *expansion*) *ratio* is defined as the ratio of the input to the output power of the compandor. A compression ratio of 2:1 is commonly used. This implies that the dynamic range of the speech has been reduced by half. The corresponding expansion ratio is then 1/2. It is necessary to specify a reference level for a compandor. Compression at other amplitudes are specified relative to this reference. The CCITT recommends 0 dBm0 signal at 800 Hz as a reference.

Companding range is defined as the input range over which the compandor operates without distortion. Typical values of companding range lie between 50 and 60 dB.

Other important parameters of a compandor are the *attack time* and the *recovery time* which are respectively the response of a compandor to speech signals applied and removed suddenly.

A parameter of importance to a system designer is the signal-to-noise ratio advantage offered by compandors. In SCPC systems, advantages of the order of 15–19 dB have been reported. The maximum available advantage is esimated as 20 dB (Freeman, 1981). Remember that the companding advantages mentioned here refer only to FM systems. [The reader should however note that there is some ambiguity in the definition of companding advantage because it is defined differently by authors (Albuerque, 1989).]

7.3 Television signals

A television camera converts a scene into electrical signals by scanning its image electronically. The smallest unit of the image determines the resolution of the picture and is known as a *pixel*. For each pixel the camera produces luminance and chrominance signal components. The chrominance signal gives the grey level which shows the relative shade of the pixel between black and white. The luminance signal gives the colour information. Scanning is performed very rapidly line-by-line, progressing from the top to the bottom. Each completed scan of an image is organized as a *frame*. Thus we have associated with television signals a *line rate* and a *frame rate*.

Black and white television became operational before colour television and therefore when colour television systems were introduced, it was made mandatory that reception of existing black and white television sets should remain unaffected by colour transmissions. To comply, colour television signals consist of *luminance* signals which are used by black and white television sets and *chrominance* signals which carry colour information for use by colour receivers. The chrominance signals consist of two components embedded in the amplitude and phase respectively of a sub-carrier. The amplitude of the colour sub-carrier gives the *saturation* of the colour, and the phase the *hue* (or shade). The colour sub-carrier is multiplexed with the luminance signal so as to cause minimal interference to the chrominance carrier.

Several standards are in use throughout the world (see CCIR Rec 568, 1982 for details). They differ in the line and frame rates, bandwidth and the colour information multiplexing technique. The most widely used systems are the NTSC and PAL systems. The NTSC system has a line rate of 525 lines/s and a frame rate of 50 Hz and is used in countries such as Japan and the USA. The PAL system consists of a line rate of 625 and a frame rate of 50 Hz. The PAL system is used in European and some Asian countries. The bandwidth of a television signal varies from 4.2 MHz (system M) to 6 MHz (systems D, K and L). Line rates of 525 or 625 lines are used by all systems.

Most of the satellite television links use frequency modulation. While discussing frequency modulation (chapter 5), it was shown that by using pre-emphasis/de-emphasis circuits an advantage in signal-to-noise ratio can be achieved. Additionally, the effect of noise perceived by a viewer does not have a uniform response. Thus a weighting filter which has a response closely following the human eye reponse is introduced for measuring noise at the receiver. This gives a net advantage, called *weighting advantage*, in the measured signal-to-noise ratio. As already mentioned, the net signal-to-noise ratio advantage (pre-emphasis/de-emphasis plus weighting) depends on the television standard used varying from 12.8 dB (system M) to 18.1 dB (systems D, K and L).

Another consideration in satellite television link design is the need to transmit the audio. Usually this is done by modulating the sound signals on a sub-carrier which is multiplexed with the television signal. The baseband television signal then consists of the composite picture and sound signals.

There is now growing awareness of the capabilities of digital television. Using coding techniques which eliminate the redundancy in picture it is possible to digitize television signals at a bit rate approaching 500 kbps. Thus a viewable quality signal is possible at RF bandwidths as low as 1 MHz. This is a vast improvement over current analog satellite television which typically requires 30 MHz. For example, digital television broadcasts regularly to mobile ship earth stations via the INMARSAT mobile satellite system using an RF bandwidth of less than 1 MHz. Another advantage of digitizing television picture is that other useful information, including the picture audio, can be readily multiplexed with the picture signal.

An area of growing interest is high definition television (HDTV) broadcasts via satellites. Development effort is underway in several countries and experimental broadcasts are regularly demonstrated. The Barcelona Olympics were broadcast on an experimental basis using European satellites. Standards have yet to be defined but it is possible that more than one standard will emerge.

High definition television has a higher resolution, implying a higher number of pixels/unit area together with a larger aspect ratio. Together, these improvements permit a much wider screen and near cinema quality picture. Some forecasts claim that HDTV will eventually replace the current television, much as colour television has replaced black and white television. Frequency allocations have not yet been assigned exclusively for HDTV. Because of the large bandwidths required it is possible that as yet relatively un-utilized bands (e.g. 20–30 GHz) will be used.

7.4 Multiplexing of baseband signals

When it is essential to establish communication links between large traffic sources such as international gateways, it is advantageous to combine traffic as a composite baseband signal. In this way all the users share the cost associated with the use of the channel.

The process of combining the baseband signals is known as *multiplexing*. The reverse process of extracting individual baseband signals is called *de-multiplexing*. Multiplexing can be performed by stacking the signals in frequency. This type of multiplexing is known as *frequency division multiplexing* (FDM). Alternatively the baseband signals are arranged in a serial time sequence called *time division multiplexing* (TDM). TDM is best suited to digital signals.

Telecommunication systems must be designed to interwork with each other – both within a nation and internationally. To enable interworking to be successful, several multiplexing standards have been developed for FDM and TDM. The CCITT has proposed both FDM and TDM standards for international communications. However, some countries choose to use their own standards. Both the FDM and the TDM multiplexing standards follow a hierarchical structure, permitting multiplexing from a few tens to thousands of telephone circuits for the FDM standard and from a few tens of kbps to over several tens of Mbps of digital data for the TDM standard. The FDM standard is applicable only to telephony signals. However, the digital standard is applicable to all types of baseband signal, provided such signals are digitized and the bit rate is compatible with the appropriate level of the selected multiplexing standard.

Frequency division multiplexing

Frequency division multiplexing of telephony channels is well illustrated by considering the CCITT multiplexing scheme. Figure 7.6 shows the CCITT multiplexing plan for obtaining the first level of multiplexing known as a *group* which multiplexes 12 telephone channels. Each telephone channel is allocated a bandwidth of 4 kHz. The actual telephone signal extends from 300 to 3400 Hz, the remaining 900 Hz being used as the guard band. Guard bands protect each telephony channel against energy spill-over from adjacent channels. Each incoming telephone channel is mixed with a carrier. The mixing process produces an upper and lower side band, out of which only the lower side band is retained through filtering. The lowest carrier frequency used is 64 kHz, each succeeding carrier being incremented in 4 kHz steps. As there are 12 telephone channels in a group, the highest carrier frequency is 108 kHz.

In the hierarchical plan, five groups are combined to produce a *supergroup* which consists of 60 telephone channels and extends from 312 to 552 kHz. Again, five supergroups may be combined to form a mastergroup comprising 300 telephone channels, and extending from 812 to 2044 kHz. Moving further up the hierarchy, three mastergroups may be combined to form a *super-mastergroup*. The super-mastergroup extends from 8516 kHz to 12 388 kHz, and consists of 900 channels. Figure 7.7 shows the CCITT FDM plan.

Although the current trend is to use digitized signals and the digital satellite multiple access technique, even now a vast number of telephone circuits are transmitted in analog form using FDM telephone signals.

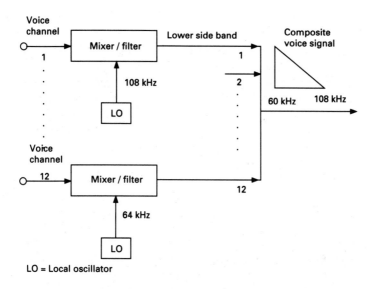

Figure 7.6   Formation of a group in CCITT multiplexing scheme.

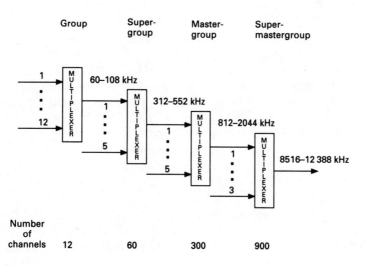

Figure 7.7   CCITT multiplexing plan.

Digital multiplexing

In digital multiplexing, data from various sources are combined and arranged as a sequential bit stream. The multiplexed data stream requires a higher bit rate as a consequence of this combination. Consider multiplexing n data sources operating at bit rates b_1, b_2, etc. as a composite bit stream b_c. Then the bit rate b_c is given as $b_c = b_1 + b_2 + \ldots + b_n + b_o$, where b_o is an overhead used for formatting and other functions. For example, when a 1200 bps and two 2400 bps data streams are combined, the multiplexed data rate is 6000 bps, plus a small overhead.

The multiplexed bits are arranged in frames. Each frame is further segmented into sub-frames (or time slots) which are allocated to various users. More than one time slot could be allocated to a user if required. Figure 7.8 shows a frame structure. When data sources operate at different bit rates, the required sampling instants for data sources differ. A timing plan is organized as part of the design to ensure proper sampling instants for each data source. Provision must also be made to account for small statistical variations in bit rate from various data sources. Other considerations include the need to impart the sampling timing information to the receiver, and provision of means for distinguishing the start of frame and time slots at the receiver. These are achieved by adding additional start-of-frame and synchronization bits to the frame, which are used at the receiver to extract the necessary information.

Multiplexers may be broadly categorized as low or high bit rate. *Low bit rate multiplexers* are used for transferring data such as facsimiles over analog telephone lines or over SCPC satellite links. Typical bit rates of such multiplexers are between 1200 and 9600 bps in steps of 1200 bps. *High bit rate multiplexers* are used for combining large amounts of information such as used on telephone trunk routes or between medium or large earth stations. To permit inter-operation between systems, several multiplexing standards have been developed. The most well known standards are the CCITT multiplexing plan and the standard developed by Bell Systems in the USA. The Bell Systems standard uses 24 64-kbps channels at the first level of hierarchy, whereas the CCITT system uses 30 64-kbps channels plus two 64-kbps channels for signalling and other system functions. The CCITT hiearchy is shown in figure 7.9. The hierarchy consists of 4 levels, beginning at 2.048 Mbps and going upwards to 139.264 Mbps.

7.5 Traffic considerations

One of the first considerations when planning a satellite system is the need to forecast the traffic handling capacity of the space segment. Similarly, to size an earth station the owners need to estimate the traffic handling re-

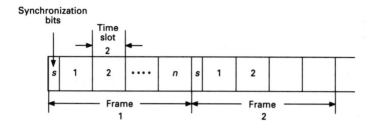

Figure 7.8   Structure of a TDM frame.

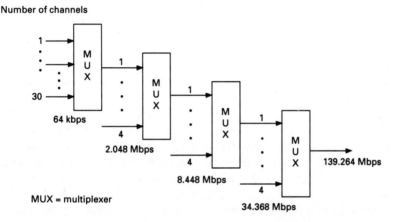

Figure 7.9   CCITT multiplexing hierarchy for digital streams.

quirement of their earth station. It is then necessary to estimate the number of RF channels required to carry the traffic to enable a system designer to estimate the desired satellite (or earth station) EIRP and bandwidth. Traffic theory is used for this purpose. We shall not attempt to delve into details of traffic engineering as it is not the purpose of this section. We shall just summarize some aspects of the subject that are useful in basic satellite system design. The interested reader should refer to the literature for details (e.g. Bear, 1980).

Consider a satellite link used to provide telephony traffic. Traffic carried by this link is given by

$$E = C_a H_t \text{ Erlangs} \tag{7.10}$$

where, C_a = average calling rate (calls/hour)
H_t = average holding time (in hours).

The unit of traffic is the *Erlang*. It has been observed that traffic in any network follows a diurnal behaviour. Most business traffic is carried during

Table 7.4   Traffic matrix in a four earth station network.*

Earth station	1	2	3	4
1	—	E_{12}	E_{13}	E_{14}
2	E_{21}	—	E_{23}	E_{24}
3	E_{31}	E_{32}	—	E_{34}
4	E_{41}	E_{42}	E_{43}	—

* The elements of the matrix show the traffic in Erlangs between earth stations. E_{mn} is the traffic in Erlangs from earth station m to earth station n. Note that for duplex traffic, the matrix elements above and below the diagonal are mirror images. (A duplex circuit consists of a two-way communication circuit, such as voice.)

working days and is characterized by a marked peak as business picks up. Therefore the network traffic (or traffic of an individual earth station) is usually sized as peak traffic carried during a working day. It is interesting to note that when a satellite serves areas covering several time zones, the peak traffic period at an earth station need not coincide with the peak traffic period of the network. This natural staggering helps to distribute satellite EIRP usage. In a fixed satellite network the traffic between various destinations is represented as a traffic matrix, as shown in table 7.4. When there are n stations in the network, there are $n(n - 1)/2$ number of paths for each direction of traffic.

Typically a network (or an earth station) serves hundreds of users. To estimate the number of channels required it is recognized that all the end users are unlikely to make calls simultaneously. Therefore it is wasteful to provide one channel for each user. Furthermore a certain degree of channel congestion is acceptable to the users. Congestion here refers to unavailability of channels. This implies that a user may obtain an 'engaged' tone for a small period of time during peak traffic period. Congestion is quantified as *grade of service* (GOS), defined as the ratio

$$\text{GOS} = C_b/C_a \qquad (7.11)$$

where C_b = number of blocked channels because of congestion
        C_a = number of call attempts made.
The GOS is often defined as a percentage figure. A 10% GOS means that out of 100 calls attempted, 10 were blocked owing to congestion.

Two models have been developed to estimate n_k, the required number of channels to serve the kth destination – the *Poisson* and the *Erlang* models. Each model provides the number of channels to meet the traffic requirement as a function of grade of service.

The *Poisson* model assumes that when a call is blocked, the user continues the attempt to make a call. The Poisson equation is given by

$$GOS = e^{-E} \sum_{n+1}^{\infty} \frac{E^n}{n!} \qquad (7.12)$$

where E = traffic level to be served

n = number of trunk lines.

The other model, known as the *Erlang-B*, is widely used in Europe and is recommended by the CCITT. This model assumes that calls arrive randomly and blocked calls are lost, i.e. the user does not attempt to make the call again. The Erlang-B equation is given by

$$GOS = \frac{E^n/n!}{\sum_{k=0}^{n} (E^k/k!)} \qquad (7.13)$$

where E and n have been defined in (7.12)

The Erlang-B equation becomes increasingly difficult to compute as the number of channels increases. Therefore look-up tables are often used (e.g. Morgan and Gordon, 1989). Gordon and Dill (1978) have proposed a method well suited for computation.

When the traffic between earth stations is low, fixed allocation is wasteful because channels remain un-utilized for long periods. To serve such thin routes it is advantageous to have a shared pool from which earth stations are assigned channels on an as-needed basis, i.e. on demand. This type of arrangement is called *demand assignment* (see chapter 8). To estimate the number of channels in a demand assignment scheme, the total traffic at the peak demand time is aggregated and the applicable traffic model used. It can be shown that the advantage of demand assignment over fixed assignment increases as the traffic between earth stations reduces. A more detailed comparison of fixed and demand assigned schemes is given in chapter 8 when discussing the frequency division multiple access scheme.

Problems

1. Write brief notes on the following:
   (a) differential pulse code modulation;
   (b) sub-band coders;
   (c) companding;
   (d) grade of service.
2. Distinguish between waveform coders and source coders as applied to the digitization of voice.

Briefly describe the principle of operation of Vocoders. Why are low bit rate speech coders being favoured for use in mobile satellite systems?

3. PCM is often used for transmission of telemetry data from satellites. Consider transmission of temperature data in the range $-40°$ to $+40°C$, taken at 1 second intervals. Suggest a bit rate for converting this data to a PCM. Assuming that there are 256 other parameters monitored and converted to PCM at the same bit rate, what is the bit rate of the multiplexed bit stream?

4. With the help of diagrams, show the digital and analog multiplexing hierarchy recommended by the CCIR/CCITT.

5. From CCIR recommendations, determine the prevalent television standard in your country and list all the parameters necessary for a satellite communication system design.

References

Albuerque, J.P.A. (1989). 'On definition of compandor advantage and its analytical calculation', *International Journal of Satellite Communications*, Vol. 7, pp 7–9.

Bayless, J.W., Campanella, S.J. and Goldberg, A.J. (1973). 'Voice signals: bit-by-bit', *IEEE Spectrum*, October, pp 28–34.

Bear, D. (1980). *Principles of Telecommunications Traffic Engineering*, Peter Peregrinus, Stevenage, Herts.

CCIR (1982), *Recommendations and Reports*, ITU, Geneva.

Flanagan, J.L., Schroeder, M.R., Atal, B.S., Crochiere, R.E., Jayant, N.S. and Tribolet, J.E. (1979). 'Speech coding', *IEEE Trans. Commun*, April, pp 710–737.

Freeman, R.L. (1981). *Telecommunication Transmission Handbook*, Wiley, New York.

Gordon, G.D. and Dill, G.D. (1978). 'Efficient computation of Erlang loss functions', *COMSAT Technical Review*, Vol. 8, No. 2, pp 353–370.

Holmes, J.N. (1982). 'A survey of methods for digitally encoding speech signals', *Radio and Electronic Engineer*, Vol. 52, No. 6, June, pp 267–276.

Morgan, W.L. and Gordon, G.D. (1989). *Communications Satellite Handbook*, Wiley, New York.

Scharf, B. (1970). 'Critical bands', in Tobias, J.V. (ed.), *Foundations of Modern Auditory Theory*, Academic Press, New York, pp 157–202.

Schwartz, M. (1987). *Information Transmission, and Noise*, McGraw-Hill, Singapore.

8 Multiple Access Techniques

This chapter addresses another important aspect of satellite communications – the techniques used to access a satellite so that the satellite spectrum and power are shared efficiently between a large number of users. A brief section projecting the future directions being pursued to improve multiple access schemes has also been included.

8.1 Introduction

A satellite is a communications node through which all types of user in the network must be interconnected as flexibly as possible. At the same time, two key resources – bandwidth and spacecraft power – must be utilized efficiently. For some applications, it may be necessary that a satellite be simultaneously accessed by hundreds of users, making accessing problems more complex. Further complications are added when factors such as a requirement for handling a mix of traffic (e.g. data and voice), traffic variations and a necessity to incorporate traffic growth are considered. Ideally a multiple access scheme must be able to optimize the following parameters (Puente et al., 1971):

1. satellite radiated power;
2. RF spectrum;
3. connectivity;
4. adaptability to traffic and network growth;
5. handling of different types of traffic;
6. economics;
7. ground station complexity;
8. secrecy (for some applications).

A single technique cannot optimize all these parameters and therefore a trade-off analysis using the applicable conditions is necessary, provided that the choice of an accessing scheme is not obvious. For example, if the application at hand is provision of communication to a large number of low-cost mobile terminals, the accessing scheme should be simple but robust so as to permit the use of low-cost mobile receivers. At the same time, a certain degree of flexibility is necessary to enable sharing of the spectrum between a large number of mobiles and to accommodate addition of mobiles to the network. Compare this with an application where a relatively few large earth stations, each carrying heavy traffic, need to be interconnected. In this case the accessing scheme can be complex and the main optimization criterion would be optimal use of the available bandwidth and satellite power rather than the need for simple earth stations.

A number of accessing schemes have evolved over the years. At the introductory phase of satellite technology, *frequency division multiple access* (FDMA) appeared to be the best candidate because of the established FDMA technology (from the terrestrial radio relay system), its simple network control requirement and the consequent low cost. The technology became widely used in all first-generation systems. This scheme, however, is inefficient with respect to both the satellite power capacity and bandwidth utilization. An improvement in FDMA was introduced by incorporating an element of flexibility in the form of a demand-assigned FDMA in which a central pool of frequencies is shared by the user on a call-by-call basis. Following increase in traffic demand, leading to a scarcity of available bandwidth, and a trend towards digital techniques, a more efficient but complex accessing scheme – *time divison multiple access* (TDMA) – was introduced. Currently the TDMA scheme is being introduced into most fixed satellite service networks used for interconnecting high traffic earth stations, although FDMA is expected to be used well into the 1990s. For some specialized applications where secrecy is vital or where a channel may suffer frequency selective fading or interference (e.g. mobile communications), *code division multiple access* (CDMA) based on spread spectrum principles was developed.

TDMA, FDMA and CDMA provide dedicated circuits for communications and are therefore suited for continuous traffic, such as voice. However, some types of data traffic, such as those originating in a computer network or a query/response system in a bank, are characterized by periods of inactivity followed by a burst of activity. Dedicated circuits for such 'bursty' traffic are inefficient in terms of channel utilization because the circuit remains idle for a significant proportion of a message session. A more efficient channel utilization could be envisaged in which a channel is shared among several users, following a certain set of rules or 'protocols' which are 'matched' to traffic characteristics. Several schemes, often referred to as *packet access schemes*, have been developed to maximize the use of a channel for data traffic.

8.2   Frequency division multiple access

In a frequency division multiplex access (FDMA) scheme, the assigned bandwidth B_T is divided into n segments which are assigned to all n earth stations in the network according to their traffic requirement. Figure 8.1 shows the time–frequency plot of an FDMA scheme. The basic principle of operation of FDMA is shown in figure 8.2 with the help of a communication route between two earth stations. In figure 8.2(a), earth station A multiplexes

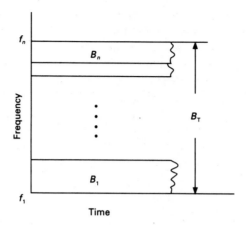

Figure 8.1   Time–frequency plot of a frequency division multiplex scheme.

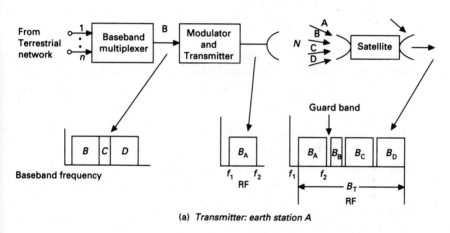

(a) *Transmitter: earth station A*

(b) *Receiver: earth station B*

Figure 8.2   Communication between earth stations A and B using FDMA: (a) transmitter; (b) receiver.

baseband data destined for various earth stations (B, C and D in this case). The multiplexed data are modulated and transmitted to the allocated frequency segment B_A. The bandwidth of a transponder is shared among n earth stations, each with a different traffic requirement – in this case there are 4. The transponder bandwidth is therefore divided into n fixed segments, with B_n allocated to the nth earth station.

Notice that there is a guard band between each segment. Guard bands reduce the bandwidth utilization efficiency, the loss being directly related to the number of accessing earth station networks. If there are N users, a total of N carriers pass through a satellite. The receiving earth station extracts the carriers containing traffic addressed to it by using an appropriate RF filter. In figure 8.2(b), earth station B filters out the frequency segment B_A and demodulates this band. The output of the demodulator consists of multiplexed telephone channels for earth stations C and D together with the channels addressed to it. A baseband filter is used to filter out the desired baseband frequency segment and finally a demultiplexer retrieves individual telephone channels and feeds them into the terrestrial network for onward transmission.

The FDMA scheme may be divided into two categories based on the traffic demands of earth stations – *multiple channel per carrier* (MCPC) and *single channel per carrier* (SCPC).

Multiple channel per carrier (MCPC)

The main elements of the MCPC scheme were illustrated in figure 8.2 and described in the previous section. In this scheme, each baseband filter in an earth station receiver corresponds to a specific transmitting station. Any change in channel capacity requires a retuning of this filter. Thus changes in traffic are difficult to implement.

The MCPC scheme may be further categorized according to the type of baseband used. A MCPC is referred to as a frequency division multiplexed/ frequency modulated/frequency division multiple access or FDM/FM/FDMA scheme when the incoming baseband signals are frequency division multiplexed. The multiplexed signal frequency modulates a carrier which is then transmitted using a FDMA scheme. Similarly a time division multiplexed/phase shift keyed/frequency division multiple access or TDM/PSK/FDMA scheme transmits a time division multiplexed digital signal using a PSK modulation into a FDMA mode.

Single channel per carrier

For certain applications, such as the provision of service to remote areas, traffic requirements are low. We know that the number of accesses required in a fully interconnected mesh network of n earth stations is $n(n-1)/2$. Consider interconnecting a network of 40 earth stations, each required to carry a small volume of traffic (e.g. 2–3 circuits during the peak demand period) spread lightly over a day. The total number of accesses required is then 780. Assigning multiple channels carriers to each earth station is wasteful of bandwidth because most of the channels remain un-utilized for a significant part of a day. For this type of application the *single channel per carrier* (SCPC) type of FDMA is used.

In the SCPC scheme, each carrier transmits a single channel. SCPC systems may be either *pre-assigned* or *demand-assigned*. In pre-assigned SCPC, a few channels (e.g. 5–10) are permanently assigned to an earth station. In the demand-assigned FDMA, a pool of frequency is shared by earth stations. When necessary, each earth station requests a channel from a pool manager. The demand-assignment concept can be extended to other types of accessing scheme. For example, in demand-assigned time division multiple access (discussed later) *time slots* are assigned on demand.

Demand-assignment permits the sharing of circuits both in the terrestrial link (i.e. from the transit centre to an earth station) and the satellite link. Therefore a reduction in cost is possible through sharing of equipment and, at the same time, the call handling capacity of each circuit is increased.

There are two ways in which the frequency pool can be managed and controlled – *distributed and centralized*.

In the *centrally controlled* frequency management scheme, channels are assigned centrally from a network control station (NCS) which maintains the frequency pool. Figure 8.3 illustrates the main elements of such a demand-assigned scheme. Requests for channel assignment are directed to the NCS on a signalling channel. The NCS then assigns a frequency from its pool using certain assignment rules. For example, the NCS may always attempt to choose the best available channel first and, if occupied, progressively move to lower quality channels until an unoccupied channel has been found. The system designer ensures that the channel quality criteria are satisfied for the worst channel in the pool. The allocation is then announced on a signalling channel known as a broadcast channel. The announcement is received by the calling and the called earth stations which then tune to the allocated channel. The communication takes place on the allocated channel. The end of call is announced by a signalling message, following which the NCS returns the channel to the common pool.

An automatic frequency control (AFC) pilot is transmitted by designated reference earth stations and all the earth stations use this reference to correct their transmission frequency. Hence drifts in satellite translation

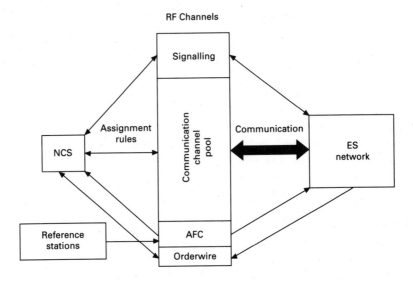

Figure 8.3   The main elements of a centrally controlled demand-assigned scheme.

frequency and frequency variations caused by the Doppler effect and the carriers retain their designated frequencies relative to each other. This feature is essential because, if uncorrected, the sum of the total frequency errors can cause carrier overlapping, as carrier bandwidths are small. Additionally, a stable receiver frequency permits the earth station receiver demodulator design to be simplified. An orderwire line is provided for communication between personnel for system management, fault diagnosis, etc.

Centrally controlled networks are simple to manage – they provide a higher usage of channels because of the availability of all information at a single point, and offer a lower connection time, while the participating earth station can use simple demand-assignment equipment. As a result they are more cost-effective. However such networks have a lower reliability and are prone to single point failure. Examples of a centrally controlled network are INMARSAT Standard A, B and M networks which provide communications to mobiles such as ships and land vehicles.

In the *distributed frequency management* scheme, each earth station can obtain a channel from the common pool on its own, following a given set of rules. The advantage of this scheme is the high reliability of the network, as the failure of a single earth station does not affect the availability of the system to other earth stations in the network. However the increased complexity in earth stations drives up their cost. When a subscriber dials a number served by another earth station in the network, the signalling information is passed by the transit centre to the demand-assignment, signalling and switching

(DASS) unit of the earth station. The DASS unit assigns a frequency pair for the call from a common pool of frequencies and broadcasts it on a *signalling* channel which is received by all the earth stations in the network. The source and the destination earth stations tune to the assigned frequency and the call takes place on this channel. At the end of the call, the frequency pair is released and returned to the common pool.

This type of demand assignment is used by INTELSAT's SPADE system (Single channel per carrier pulse code modulation multiple access demand assignment equipment)(Puente *et al.*, 1971). In the SPADE system the signalling channel is a time division multiplexed access (TDMA) channel (TDMA is discussed in the next section). The channel assigns 1 ms slots to up to 48 earth stations. The first burst is the reference burst followed by 49 slots of 1 ms each, permitting the participation of up to 49 earth stations. Each earth station receives this channel and constantly updates its frequency assignment table according to the assignment message contained in each burst. When a request for a channel is received by an earth station, the DASS unit assigns a frequency pair from its latest frequency table. This assignment is transmitted on the signalling channel. If the frequency becomes assigned to another earth station before the source earth station has received its own broadcast, the earth station assigns another frequency pair. The process continues until a successful assignment is completed. The connection for the call is then made by the source and the destination earth station. At the end of the call a channel-release message is announced on the signalling channel and all the earth stations then re-enter the released channel in the list of available channels.

It is possible to have a *hybrid* frequency management scheme in which the network provides a combination of distributed and centrally controlled frequency management functions, bringing together the advantages of each.

Design considerations

The design of an FDMA system requires careful evaluation and optimization of several parameters (Dicks and Brown, 1974). To begin with, the major sources of impairments are summarized below. Areas where there are differences in MCPC and SCPC have been identified. This is followed by a discussion on transponder utilization using FDMA. A trade-off analysis between demand and pre-assigned FDMA concludes the section.

(i) *Impairements caused by satellite high-power amplifier*
In chapter 4 we saw that high-power amplifiers (HPAs) used in satellites, and in particular TWTs, exhibit increasing non-linearity as the output level approaches saturation. This non-linearity causes inter-modulation noise whenever a number of carriers pass through the HPA. The FDMA scheme is

therefore susceptible to inter-modulation noise caused by amplitude and phase non-linearity of the HPA. To minimize the effect of inter-modulation, the drive level of the final stage of the satellite transmitter is reduced so that it begins to operate in a linear region. The transmitter is then said to be '*backed off*'. As a result, the full power capability of the amplifier is not utilized, causing a reduction in capacity.

Another consideration is the amplitude–frequency response characteristic of the TWT. A slope in this response causes variations in the frequency of incoming signals to appear as amplitude variations. Such amplitude variations introduce a phase modulation to other carriers sharing the TWT. Information from one carrier is thus coupled to other carriers (and vice versa), causing an intelligible cross-talk between carriers. The cross-talk component is minimized by using TWTs which minimize AM–PM conversion. The CCITT recommendations specify the permissible cross-talk to be better than 58 dB. This requirement is readily met by TWTs in use now.

(ii) *Other impairments*

To maximize the frequency utilization, adjacent carriers are brought as close to each other as possible. The spectrum overlap known as *adjacent channel interference* causes an increase in noise. When the adjacent carrier has a smaller amplitude than the desired carrier, the interference is usually treated as an equivalent increase in noise floor. This type of noise is called *convolution noise*. When the adjacent carrier has a larger level than the desired carrier, the adjacent carrier interference occurs as clicks for voice channels or as error bursts for data channels. This type of noise is generally dominant and categorized as *impulse noise*. Several mathematical models have been proposed in the literature to quantify impulse noise (e.g. Wache, 1971). Adjacent carrier interference is best minimized by providing suitable guard bands. A guard band of 10% of occupied bandwidth generally provides adequate protection.

Inter-modulation noise from *adjacent* transponders carrying FDMA traffic can fall within the desired band. This interference is minimized by filtering the out-of-band inter-modulation in each transponder. The out-of-band rejection property of such filters is a function of the number of sections of filter used, and therefore a mass penalty has to be paid to increase the rejection.

Phase non-linearity can be introduced when a carrier located at band edges arrives at the output after travelling through the desired and an adjacent transponder. This happens when the filter response of the adjacent transponder cannot provide adequate isolation at the band edges. The resulting impairment causes group delay distortions. This type of noise can be minimized by optimizing the transponder guard band and increasing the out-of-band attenuation. The trade-off is between a loss in bandwidth and increase in complexity (and hence its weight and cost) of the output filter.

Another source of distortion is the group delay associated with each filter in the transmission path – i.e. the earth station transmit/receive chain and the satellite transponder. The problem is much less pronounced for SCPC, which uses narrow channels. The net effect of group delay is an increase in noise. The magnitude of group delay increases rapidly at the band edges and therefore the carriers occupying frequencies close to band edges are susceptible to group delay distortion.

Transponder utilization

There are two major factors which limit the number of FDMA accesses through a transponder:

(a) an increase in inter-modulation noise with an increase in the number of accesses; and
(b) for MCPC, lower spectrum utilization efficiency of smaller carriers (i.e. carriers carrying smaller numbers of traffic channels).

(a) *Inter-modulation noise*
We have already discussed the inter-modulation characteristics of a satellite in chapter 4 and in the preceding section.

(b) *MCPC spectrum utilization efficiency*
The reduction in spectrum utilization efficiency is attributed to the channel loading characteristics of FDMA given as equation set (5.20) (see chapter 5), which favours larger numbers of users.

The spectrum utilization efficiency, η_B, for various carrier sizes can be compared by calculating the RF bandwidth/channel for each carrier size. Table 8.1 shows the number of channels for various standard carrier bandwidths used by INTELSAT, together with RF bandwidth/channel as a function of carrier-to-noise ratio (Dicks and Brown, 1974). It can be seen that bandwidth utilization increases progressively as the carrier size increases. For a carrier-to-noise ratio of 13 dB the RF bandwidth/channel reduces from ~104 kHz to ~76 kHz as carrier size is increased from 2.5 MHz to 10.0 MHz. The table also illustrates the bandwidth versus power trade-off. We note that for a given bandwidth the capacity increases with carrier-to-noise ratio. Taking 2.5 MHz as an example we note that the RF bandwidth/channel reduces from ~104 kHz at a carrier-to-noise ratio of 13 dB to ~42 kHz at a carrier-to-noise ratio of 20 dB.

Table 8.2 and figure 8.4, obtained through operational experience, show the utilization of a 36 MHz transponder (Dicks and Brown, 1974). It is seen that the channel capacity of the transponder reduces significantly as the number of accesses increase. For example, when the number of accesses increases to 8,

Table 8.1   Number of available channels (N) and RF efficiency η_B, (RF bandwidth/channel in kHz), as a function of carrier-to-noise ratio for standard INTELSAT carrier sizes (Dicks and Brown, 1974).

| Carrier bandwidth (MHz) | Carrier-to-noise ratio (dB) | | | | | |
| | 13 | | 15 | | 20 | |
	N	η_B	N	η_B	N	η_B
2.5	24	104.2	36	69.4	60	41.6
5.0	60	83.3	96	52.1	132	37.9
7.5	96	78.1	132	56.8	192	35.1
10.0	132	75.8	192	52	252	39.7

Table 8.2   Reduction in a 36 MHz transponder capacity with an increase in the number of accesses (Puente *et al.*, 1971). 100% corresponds to 900 FDM voice channels.

Number of accesses	Approximate channel capacity relative to a single carrier access
1	100
2	90
4	60
8	50
14	40

the channel utilization reduces by 50% relative to a single access. Some capacity is also lost because the assigned channel blocks are not always filled with traffic, leading to a further reduction in transponder capacity utilization.

Now consider the capacity of a demand-assigned SCPC system – SPADE for example. The SPADE system uses 45 kHz spaced carriers, providing a total of 800 carriers (400 duplex channels). Note that the SPADE system provides a capacity approaching that of a single access. This improvement is attributed to the following reasons. Each SPADE SCPC carrier is voice-activated, i.e. carriers are transmitted only in the presence of voice. It has been shown that in a large number of conversations only 40% of speech activity is present at any given time (Byrov *et al.*, 1968). This permits an increase in the utilization of the transponder by a factor of 2.5. In addition, random voice-activated loading of the transponder reduces the inter-modulation noise in the worst channel by approximately 3 dB (McClure, 1970).

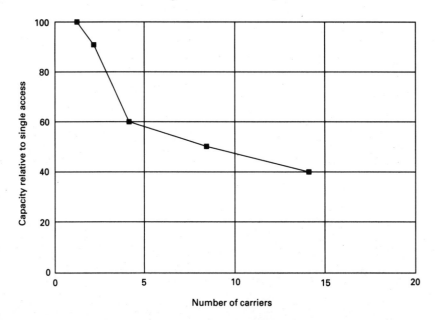

Figure 8.4   Transponder capacity as a function of number of frequency division multiple accesses (data from Dicks and Brown, 1974).

Demand and pre-assigned FDMA – trade-off analysis

It was mentioned above that demand-assigned SCPC provides a cost-effective solution to thin traffic routes. A system designer needs a quantitative assessment of the improvement offered by demand-assigned relative to pre-assigned access – in particular when the offered traffic cannot be easily quantified as 'thin'. The improvement offered by the demand-assigned scheme can be obtained by comparing the number of pre-assigned two-way circuits with the number of demand-assigned circuits for a given traffic load. An improvement factor, I, as defined below, can be used for such a comparison:

$$I = n_{PA}/n_{DA} \tag{8.1}$$

where n_{PA} and n_{DA} are respectively the number of pre-assigned and demand-assigned circuits used to provide the same grade of service.

Studies have been done to estimate the improvement in the number of circuits of a fully variable demand-assignment scheme relative to a pre-assigned access scheme on terrestrial access circuits and in the satellite link (Dill, 1972). A model capable of providing a cost comparison between pre-assigned and various alternatives of demand assignment has also been proposed (Laborde, 1985). The details of these studies will not be given here

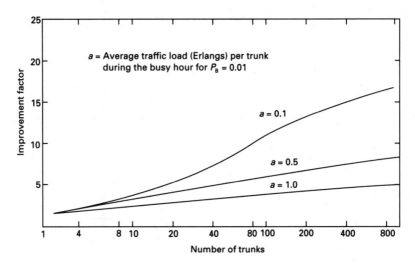

Figure 8.5   Improvement factor of demand-assigned FDMA as a function
of average traffic load per trunk and number of trunks (Dill,
1972).

and interested readers are referred to the literature.

Figure 8.5 (Dill, 1972) shows a plot of the number of trunks and the
improvement factor for various offered traffic per trunk, applicable to both
terrestrial and satellite access. *Trunk* is defined here as a collection of circuits grouped together to carry traffic between two locations. It is observed
that the improvement is larger as the traffic offered/trunk reduces and the
number of earth stations in the network increases. For example, the use of a
demand-assigned scheme in a network of 40 earth stations with traffic of
0.1 Erlang/destination provides a reduction in the number of circuits by a
factor of 8 relative to a pre-assigned scheme. This leads to a reduction in
termination equipment, transmission requirements, installation costs and
operation/maintenance costs.

Summary of salient features

We have discussed the main features of FDMA in the above sections. In
this section we summarize the main advantages and disadvantages of FDMA.
The advantages of FDMA are:

1. Uses existing hardware hence technology is mature and cost is low.
2. There is no need for network timing.
3. There is no restriction regarding the type of baseband (voice or data) or
   type of modulation (e.g. FM or PSK).

The disadvantages are:

1. Inter-modulation noise in the transponder leads to interference with other links sharing the TWT, and consequently reduces the satellite capacity.
2. Lack of flexibility in channel allocation, since for each change in allocation the receiver filters need to be retuned. This disadvantage applies to the MCPC type of FDMA. Demand-assigned SCPC removes this limitation.
3. Requires uplink power control to maintain the link quality.
4. In a mix of traffic containing strong and weak carriers, especially in MCPC, the weak carriers tend to be suppressed.

8.3 Time division multiple access

We have seen that in the MCPC frequency division multiple access scheme there is a significant reduction in capacity because of the necessity to back off transponders to minimize the inter-modulation noise.

We have noted that the full capacity of a transponder can be realized by a single carrier access. In the *time division multiple access (TDMA)* scheme only a single carrier is allowed access to the transponder (or in some cases, a specific segment of the transponder) at any given time. However, to allow all the users access to the satellite, the transponder is time-shared between users. Each user is allocated a specific time slot for transmission. Thus transmissions arrive at the satellite in a sequence of non-overlapping bursts. The capacity of the system is increased considerably as only a single carrier is present at any given time. However, the TDMA scheme adds both hardware and network complexity. The messages need to be stored, compressed and transmitted during one (or more) specific time slots. Also, the receiving earth stations must be able to demodulate all the bursts in a very short time (~ milliseconds). At the network level, all the transmissions must be synchronized to avoid collision between bursts. Because of the need to compress, store and process information, a digital mode of transmission is best suited.

Figure 8.6 shows a time–frequency plot of the scheme. Each earth station has full access to a transponder during its allocated time slot. The transmissions are organized into frames, an earth station transmitting in its designated time slot in every frame. Each earth station receives all the bursts and removes data addressed to it. The organization of time slots in a frame is shown in figure 8.7. Every frame contains at least one reference burst transmitted by a station, designated for synchronization purposes. Each time slot is separated from the adjacent time slot by a guard time. The sizes of the time slots depend on the traffic requirement of the station. The size of the slot can be re-allocated in the case of any change in traffic demand, thereby increasing the flexibility in handling changing traffic demands. This change

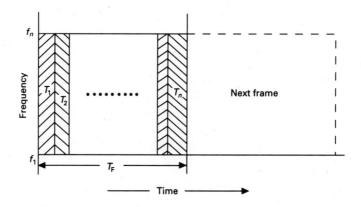

Figure 8.6 Time–frequency plot of TDMA. T_n is the alloted time of the nth user. T_F is the frame time.

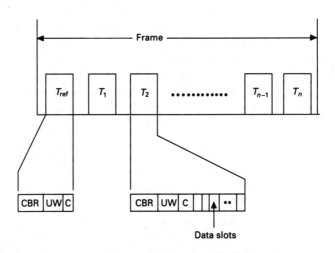

Figure 8.7 Organization of time slots in a frame. T_{ref} is the reference burst. T_n is the allocated time slot of the nth user.

in slot dimension can be achieved with minimum alteration in hardware. (Note that in the MCPC FDMA scheme, any change to traffic requirement requires a significant change to hardware.) To improve the flexibility further, TDMA can be operated on a demand-assigned basis in which slots are allocated to users on demand.

Figure 8.7 includes details of each burst in a frame. The reference burst consists of three parts. The *carrier and bit time recovery* (CBR) slot is used by receivers for recovering the carrier and bit time essential for coherent demodulation. The *unique word* (UW) is used for burst synchronization. This is achieved by correlating a stored replica of the unique word with the received bits. A correlation peak indicates the start of a new frame. Careful optimization of the unique-word structure minimizes the possibility of missing a unique word caused by bit error or of wrongly flagging the reception of a unique word from a data string. *Control bits* (C) contain information such as station identification and engineering service messages for network management. The bursts from traffic-carrying earth stations consist of CBR, UW and C, collectively called the *preamble*, followed by data bursts containing the traffic bits, together with the address of the destination earth stations.

For successful operation of the system it is essential that each terminal transmits its burst in its allocated time slot. This is achieved by network-wide synchronization. Constant motion of the satellite caused by various perturbations results in variations in satellite range. Existing satellites are kept within a window of ±0.1° in longitude and 2–3° in inclination of their nominal position. This results in an uncertainty window of several tens of kilometres, which is equivalent to an uncertainty of several hundreds of microseconds in time. For example, an uncertainty of ±75 km is equivalent to 500 microseconds in time. A guard band of at least 500 microseconds would be necessary to avoid possible collision of adjacent bursts. Such a large guard time is wasteful of satellite resources as such periods remain un-utilized. To minimize the guard time, it becomes necessary to synchronize each burst on a dynamic basis. The achievable accuracy depends on the synchronization technique (see the next section).

Another associated problem is related to the fact that the frame time depends on the earth station–satellite range, which varies with time. This means that earth stations perceive shorter frame times. As the transmission bit rate is an integral multiple of frame rate, the net result is that the bit rate of each station differs slightly. Each earth station receives traffic data from various earth stations and sends them to the fixed network after multiplexing. To resolve the problem of multiplexing slightly differing bit rates, extra bits are added, where necessary, to maintain a uniform bit rate for the multiplexed stream.

There are two methods of TDMA synchronization – open-loop and closed-loop.

Open-loop synchronization

In an open-loop synchronization each earth station maintains an accurate clock, independently of other earth stations in the network. Open-loop synchronization is relatively simple to implement and hence is useful for simple ground stations. The uncertainty is reduced by using an agreed satellite motion model. It is possible to maintain guard bands of the order of 100 microseconds by using this technique.

Close-loop synchronization

In this method, continuous adjustments are made to the burst positions based on real-time measurements. Figure 8.8 shows a method of closed-loop synchronization. An earth station, say i, when beginning transmission, must first acquire its assigned time slot. The initial burst is transmitted on an estimated position of the burst position, T_m. The burst is transmitted at a low power to avoid interference to adjacent slots in case of a gross error in the estimate. The received burst is compared with the desired position T_a of the slot and a correction of $(T_a - T_m)$ applied to the initial estimate in the next transmission, to acquire the correct time seat.

Alternatively, a carrier modulated with a pseudo-random signal may be transmitted. This type of signal has a noise-like property, and when transmitted at low power does not affect the other bursts. The receiver consists of a correlation receiver to obtain the burst position. A correction equal to the difference between the received and the desired burst positions is then applied to the original burst position. After entering the network the process of measurement/correction is repeated at frequent intervals (~1 s) so that the burst position errors remain within tolerable limits.

In another synchronization scheme a reference station observes burst positions of each earth station in the network and provides feedback to each station to apply the necessary correction. This type of synchronization is useful in a multiple beam system when the transmitting earth station is unable to receive its own transmission.

Frame efficiency

The frame efficiency η_f in a TDMA system is defined as the ratio of time devoted for useful transmission to the total frame length:

$$\eta_f = \left(1 - \frac{\Sigma t_i}{T_F} \right) \tag{8.2}$$

where t_i is the sum of all guard times and preamble including the reference burst and T_F is the frame time.

When considering voice channels, a sampling period of 125 microseconds is a well-accepted standard, and therefore a frame period which is a multiple of 125 microseconds is well suited. From (8.2) we note that the frame efficiency increases when the frame length is increased. A compromise can be exercised between the storage capability in the earth stations and the frame length. Frame efficiencies of the order of 0.9 are possible at present.

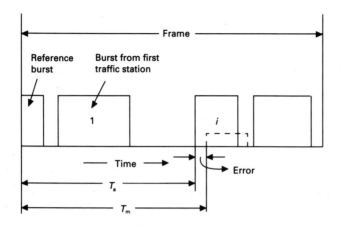

Figure 8.8 A method of close-loop synchronization. T_a is the desired slot position. T_m is the measured position. A correction is applied to cancel the error.

The voice channel capacity of a TDMA system can be obtained as

$$N_c = (R_s - kn/T_f)/R_c \qquad (8.3)$$

where R_s = TDMA system bit rate
  k = number of bits in each preamble
  n = number of bursts/frame
  T_f = frame time
  R_c = voice channel bit rate.

One of the important considerations in a TDMA is the performance of the demodulator. It is essential that the demodulator be capable of very quick carrier and bit time recovery together with accurate demodulation. The performance of such modulators is often given in terms of bit error rate as a function of time from the beginning of the burst. The performance generally improves with time and therefore at the preamble stage the bit error rate is relatively high (e.g. 5×10^{-4}) but improves for the traffic part of the burst (e.g. 1×10^{-4}).

An earth station for carrying TDMA traffic is described in chapter 10 (section 10.4).

Transponder utilization

Transponder utilization in the case of TDMA depends on the satellite EIRP, the G/T values of the receiving earth stations and the efficiency of the modulation scheme (Gabbard and Kaul, 1974). The utilization expressed as the

transmission bit rate can be either EIRP or bandwidth limited.
By rearranging the transmission equation, we obtain

$$C/N_0 = \text{EIRP} - L + G/T - k - M \qquad (8.4)$$

We know that

$$C = E_b R$$

where E_b = energy/bit
 R = bit rate.
Substituting in (8.4), the bit rate R_p for a power limited link is

$$R_p = \text{EIRP} - L + G/T - k - E_b/N_0 - M \qquad (8.5)$$

Equation (8.5) provides the bit rate in a power-limited case for a given G/T and E_b/N_0.

When we have sufficient EIRP, the maximum permissible bit rate is governed by the available transponder bandwidth. Thus a more spectrally efficient modulation scheme provides a higher bit rate:

$$R_b = B\eta \qquad (8.6)$$

where R_b = maximum bit rate in a bandwidth limited case
 B = transponder bandwidth
 η = modulation efficiency (bits/Hz).
Thus using equations (8.4) and (8.5) it is possible to determine if a system is power or bandwidth limited.

Summary of salient features

The advantages of TDMA are:

1. Maximum use can be made of the available satellite power since inter-modulation noise is minimal.
2. Uplink power control is not required.
3. Transmission plans are easier to construct and modify. Capacity management is simple and flexible.
4. The digital format of TDMA permits utilization of all the advantages of digital techniques. For example, techniques such as digital speech interpolation, source and channel coding, etc. can be incorporated.

The disadvantages of TDMA are:

1. It requires network-wide timing synchronization hence it is relatively complex.
2. Analog signals must be converted to digital form.
3. Interface with analog terrestrial plant is expensive.

8.4 FDMA/TDMA operation in a multiple beam environment

Consider a multiple spot beam system in which connectivity between all the n spot beams is desired. Here, the routing of signals to the appropriate beam can be achieved by the use of a frequency-to-beam correspondence. The passband of the satellite is segmented into sub-bands. Each band provides a unique route between two spot beams. Thus if an earth station in beam 1 wishes to transmit to earth stations in all the n spot beams, it chooses transponder 1 to transmit to earth stations in spot beam 1 and transponder n to transmit in spot beam n.

Transponders can be accessed in the FDMA or the TDMA mode. In the TDMA scheme, earth stations switch between desired beams, transmitting in their designated slot in each corresponding transponder. This is often known as *transponder hopping* because earth stations have to hop between transponders to route traffic to the desired spot beam. Each beam in this configuration has n transponders. For an n spot beam system there must be n^2 transponders.

To incorporate changes in the traffic pattern, some flexibility in altering the frequency bands is desirable. For example, it may turn out that during the lifetime of the satellite the growth in one spot beam traffic is larger than estimated, whereas another spot beam remains under-utilized. The use of switchable filters on the satellite provides one possible solution. The filters can be switched as desired so as to change the available bandwidth to each beam. This configuration has been used in INMARSAT-3 satellites.

The configuration mentioned above provides a practical solution to accommodate routing when the number of spot beams is low. However the number of transponders begins to get unwieldy as n increases. The use of *satellite switched* TDMA provides a good solution by incorporating a certain degree of on-board processing. In this technique, a programmable switch located on the satellite routes bursts to spot beams according to a set plan. An earth station can direct its transmission to any spot by transmitting in the appropriate time slot. Transmissions directed to a given beam are arranged in non-overlapping time slots. Figures 8.9(a) to (c) show the concept. Figure 8.9(a) shows the main elements of a satellite-switched TDMA. Figure 8.9(b) shows the input to the satellite-switched matrix from various beams, and figure 8.9(c) the bursts arranged in the switch according to the destination beam. Here, each beam can be served by a single transponder and the total transponder requirement is reduced to n. Earth station complexity is also reduced because only a single up/down converter chain is necessary. Further, the total (i.e. both ways) bandwidth requirement of this scheme is $2B_t$ compared with the $2nB_t$ required in the conventional scheme. This technique is in use on INTELSAT VI satellites.

Speech interpolation

The channel capacity of a demand-assignment SCPC is increased by taking advantage of pauses in speech. Time-assigned speech interpolation (TASI) for analog signals, and digital speech interpolation (DSI), its digital implementation (Campanella, 1976), also utilize this speech property but in a different manner. In this technique a channel in use by user A is allocated to another user whenever a speech pause is detected during a conversation of user A. A new channel is subsequently allocated to the user A when his next speech spurt begins. The resulting improvement in channel utilization, known as the DSI gain, is given as the ratio N_t/N_s where N_t is the number of incoming terrestrial channels at an earth station and N_s the number of transmission channels. DSI implementation can improve the channel usage by a factor of 2. With a well-designed DSI system the percentage of speech lost while waiting for reassignment, called freeze-out fraction, is less than 0.5%.

8.5 Code division multiple access

Code division multiple access (CDMA) is based on the use of a modulation technique known as spread spectrum. In direct contrast to a FDMA system where an attempt is made to minimize the transmitted bandwidth, in a code division multiple access scheme all users transmit signals simultaneously on the multiple access channel. Each user employs a spread spectrum modulation. In this modulation scheme the message signal is spread over a wide band by multiplying it with a noise-like or *pseudo-random* spreading signal. To demodulate the signal the receiver cross-correlates the received signal with an exact replica of the spreading function. The cross-correlation produces a maximum only when the codes are matched. Otherwise the result produces a very low value. Each user is assigned a unique code, thereby all the users can coexist even though each user is interfered by transmissions from others. Such an accessing scheme is called *code division multiple access*. The carrier-to-interference ratio is determined by the ratio of the peak auto-correlation of the code to the sum of cross-correlation noise produced by all simultaneous users. Therefore codes are chosen to exhibit very low cross-correlation with each other. Although CDMA has not been used extensively to date in civilian applications, there are a number of attractive features which could be exploited. Some of these advantages are:

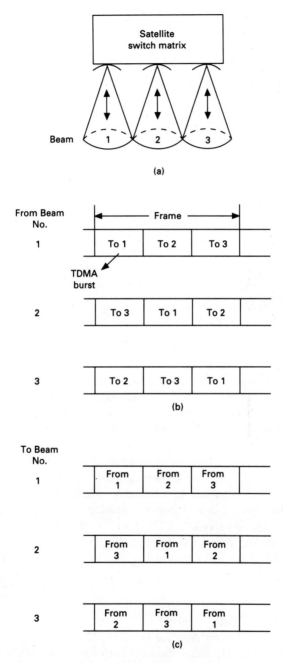

Figure 8.9 (a) A schematic of the switching elements of a satellite-switched TDMA. (b) Input to satellite switch matrix. (c) Output from switch matrix.

1. CDMA is highly resistant to interference and therefore satellite spacings can be reduced considerably without causing unacceptable degradation in received signal quality.
2. Spread spectrum systems are resistant to multipath noise which is commonly experienced by mobile terminals.
3. Small antennas can be employed without any problem of interference from adjacent satellites.
4. This technique offers a highly secure form of communication.

There are several ways of implementing a CDMA system. The most commonly used are the use of:

(a) a direct sequence spread spectrum;
(b) a frequency hopped spread spectrum.

A CDMA system utilizes the properties of pseudo-random sequences extensively. Therefore in order to understand the principle of this technique fully it is useful to review briefly the main properties of a pseudo-random sequence.

Pseudo-random sequences

A pseudo-random sequence is a random sequence of signals within a specified time period T_r, at the end of which the sequence repeats itself. Therefore such a sequence is not truly random and hence is termed as 'pseudo' random. Shift registers with feedback are often used for its digital implementation (see Golomb, 1967; Utlaut, 1978; Dixon, 1984). When an m-bit shift register is used, the maximum possible length of code is $p = (2^m - 1)$. Such maximum length codes are known as maximum-length linear shift register sequences. Each sequence has randomly placed (2^{m-1}) ones and $(2^{m-1} - 1)$ zeros, which results in a sequence appearing to be random. The *upper limit* on the number of users in a CDMA system depends on the size m of the shift register. The maximum number, N_{max}, of maximum-length linear codes using a shift register of size m is given by (Golomb, 1967)

$$N_{max} = [\phi(2^m - 1)]/m \tag{8.7}$$

where $\phi(2^m - 1)$ is a Euler number.
For example, table 8.3 gives N_{max} for m of 14, 15 and 16.
Figures 8.10(a) and (b) show the auto-correlation function and the envelope of the spectrum of the signal. The auto-correlation function peaks periodically every $2^m - 1$ bits with very low magnitude (average of $-1/p$, where p is the number of bits in the sequence) at other offsets. The power

Table 8.3 Maximum number of pseudo-random sequences possible with an m-stage shift register.

m	N_{max}
14	756
15	1800
16	2048

spectral density function of the pseudo-random code is given by

$$\delta(\omega) = v^2 \frac{(p + 1)}{p^2} \left[\frac{\sin(\omega t_0/2)}{\omega t_0/2}\right]^2 \sum_{\substack{n=-\infty \\ n \neq 0}}^{\infty} \delta\left[\omega - \frac{2\pi n}{p t_0}\right] + \frac{v^2}{p^2}\delta(\omega) \quad (8.8)$$

where p = number of bits in the sequence
t_0 = period of one digit
v = pulse height in volts
$\delta(\cdot)$ = impulse function.

The spectral density is a line spectrum, spaced $(1/p)$ times the pulse repetition frequency, with a $(\sin(x)/x)^2$ envelope and a scale factor inversely proportional to p. The final term $[(v^2/p^2)\delta(\omega)]$ is a very small DC term, arising because the sequence is not truly random – there being always a difference of 1 between the number of '0's and '1's, and represents the carrier power at the centre frequency.

The spread in spectrum increases with an increase in bit rate. The spread between nulls is R_c (where R_c = code rate). Typically, the RF bandwidth used is 100–1000 times the information rate. To maximize the number of users in a multiple access environment it is necessary that the cross-correlation between codes be as small as possible. As noted above, the number of maximum-length linear codes satisfying this requirement is limited. However, several codes having a very low cross-correlation property and all with period 2^{m-1} have been devised (Sarvate and Pursley, 1980).

Direct sequence spread spectrum

Figure 8.11(a) illustrates the principle of code division multiple access using a direct sequence spread spectrum technique. The information stream $m_1(t)$ at a bit rate of b_1 bits/s is modulated and the modulated signal $s_1(t)$ multiplied by a spreading function $g_1(t)$ (see figure 8.11a). The spreading function $g_1(t)$ is a pseudo-random code with a bit rate B_s significantly larger than the information bit rate. The code length and the bit rate depend on the application. In another type of implementation the spreading function is applied at the baseband.

Other users in the network transmit on the same channel but each user has a unique code. The received signal consists of the desired signal together with interference caused by other users sharing the channel and inherent system noise comprising thermal and inter-modulation noise. Figure 8.12(a) and 8.12(b) show representative spectra of the transmitted and received spectrum. Figure 8.12(b) also shows a narrow-band interferer. Thus the received signal may be given as

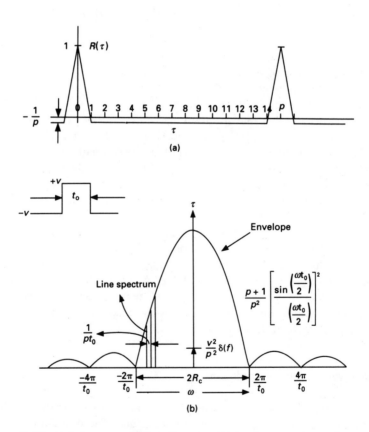

Figure 8.10 (a) Autocorrelation function $R(\tau)$ of a pseudo-random code generator. (b) Power spectrum $S(\omega)$ of a pseudo-random code generator.

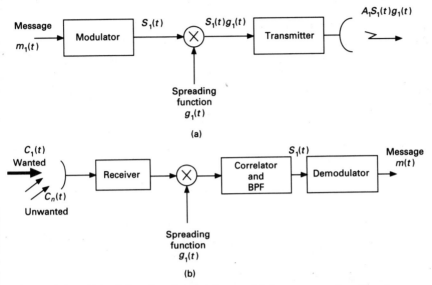

(a)

(b)

Figure 8.11   Principle of code division multiple access using direct
sequence spread spectrum: (a) transmitter; (b) receiver.

$$R_x(t) = C_1(t) + C_2(t) + \ldots C_n(t) + n(t) \tag{8.9}$$

where $C_n(t)$ is the received signal from the nth transmitter and $n(t)$ is the
system noise.

In the receiver (see figure 8.11(b)), the composite signal is correlated
with a replica of the transmitted code. The code generated at the receiver
must be synchronized with the transmitted code to provide well-defined auto-
correlation peaks. The synchronization technique depends on factors such as
the speed at which synchronization must be acquired, the receiver sensitivity and the complexity. Figure 8.12(c) shows the output of the correlator.
The net effect of the correlation is to despread all the interfering signal and
peak the desired carrier. The correlation process is followed by a bandpass
filter, used to reject noise components outside the message bandwidth. Note
that if there is a narrow-band interfering carrier, the correlation process spreads
the carrier bandwidth in the same manner as the message at the transmitter
end. Hence at the output of the correlator the power spectral density of the
narrow band is reduced by B_c/B_i, where B_c and B_i are the RF bandwidth of
code and interfering signals respectively.

The *processing gain* of the spread spectrum signal can be approximated by

$$G_p = (C_o/N_o)/(C_i/N_i) \tag{8.10}$$

where C_i and C_o are the input and output signals of the correlator respectively

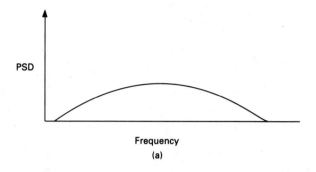

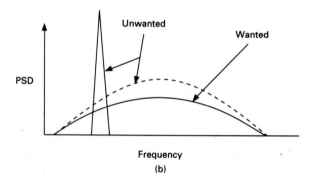

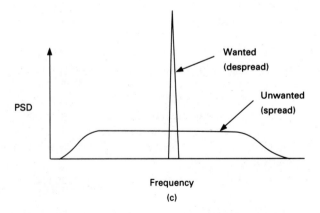

Figure 8.12   Representative power spectral densities of a code division multiple access scheme: (a) transmitter output; (b) receiver input; (c) correlator output.

$N_i \quad = I(f)B_c$

$I(f) = $ noise power spectral density

$B_c \quad = $ occupied channel bandwidth

$N_o \quad = I(f)B_m$

$B_m \quad = $ occupied message bandwidth.

Occupied channel bandwidth is defined here as the spectrum between nulls of the spectrum, as shown in figure 8.10(b).

Substituting for N_i and N_o in equation (8.10), the processing gain is given by

$$G_p = B_c/B_m \qquad (8.11)$$

From figure 8.10(b) we note that $B_c = R_c$ (channel bit rate), assuming $B_m = R_m$ (message bit rate). Then substituting in (8.11):

$$G_p = \frac{R_c}{R_m} \qquad (8.12)$$

In practice, the processing gain is reduced because of losses in the correlation process caused by factors such as imperfect cross-correlation. Further, it should be noted that by making processing gain arbitrarily large, the system becomes thermal noise limited and therefore there may not be much advantage in increasing the processing gain beyond a certain limit. The tolerance of a spread spectrum signal to interference is given in terms of an *interference margin*. Interference margin (M_i) gives the tolerable level of interference above the desired carrier-to-noise ratio and may be defined as

$$M_i = G_p - L - C_o/N_o \qquad (8.13)$$

where $G_p \quad = $ processing gain

$L \quad = $ implementation loss

$C_o/N_o \quad = $ desired carrier-to-noise ratio at the correlator output.

Example 8.1

Consider a spread spectrum system with a processing gain of 1000 and an implementation loss of 2 dB. The required carrier-to-noise ratio at the demodulator input is 7 dB. Determine the interference which can be tolerated by the system.

Solution

The tolerance of the system to interference is determined by the interference margin given by equation (8.13). Here

$$G \quad = 1000 \text{ or } 30 \text{ dB}$$

$$L \quad = 2 \text{ dB}$$

$$C_i/N_i = 7 \text{ dB}.$$

Therefore

$$M = 30 - 2 - 7$$

$$= 21 \text{ dB}$$

This implies that the total interference permitted to enter the system is up to 21 dB above carrier power. This interference may be suitably apportioned between other users and interference entering the system from other sources such as other satellite and terrestrial systems operating in the same band.

Frequency hopped spread spectrum

There is another technique of implementing a spread spectrum system. This technique, known as *frequency hopping*, is shown in figure 8.13. Here, the spreading function is used to alter the transmission frequency in discrete steps (Δf) using a frequency synthesizer. At the receiver, an identical code synchronized to the transmission is used to alter the frequency of the local oscillator (figure 8.13b). As a result of the mixing process, the frequency hopping element is removed. The mixer output is bandpass filtered to remove the undesirable frequency product generated by the non-linearity of the mixer and other system noise. The resulting signal is a fixed IF which is then fed into the demodulator for retrieving the message signal. The spectrum B_{rf} of the transmitted signal is given by

$$B_{rf} = (2^n - 1)\Delta f \tag{8.14}$$

where　n　=　number of stages in code generator shift register
　　　　Δf　=　unit frequency increment
　　　　B_{rf}　$\gg$　B_i (information bandwidth).
　　The processing gain of a frequency hopped system is given by

$$G_p \quad = B_{rf}/B_i$$

$$= (2^n - 1)\Delta f/\Delta f \tag{8.15}$$

assuming that $B_i = \Delta f$.
　　The code rate of a frequency hopped system can be much lower than the code rate of a direct sequence type of spread spectrum system.

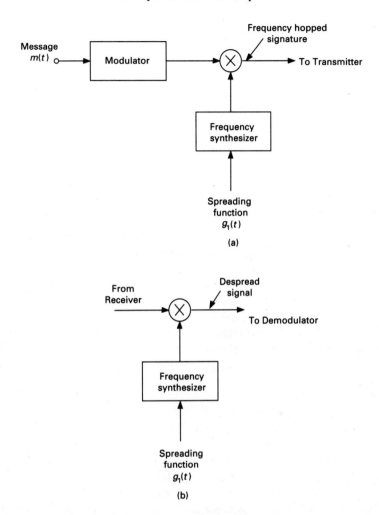

Figure 8.13 Principle of frequency hopped spread spectrum scheme: (a) transmitter; (b) receiver.

Interference mechanisms in frequency hopped systems are different from those in the direct code spread spectrum system. Interference in the frequency hopped spread spectrum is caused when an unwanted signal appears within the passband of the desired signal. This can occur under the following conditions:

1. The transmission of other users of the multi-access channel falls within the receiver passband. The probability of occurrence of this type of interference can be minimized by choosing codes with low cross-correlation properties. When an interfering source is narrow band, interference

occurs only when the interfering signal falls within the passband of the instantaneous hopped frequency of the wanted signal.

2. The inter-modulation products resulting from non-linearities in the channel fall within the receiver passband.

When the hopping rate is much larger than the information rate, interference is noise-like. However when the hopping rate is small relative to the information rate, interference tends to be coherent but the occurrence is intermittent. The net result is that some hops suffer large signal degradation followed by a long period of interference-free transmission, and the probability of occurrence of interference reduces as the code length is increased.

The maximum frequency hopping rate is limited by the capability of the frequency synthesizer. One of the main problems in a frequency hopped system is the need to meet the phase noise characteristics in each frequency hop.

Spread spectrum techniques were mainly developed to permit operation under interfering environments and very low carrier-to-noise levels. In the present satellite communication environment the spread spectrum technique can be used for a number of applications. In military applications it can be used to provide counter-measures against jamming. Another possible and interesting application is the addition of extra capacity by overlaying spread spectrum signals on existing narrow-band users – provided this is agreeable to the existing users. It has been shown that the spread spectrum technique offers capacity advantage when the duty cycle of users is low.

Capacity of spread spectrum systems

A system designer is interested in estimating the capacity of a CDMA in comparison with other multiple access techniques. The CDMA capacity can be determined by estimating the maximum number of users that can access the channel simultaneously without exceeding the specified carrier-to-interference noise ratio at the receiver demodulator. Note that resistance to interference can be increased by coding the information with a suitable FEC coding technique (see chapter 6), giving improved capacity. Closed-form mathematical expressions have been developed in the literature under various assumptions (e.g. Ha, 1985; Johannsen, 1988). Here we shall take a simplified approach to illustrate the technique (Utlaut, 1978). We have seen that the interference margin of a spread spectrum signal is given by

$$M = G_p - L - C_o/I_o \qquad (8.16)$$

Substituting for G_p and rearranging, we obtain the maximal interference, I_{im}, at the receiver input as

$$I_{im} = (C_i + M + L) \quad dB \tag{8.17}$$

To obtain the maximum number of users that can share the channel we integrate the interfering power from each user since

$$10 \log \sum_{n=1}^{n} i_n = I_{im} \quad dB \tag{8.18}$$

where i_n = interfering power (in watts) from each active user. Assuming for simplicity equal power from each source:

$$10 \log (ni_n) = M + L + C_i$$

$$10 \log(n) = M + L \tag{8.19}$$

(note: $10 \log i_n = C_i$)

$$= G_p - \left(\frac{C_o}{I_o}\right)$$

Finally, it is worth noting that spread spectrum systems have an inherent capability to accommodate traffic growth without any change to the network configuration, provided that the system design has been dimensioned for this growth at the outset, recalling that here growth in customer base causes an increase in interference. Another interesting feature is that, since a user can access the channel without any delay, the grade of service in a spread spectrum system is always good. It may happen that during the peak traffic period the interference levels exceed the limit. This increase is perceived by the user as a degradation in signal quality. The users can then decide whether to accept the degraded quality or wait until the quality improves. While this may or may not be an acceptable solution, it is worth noting that the degradation is gradual and the circuits are readily available to the users at all times.

8.6 Access protocols for data traffic

The fixed and demand-assigned schemes discussed above were optimized mainly for speech or continuous streams of data. The characteristics of most data traffic are different and this feature influences the selection of multiple access for data traffic. Examples of data traffic are: the transfer of large volumes of data; a request for channel allocation in a demand-assigned system; a query–response system, etc. Unlike speech which requires several minutes of continuous connection, data messages are characterized by bursts of high activity. For instance, in a query–response system a carriage return sends a burst of data within a few milliseconds, followed by a relatively long pause while the user awaits a reply. Some types of data messages such as a large data file transfer, however, do tend to be continuous. Another important difference between data and voice traffic is the larger tolerance of data traffic to message delivery delays. A conversation becomes incoherent when time delays of over 400 ms occur but in data communication applications, e.g. a query–response system, the user is more tolerant to delays.

When a network consists of a large number of data users, the throughput of the multiple access channel can be increased by allocating capacity on a *per message* or *packet basis*. The increase occurs because of the averaging effect of the channel. These accessing schemes can be categorized on the basis of degree of network coordination. One extreme is fully random access in which a terminal is allowed transmission at any time without coordination with other users. The penalty incurred for the offered simplicity is an increased probability of message loss through collision. The other extreme is fully scheduled access in which a channel is reserved by a user for the duration of a message. The increased reliability in message transfer is offered at the expense of increased terminal and network control complexities. We can accordingly categorize accessing scheme for data traffic as follows (Lam, 1977):

(a) channel reservation schemes;
(b) contention protocols;
(c) packet reservation protocols.

We shall see that most accessing schemes attempt to minimize the conflict between users who require access to a channel at the same time by the use of certain rules or 'protocols'. Hence multiple access schemes for data traffic are often called multiple access *protocols*.

In assessing the performance of multiple access schemes for speech, we used channel capacity and satellite EIRP as the main criteria. The performance of accessing schemes for data traffic is measured as channel *throughput*. Throughput is defined as the number of messages successfully transmitted per unit time. In the definition of throughput the *actual* transmitted blocks are considered, i.e. all the overheads such as bits required for coding, synchronization etc. are included in the packet. If P is the probability of successful transmission of a packet and G is the offered load, then throughput S is given by

$$S = GP \qquad (8.20)$$

To assess the multiple access performance quantitatively, the traffic environment must be modelled precisely. The environment is assumed to contain an infinite number of traffic sources with each source modelled as a point process, the message arrival instant being the point of interest (Lam, 1977). Each message is segmented into blocks called *packets*. Each message type is characterized by an *inter-arrival time* and permissible *delay* in source–destination message delivery. Several types of message can originate at a traffic source.

The burstiness of data traffic can be measured as the peak-to-average data rate. Lam (Lam, 1977) proposes the following definition of a bursty factor:

$$\beta = \delta/T \tag{8.21}$$

where δ = average message delay constraint
T = average inter-arrival time between messages.
Bursty factor gives an upper bound on the duty cycle of a traffic source, i.e.

$$S \leq \beta$$

For example, consider a message with an average delay constraint of 3 seconds. If such messages arrive every second the measure of bursty factor is 3. Now, if the arrival rate is changed to 100 seconds the bursty factor is reduced to 0.03. This latter traffic is then said to be more bursty since it is characterized by spurts of activity followed by longer periods of inactivity. We note that, as β reduces, traffic gets more bursty.

It should be emphasized here that in evaluating the applicability of a protocol for an application, in addition to the throughput offered by a protocol, factors such as terminal and network complexity and cost are equally important.

The main features of some well-known data accessing schemes are described next (Lam, 1977). The discussion is focused solely on the performance evaluation of the protocols, i.e. error-free transmissions are assumed and loss of packets is assumed to occur only when packets collide.

A. Channel reservation

By channel reservation scheme we mean schemes where physical channels are reserved for the duration of each message transmission. The channels may be pre-assigned or demand-assigned for a session. Conventional access techniques such as FDMA or TDMA are included in this category. For messages characterized by long continuous transmissions, such as voice, the time required to set up a channel connection (typically, a few seconds) is an insignificant part of the message time. But for bursty traffic, channel connection time could become comparable to message time.

The upper bounds on channel throughput S for a pre-assigned channel is given by

$$S \leq \beta \tag{8.22}$$

where β is the bursty factor given by (8.21).

The upper bounds on channel throughput S for a demand-assigned channel is given by

$$S \leq (\delta - t_a)/(\delta + t_d) \tag{8.23}$$

where t_a = allocation time

t_d = de-allocation time

δ = average message delay constraint.

Figure 8.14 (Lam, 1977) shows plots of upper bound of channel throughput for fixed and demand-assigned schemes as a function of message inter-arrival delay and the delay constraint.

It is seen that for a fixed assigned scheme the throughput approaches the maximum ($S = 1$) as the traffic begins to become less bursty ($\beta \rightarrow 1$) in accordance with equation (8.22). When the traffic gets increasingly bursty ($\beta \rightarrow 0$) the throughput of the fixed assigned scheme decreases. This is expected because with an increase in message inter-arrival delay (for a given δ) the channel remains idle for longer periods.

The throughput for the demand-assigned scheme is solely dependent on the channel allocation and de-allocation time, assuming the channel remains un-utilized during these periods. The inter-arrival message delay does not influence the throughput of the demand-assigned scheme, as the channel is reserved on a per message basis. The throughput is however affected by the delay constraint on the message, throughput increasing as average message delay requirement increases. It should be noted that, in practice, additional delay is introduced in the demand-assigned reservation process.

B. Contention protocols

In contention protocols, each user accesses a multiple-access channel without attempting to coordinate with other users in the network. Hence some packets are lost as a result of collision. Such packets must be re-transmitted. Varying degrees of discipline can be incorporated into such protocols to minimize the probability of collision, at the expense of receiver complexity.

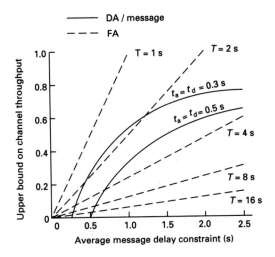

Figure 8.14   Upper bounds of channel throughput for fixed and demand-assigned schemes (Lam, 1977).

ALOHA schemes

ALOHA is the best known random access protocol (Abramson, 1970). In this scheme there is no coordination between users. Each user transmits a data packet as soon as it is received. If the packet is lost owing to collision, it is re-transmitted after a random interval to avoid a further collision. Randomization of re-transmissions is essential in establishing the stability of this scheme. A collision can be sensed by each earth station by listening to its own burst or awaiting the acknowledgement of receipt.

The throughput of ALOHA has been derived for single packet messages and an infinite user population (Abramson, 1970), assuming that the traffic generation at each terminal follows a Poisson process:

$$S = Ge^{-2G} \tag{8.24}$$

where G is offered traffic on the channel including new transmissions and retransmissions, and measured in packets/packet time. The maximum throughput is $1/2e$ or 18.4%, corresponding to an offered load of 0.5 packet/packet time.

The ALOHA protocol is a low-maintenance network capable of supporting simple terminals. Terminals may enter the network without any coordination.

Figure 8.15(a) shows the conditions under which collision-free operation of an ALOHA scheme takes place. It can be seen that a packet is vulnerable

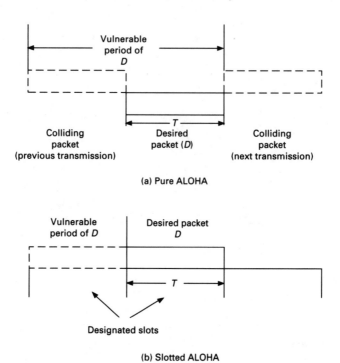

Figure 8.15   Period when packets are vulnerable to collision from other
packet transmissions: (a) pure ALOHA; (b) slotted ALOHA.

to collision for a total period of $2T$ seconds, T being the packet duration.
When the packet transmissions are confined to specific time slots, the vulnerable period of a packet collision is reduced, being restricted to transmissions occurring only during the slot. If a packet arrives before the start of a
slot, its transmission is delayed until the beginning of the next slot. Thus
the vulnerability of packets to collision (see figure 8.15b) is reduced to
half, resulting in an improved throughput. This improved version of ALOHA
is known as *slotted ALOHA*. Under identical assumptions as pure ALOHA,
the throughput of the slotted ALOHA scheme is given by

$$S = Ge^{-G} \tag{8.25}$$

The maximum throughput or *channel capacity* of this scheme occurs at a
traffic load of 1 packet/packet time and equals 1/e or 36.8%.

Reservation ALOHA

The ALOHA and slotted ALOHA schemes are best suited for an application consisting of a large population of very bursty users with very short messages. (Note that derivation of throughput assumed single packet messages.) When users have long messages for transmission, a modified version of ALOHA, known as reservation-ALOHA (Crowther, 1973), offers improvement. The transmission is organized into frames. Each frame is segmented into slots which are available to users on a contention basis as in slotted ALOHA. Whenever a user achieves successful transmission, the slot is reserved and no other user contends for it. If the user stops transmission, the slot is available for contention in the next frame. The capacity can be further increased if the user announces the end of use in its last message – then users do not have to wait for an occupied slot to become empty, indicating the end of use. Since a slot remains reserved for a user until released, the channel is able to support continuous traffic when necessary. The throughput of a channel is given by (Lam, 1977)

$$S_{RA} = S_{SA}/(S_{SA} + 1/L) \qquad (8.26)$$

where S_{SA} = slotted ALOHA throughput
L = average number of packets transmitted before a user releases a captured slot.
The throughput (or capacity) for an infinite population model is given by

$$C = 1/(1 + e/L) \qquad (8.27)$$

Thus $1/(1 + e) \leqslant C \leqslant 1$, for L ranging from 1 to infinity. When the end of message is announced $1/e \leqslant C \leqslant 1$, for L ranging from 1 to infinity.

Although simple to implement, ALOHA schemes tend to become *unstable* as traffic increases. In the limiting case, the transmitted packets keep colliding with each other and all the packets are lost. Figure 8.16 (Li, 1987) shows the throughput versus time delay for the pure and slotted ALOHA schemes.

Packet reservation

Channel reservation schemes were discussed earlier in the section, where a *channel* (e.g. a frequency channel in the FDMA SCPC scheme or a time slot in a TDMA scheme) is reserved for the duration of a message (on demand). In packet reservation schemes, slots are reserved on demand for transmission of *packets*. Reservation requests are sent on a separate channel, which could be fixed assigned, or use a contention access scheme such as ALOHA.

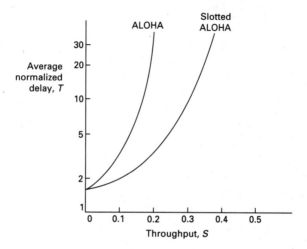

Figure 8.16 Delay–throughput relationship for ALOHA protocols (adapted from Li, 1987).

Reservation access schemes provide a much higher throughput than contention protocol since losses due to contention are eliminated. Instead, the contention problem is shifted to the reservation channel, if a contention protocol is used. It is desirable that the reservation scheme be simple. The advantage gained by using such a scheme is at the expense of an additional message delivery delay incurred during the reservation process. Delay incurred in the reservation process is about twice the propagation delay, $\sim 0.54S$. The mechanisms involved in packet reservation schemes are: *reservation, queue management* and *recovery* in case of a packet loss caused by erroneous access.

The reservation channels can be fixed assigned time slots or a contention scheme such as slotted ALOHA. The reservation packets are typically a fraction of the size of message packets, and hence very bursty, and therefore contention protocols are well suited for the purpose.

Queue management can be done either centrally or through distributed control. In centrally controlled queue systems, a network control centre assigns the transmission capacity, based on a centrally managed queue. Assignments are made on a designated part of the channel. In distributed queue management systems, each user maintains the status of the global queue and decides when to transmit. The correct transmission time can only be obtained when a user is synchronized to the queue – otherwise a collision occurs. Several types of protocol have been proposed for the purpose (e.g. Roberts, 1973). To accomplish error recovery, users sensing a loss of packet re-acquire the queue status and repeat the transaction.

Table 8.4   Preferred protocols for various types of traffic (Lam, 1979).

Type of traffic	*Suitable protocol*
Non-bursty	Fixed assigned (e.g. TDMA, FDMA)
Bursty with short messages	ALOHA, slotted-ALOHA
Bursty with long messages and large number of users	Reservation protocols with contention channel for reservation
Bursty users with long messages but small number of users	Reservation protocols with fixed assigned reservation channel

Reservation TDMA

Reservation TDMA is useful when large volumes of data need to be transferred. In this scheme, users first reserve a slot by means of a reservation slot. The main traffic is then transmitted into the allotted traffic slot. The assignment may be done either by a central controller or by a suitable distributed algorithm.

Choice of data access protocol

The selection of a protocol for data access depends mainly on the type of application. Some of the considerations are:

1. message delay and throughput of the channel;
2. cost of earth stations;
3. network complexity.

In the preceding section we have mainly discussed the first criterion. Based on this, table 8.4 (Lam, 1979) shows the preferred multiple access scheme for various types of traffic. In practice, however, the cost of earth stations and the network complexity may be the real considerations. Even though a theoretical analysis may indicate a specific protocol as highly efficient, the application must be able to earn enough revenue to justify the use of an earth station. It may well be that the cost of a dedicated earth station is too high to permit the use of very bursty traffic sources. In such cases it may be more economical to concentrate several traffic sources at a single point to make the use of an earth station economical.

When the choice is not obvious, an analysis involving the applicable traffic model, the number of users and other techo-economic factors must be done.

In the discussion above only a few representative protocols have been described to illustrate the principles. The reader should be aware that several types of protocol have been developed, with each type tailored to a specific traffic environment (e.g. Li and Yan, 1984).

8.7 Multiple access examples

In selecting a multiple access scheme, an attempt is made to match the multiple access channel to the network traffic and channel characteristics, economic constraints and state of the art. It is difficult to satisfy all the criteria when several types of traffic coexist – for example, speech and data. Since these characteristics cannot be satisfied with a single scheme, combinations of accessing schemes are sometimes used to optimize a network.

We shall illustrate two examples here – one in the mobile satellite service and the other in the fixed satellite service: the INMARSAT network for the MSS and the INTELSAT network for the FSS.

Mobile satellite service

Let us first summarize the main features of a mobile communication system which influence the choice of multiple access scheme:

1. Traffic for mobile service is characterized by a large number of widely dispersed users each with low traffic demand.
2. There is a need to accommodate growth in the user base. For example, a typical growth figure for the INMARSAT network in 1992 was in the range of several hundreds of users per month. The present user base in the network is about 20 000, and predictions show that the number will rise to well over 100 000 in the last part of the decade.
3. The need to operate in a multipath propagation environment.
4. The need to keep terminals simple and low-cost, hence well established technology is preferred.
5. The need to serve mobiles in aeronautical, land and maritime environments.
6. The need to conserve bandwidth.

Fixed assignment for this type of environment is wasteful of bandwidth and too restrictive for growth. Demand-assigned schemes – both in the time and frequency domains – provide higher channel usage through sharing and are inherently flexible to accommodate traffic growth, and are therefore well suited for the application. Another possibility is the use of the spread spectrum system which provides a natural resistance to noise and multipath together with a capability to accommodate growth and a high grade of service at all times. The spread spectrum has been used in an experimental low-bit rate mobile communication system in Europe. But the spread spectrum technology is relatively new and evolving even in the 1990s, and therefore was not considered favourably in the INMARSAT network which became operational in the early 1980s. With this background, we shall consider the accessing scheme for the *INMARSAT* network.

The INMARSAT network provides worldwide voice and data communication for various types of terminals mounted on ships, aircraft and land mobiles. As we have noted in chapter 3, each of these environments is different, both from propagation considerations and terminal characteristics. Moreover, there are differences in traffic characteristics. For example, currently, data transmissions are more commonly used by aircraft (to provide safety-related information) and land mobiles, whereas voice communication is the main type of traffic from ships, although a gradual shift towards facsimile and data transmissions is emerging. Additionally, these traffic patterns may change in future. Accordingly, the INMARSAT network has provisions to accommodate these varying requirements by providing a mix of multiple access schemes.

The INMARSAT-A system is a first-generation system which provides analog voice transmission and telex to ships. The INMARSAT-B system is an improved implementation of the Standard-A system using digital techniques with a capability of providing digitized telephony, telex, facsimile and data. The INMARSAT-M system has been designed to operate with smaller terminals in the maritime and land environment, with a capability to support medium-bit rate telephony, data and facsimile services. INMARSAT-C is a digital store-and-forward low-bit rate system (600 bps) to provide communication to low-cost maritime and land mobile terminals. The INMARSAT-Aero system provides telephony, data and facsimile services to the aeronautical community. Here we shall only discuss, briefly, the INMARSAT-A system.

The accessing schemes are different in the forward and return directions for telex traffic. This is so because shore to ship traffic is concentrated in a few LESs whereas ship-to-shore traffic originates in widely dispersed areas.

The INMARSAT-A network consists of a network control station (NCS) which provides central control of the network, several fixed land earth stations (LESs) distributed all over the Earth, which serve as gateways to adjacent geographical regions, and the ship earth stations. The NCS broadcasts network-related information on a *time division multiplexed* broadcast channel. Call assignment (or rejection) messages are carried on this channel. All ship earth stations (SESs) and LESs stay tuned to the NCS broadcast channel to receive the necessary network information. All the channel assignment requests are made to the NCS. Ship earth stations use a *pure ALOHA* channel for sending requests. Pure ALOHA is suited for this type of message as request messages are very bursty. There are two request channels in the network. The NCS to LES communication takes place via the broadcast channel and *pre-assigned* time division multiplexed (TDMs) broadcast channels (LES-NCS). Voice communication between ship and shore takes place using *demand assigned single channel per carrier*. For each voice channel request, the NCS assigns a voice channel pair from a pool of channels and broadcasts the assignment on the broadcast channel. The requesting ship and destination LES (or vice versa) then tune to the assigned frequency and begin communication. At the end of the call the NCS returns the channel to the pool. Telex messages in the ship-to-shore direction take place via *demand assigned time division multiple access*. In this case the request is sent to the NCS but the assignment is made by the LES with which the terminal wishes to communicate. Each LES manages its own pool of time slots. The assignment is made as a time slot assignment. Thereafter, the communication takes place on the assigned time slot for the duration of the call. At the end of the call the time slot is returned to the LES pool. Telex communication in the shore-to-ship direction is made on the LES TDM channel, wherein the telex messages are multiplexed with other types of information.

Fixed satellite service

The accessing scheme for fixed satellite services depends on the type of service and the traffic density between communicating points. For example, a random access scheme may be suitable in a VSAT network whereas a high-bit rate TDMA may be a better choice for connecting two large earth stations.

It is interesting to examine the way in which accessing schemes in INTELSAT networks developed to meet the growing and changing demand over the years. During the initial years, fixed assigned *frequency division multiple access* was a natural choice because of its simplicity and the state of the technology. With the addition of earth stations having low traffic, INTELSAT introduced the SCPC system – both *fixed* and *demand-assigned* schemes in the form of *SPADE* (discussed in a previous section) were introduced to serve thin routes and their use has continued into the 1990s. With an increase in demand, increased availability of digital hardware and a growing need for bandwidth and network flexibility, *time division multiple access* was introduced in the late 1970s. *Digital speech interpolation* was introduced to increase the capacity further. Multiple spot beams were introduced to enhance system capacity. This led to the need for a large number of transponders on satellites to provide the necessary interconnections. To economize on the space segment, *satellite switched TDMA* was introduced. With regard to video transmission, during the initial years one television channel was used per transponder. This has now been increased to two television

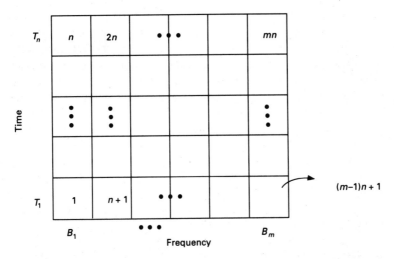

B_i = ith frequency segment

T_i = ith time slot

Figure 8.17   Two-dimensional grid of a hybrid FDMA–TDMA scheme. Assuming a pre-assigned grid, a total of mn users may be accommodated where m = number of frequency segments and n = number of time slots per frequency segment.

channels/transponder. At present, the INTELSAT network consists of a mix of all these schemes.

8.8   Future trends

In future, the basic multiple access schemes may not be able to meet the increasing demands of satellite resource sharing with the desired flexibility and cost. The maximum number of interconnections in a full mesh network of n earth stations is given as $n(n - 1)/2$. Therefore if a base of 100 customers (e.g. VSAT users) needs simultaneous interconnections, the number of accesses totals 4950. The basic multiple access schemes are not well suited to provide such a large number of simultaneous accesses efficiently. However, it is felt that the financial advantages which this type of network could potentially provide may warrant more novel solutions (Pelton and Wu, 1987).

Several hybrid schemes can be envisaged. As seen in the above discussions, the available multiple access resources are frequency, time, space and code. It is possible to combine the time and frequency domains to form a two-dimensional grid as shown in figure 8.17. In this scheme the available bandwidth is segmented into sub-blocks and each sub-block is then shared in a TDMA mode. Several-fold increases in access are possible using this scheme, although at the expense of a more efficient modulation scheme – realizing that bandwidth segmentation requires compression in bandwidth. Similarly a three-dimensional plane can be visualized by introducing CDMA in each sub-band, increasing the number of accesses further. The interested reader should refer to the literature (e.g. Wu, 1984) for further persual of the topic.

Problems

1. With the help of diagrams, discuss the principle of operation of the following multiple access schemes:
   (a) frequency division multiple access;
   (b) code division multiple scheme;
   (c) various ALOHA schemes.
   Mention the advantages and limitations of each.
2. Suggest possible satellite multiple accessing techniques for the following applications, giving reasons for your choice:
   (a) an international FSS network;
   (b) an MSS network.
3. Discuss the suitability of each multiple access scheme for the return link of a VSAT network. Suggest a possible solution.
4. Describe the operation of a pre-assigned TDMA system. Illustrate a typical TDMA frame structure. Mention the advantages of TDMA over FDMA.
5. (a) What do you understand by (i) a bandwidth limited system, (ii) a power-limited system?
   (b) Determine the TDMA transmission capacity of a transponder with the following characteristics:

   Transponder bandwidth (B_T) = 72 MHz
   Modulation = QPSK
   B_t/R_b = 1.2
   where R_b is the symbol rate.

   (c) Determine the TDMA capacity of a 4 GHz link with the following characteristics:

   Satellite EIRP = 22.5 dBW
   Path loss = 197 dB
   G/T of earth station = 30 dB/K
   Total link margin = 5 dB
   Required E_b/N_o = 8.5 dB.

(d) What is the voice channel capacity of a TDMA system with the following characteristics?

Transmission bit rate = 60 Mbps
Voice channel bit rate = 64 kbps
Number of bursts/frame = 10
Number of bits in each preamble = 150
Frame time (microseconds) = 750.

References

Abramson, N. (1970). 'The ALOHA system – another alternative for computer communications', *AFIPS Conf. Proc.*, Vol. 42, AFIPS Press, Montvale, New Jersey, pp 281–285.

Byrov, V.L., Borokov, V.A. and Khomutov S.M. (1968). 'Improving channel capacity of satellite communication systems with multistation access and frequency multiplexing', *Telecommunication Radio Eng.*, Vol. 22, January 1.

Campanella, S.J. (1976). 'Digital speech interpolation', *COMSAT Technical Review*, Vol. 6, Spring, pp 127–158.

Crowther, W. (1973). 'A system for broadcast communications: reservation-ALOHA', *Proc. 6th HICSS*, University of Hawai, Honolulu, January.

Dicks, J.L. and Brown, P. Jr. (1974). 'Frequency division multiple access for satellite communication systems', *IEEE Electronic and Aerospace System Convention (EASCON)*, October 7–9, pp 167–178.

Dill, G.D. (1972). 'Comparison of circuit call capacity of demand-assignment and pre-assignment operation', *COMSAT Technical Review*, Vol. 2, No. 1, Spring.

Dixon, R.C. (1984). *Spread Spectrum Systems*, Wiley-Interscience, New York.

Gabbard, O.G. and Kaul, P. (1974). 'Time division multiple access', *IEEE Electronic and Aerospace System Convention (EASCON)*, October 7–9, pp 179–184.

Golomb, S.W. (1967). *Shift Register Sequence*, Holden-Day, San Francisco, California.

Ha, T. (1985). 'Spread spectrum for low cost satellite services', *International Journal of Satellite Communications*, Vol. 3, pp 287–293.

Johannsen, K.G. (1988). 'CDMA versus FDMA channel capacity in mobile satellite communication', *International Journal of Satellite Communications*, Vol. 6, pp 29–39.

Laborde, E. (1985). 'Optimization of demand assigned SCPC satellite networks', *Space Communication and Broadcasting*, Vol. 3, pp 201–206.

Lam, S.S. (1977). 'Satellite multiaccess schemes for data traffic', *Int. Conf. Communications*, June 12–15, Vol. III, pp 37.1–19 to 37.1–24.

Lam, S. (1979). 'Satellite packet communication – multiple access protocols and performance', *IEEE Trans. Commun.*, Vol. COM-27, No. 10, October, pp 1456–1466.

Li, V.O.K. (1987). 'Multiple access communications networks', *IEEE Communications Magazine*, Vol. 25, No. 6, pp 41–47.

Li, V.O.K. and Yan, T.Y. (1984). 'An integrated voice and data multiple-access scheme for a land mobile satellite system', *Proc. IEEE*, Vol. 72, No. 11, November, pp 1611–1619.

McClure, R.B. (1970). 'Analysis of intermodulation distortion in an FDMA satellite communication system with a bandwidth constraint', Presented at the *IEEE Int. Conf. Communications*, San Francisco, California, June 8–10.

Pelton, J.N. and Wu, W.W. (1987). 'The challenge of 21st century satellite communications: INTELSAT enters the second millennium', *IEEE Journal on Selected Areas in Communications*, Vol. SAC-5, No. 4, May, pp 571–591.

Puente, J.G., Schmidt, W.G. and Werth, A.M. (1971). 'Multiple-access techniques for commercial satellites', *Proc. IEEE*, Vol. 59, No. 2, pp 218–229.

Roberts, L. (1973). 'Dynamic allocation of satellite capacity through packet reservation', *AFIPS Conf. Proc.*, Vol. 42, AFIPS Press, Montvale, New Jersey, pp 711–716.

Sarvate, D.V. and Pursley, M.B. (1980). 'Crosscorrelation properties of pseudo random and related sequences', *Proc. IEEE*, Vol. 68, pp 593–619.

Utlaut, W.F. (1978). 'Spread-spectrum principles and possible application to spectrum utilization and allocation', *Telecommunication Journal*, Vol. 45, No. 1, pp 20–32.

Wache, W. (1971). 'Analysis of adjacent channel interference in a multi-carrier FM communication system', *COMSAT Technical Review*, Vol. 1, No. 1, Fall.

Wu, W.W. (1984). *Elements of Digital Satellite Communications, Vol. 1 – System Alternatives, Analyses and Optimization*, Computer Science Press, Rockville, Maryland.

9 Communication Satellites

9.1 Introduction

A communication satellite is required to provide service reliably within a given geographical area throughout its lifetime. The design is governed by the communication capacity of the satellite, the physical environment in which it operates and the state of the technology.

The preferred size, complexity and cost of the ground stations together with the required traffic to be served by the satellite define the power and bandwidth requirements of the satellite. For example, to provide communication to small portable terminals requires high-powered large satellites.

The geographical area to be served by the satellite is another important consideration. A satellite is most efficient when its transmissions are focused within the desired geographical area. An additional advantage is the reduction in emissions outside the service area, thereby minimizing the interference to other systems and permitting more efficient spectrum usage. The geographical areas are usually irregularly shaped and therefore the satellite antenna patterns must be shaped accordingly to fit the coverage area as closely as practical. Thus there is a trade-off between the advantage gained in using a shaped spot beam and the complexity of a satellite's antenna system.

A geostationary satellite is subjected to harsh environmental conditions such as shocks and vibrations during the launch, vacuum, large temperature variations and the effect of small particles present in space. Considering that a spacecraft is very difficult to access after launch, the design must take into account both the short- and long-term effects so that the reliability of the spacecraft is acceptable throughout its lifetime.

Several types of forces act on a geostationary satellite causing it to drift away from the assigned orbital location, and hence its orientation with respect to Earth is subject to change. These extraneous forces must be compensated by a control system. The spacecraft must also be capable of generating the necessary electrical power for its proper functioning.

The design of a satellite begins with a synthesis of a baseline spacecraft design, meeting all technical requirements such as EIRP and coverage. The synthesis process provides useful parameters such as the size and weight of the spacecraft. This is followed by optimization of this basic design, taking into consideration technical constraints and costs. For example, the assumed spot beam in the initial design may be too complex to implement, increasing the cost beyond the target, and therefore the constraints on coverage area may have to be relaxed.

The subsequent development programme of the spacecraft follows a sequence of distinct phases until it is ready for launch. At each stage, changes to the design may be introduced if the original design is too difficult to implement without significantly impacting the schedule, or the implementation costs become prohibitive.

This chapter begins with a description of the main considerations necessary in the design of a geostationary satellite. This is followed by a description of the main features of each sub-system. A synthesis approach useful for estimating power and mass during the baseline design phase is introduced next. The chapter concludes with a brief description of a spacecraft's development phases up to its launch.

Throughout, no distinction is made between the use of the word 'satellite' and 'spacecraft', although it may be argued that 'spacecraft' is a more appropriate description of a man-made vehicle launched into space.

9.2 Design considerations

The design and configuration of a spacecraft is dictated by its mission goals. For instance, the mission goal of a scientific satellite is gathering scientific data. Therefore the size/complexity of the spacecraft and ground stations are dictated solely by the need to obtain accurate data, commercial interests being less significant.

Communication considerations

For a telecommunication satellite the main considerations are:

1. type of service to be provided (e.g. mobile communication, direct-to-home broadcasts);
2. communication capacity (transponder bandwidth and satellite EIRP);
3. coverage area;
4. technological limitations.

Initially, the basic specifications are laid out for the satellite, based on the given communication requirement. For a domestic fixed satellite service it may be the EIRP per carrier, number of carriers and coverage area. Similarly for a direct broadcast satellite the number of television channels and coverage area are specified. The task of the system designer is then to develop an optimal spacecraft configuration to meet the needs within the specified costs and technical constraints. In addition, the recommended configuration from a manufacturer is often based on factors such as previous experience and in-house capability. Thus for the same set of requirements, different types of configuration are often proposed. For example, the same communication requirements can be equally well supported either by a spin-stabilized or a three-axis-stabilized satellite.

Environmental conditions

A spacecraft is designed to operate reliably in all types of environments encountered during its mission – beginning from the launch to the in-orbit deployment, and extending throughout its operational phase. The stresses encountered are mechanical, thermal and radiative. The most significant stresses needing consideration are briefly discussed below.

(i) Zero gravity

At the geostationary orbit, gravitational force is negligible giving rise to 'zero-gravity' effects. One of the major effects of the zero-gravity condition is the difficulty of liquid fuel flow, and therefore external provision has to be made to force liquids to flow. On the beneficial side, the absence of gravity facilitates operation of the deployment mechanisms used for stowing antennas and solar panels during launch.

(ii) Atmospheric pressure and temperature

At the geostationary altitude the atmospheric pressure is extremely low – of the order of 10^{-17} torr. The effect of such low pressures is to make thermal conduction negligible and increase friction between surfaces. Special materials are therefore required for the lubrication of moving parts on the spacecraft (e.g. the bearings used in the stabilization system). Fortunately the pressures inside a spacecraft are several orders of magnitude higher because of outgassing of electronic components, and hence the conditions for the electronic sub-systems housed inside a spacecraft are more manageable than those outside.

The temperature of a spacecraft is mainly affected by heat from the Sun and heat generated by various spacecraft sub-systems. The excursion in the external temperature may range from 330–350 K in the presence of sunlight to 95–120 K in eclipse conditions.

(iii) Space particles

Several types of particle exist in space. These are cosmic rays, protons, electrons and meteoroids. The intensity of particles emitted from the Sun depends mainly on the solar activity. The main effect of bombardment by particles on a satellite is to cause a degradation in the solar cells and certain solid-state components within the satellite. The effect of meteoroids is negligible on geostationary satellites.

(iv) Magnetic fields

The magnitude of the Earth's magnetic field is very weak at the geostationary orbit, ~1/300 relative to its value at the Earth's surface. The effect of the magnetic field can be compensated by the use of a large coil, if considered necessary.

The Earth's magnetic field deflects charged particles which are trapped in the region surrounding it. This region is called the Van Allen belt. The most intensely charged layer around the equator lies below $4R_e$, where R_e is the Earth's radius, and therefore does not affect a geostationary satellite under normal operation. Satellites pass through the Van Allen belt during orbit-raising. Also, satellites in a highly elliptic Molniya type orbit could spend a significant time in the radiation belt if the perigee of the elliptical orbit is below the Van Allen belt. The electric charges affect electronic components and special manufacturing techniques need be used to harden the electronic components against radiation.

(v) Other considerations

There are a number of other considerations peculiar to a satellite in a geostationary orbit.

Because of the eccentricity in the Earth's orbit (~0.0167) there is a variation in the distance of the Earth from the Sun throughout the year, resulting in a variation of ±3.34 % in the received radiation intensity from the Sun. This fluctuation cause a variation in the DC generation capability of a satellite and must be taken into account in the design of the satellite's power system.

Additionally, the spacecraft power system must be designed to account for the loss of power from the solar cells during eclipses, gradual degradation in the efficiency of solar cells occurring over the years, and change in the direction of sun-rays due to relative movement between the Earth and the Sun (see section 2.6).

Geostationary satellites are affected by a number of external forces. Chapter 2 discussed and quantified the major natural effects which cause a satellite to deviate gradually from the ideal orbit. In addition, solar radiation incident on the surface of a satellite generates pressure, causing a gradual change in the eccentricity of the orbit. There are other short-term perturbations on the satellite caused by the movement of mechanical parts and fuel within it. A further consideration is the need to maintain the antennas pointing accurately at the Earth. Note that the satellite rotates a full 360° along its north–south axis over a sidereal day. The *net effect* of these forces on the satellite is to cause a gradual drift from its nominal position together with short-term variations in pointing towards the Earth. These perturbations are compensated by an on-board orbit and control system.

9.3 Lifetime and reliability

Lifetime

The useful lifetime of a geostationary satellite is determined by the maximum acceptable deviation in inclination and orbital location together with the reliability of the satellite's critical sub-systems (see also section 2.5, *Atmospheric Drag*, for a general discussion on satellite lifetime).

We shall see later that a satellite is maintained in its orbital location by firing thrusters regularly, using stored fuel. Hence it is possible to increase the operational lifetime of a satellite by the following means (see also chapter 11):

(a) increasing the fuel capacity;
(b) saving fuel by accepting orbital deviation to the maximal extent possible.

Both these techniques are used when applicable. There is a practical limit to a satellite's fuel storage capability. A more efficient alternative is to improve the efficiency of the fuel used for propulsion. Useful satellite lifetimes have increased from 7–10 years (1980s) to 12–15 years (1990s) as a direct result of improved fuel efficiency.

Reliability

The overall reliability of a satellite is mainly governed by the reliability of its critical spacecraft components. Reliability is improved by employing redundancy in the critical sub-systems and in components such as travelling wave tube amplifiers known to have a limited lifetime. However it is not possible to duplicate certain critical components such as antenna reflectors and bearing assemblies. The reliability of a spacecraft is calculated by means of a reliability model. The basic concepts used in reliability modelling are developed here.

Reliability is defined as the probability that a given component or system performs its functions, as desired, within a specified time t. The reliability R of a component can be represented by

$$R = e^{-\int_0^t \lambda \, dt} \tag{9.1}$$

where λ = failure rate of the component.

The unit of λ for satellite components is often specified as FIT, the number of failures in 10^9 hours (or sometimes in 10^6 hours).

Most components follow the failure mode as shown in figure 9.1. Three

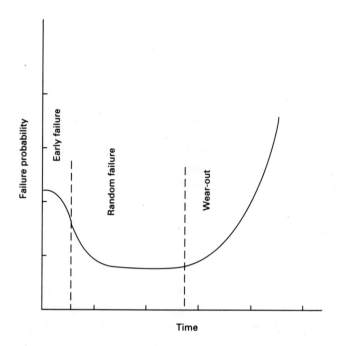

Figure 9.1   Failure modes of components.

regions can be identified – an early high failure rate region attributed to manufacturing faults, defects in materials, etc.; a region of low failure attributed to random component failures; followed by a region of high failure rate attributed to component wear-out.

In a satellite system, early failures are eliminated to a large extent during testing and burn-in. The main aim is to minimize the random failures which occur during the operational phase of the satellite by using reliability engineering techniques. The beginning of wear-out failure can best be delayed by improving the manufacturing technique and the type of material used.

In the analysis of reliability, it is usual to assume that failure rate is constant over time. Then the reliability can be expressed as

$$R = e^{-\lambda t}$$
$$\quad = e^{-t/m} \qquad (9.2)$$

where $m = \dfrac{1}{\lambda}$

$\quad\quad\quad = $ mean time between failures.

When several components or sub-systems are connected in series, the overall reliability is

$$R_s = R_1 R_2 \ldots R_n \tag{9.3}$$

where R_i is the reliability of the ith component. In terms of the failure rate:

$$R_s = e^{-(\lambda_1 + \lambda_2 + \ldots \lambda_n)t} \tag{9.4}$$

The overall reliability is improved by employing redundancy, wherein the less reliable and critical components are duplicated. If a failure occurs in the operational unit, the standby unit takes over its function. In developing a reliability model for such a configuration the redundant elements are considered to be in parallel. Parallel redundancy is useful when the reliability of an individual sub-system is high. Consider a system of i parallel elements in which the reliability of each element is independent of all others. If Q_i is the unreliability of the ith parallel element, the probability that all units will fail is the product of the individual unreliabilities:

$$Q_s = Q_1 Q_2 \ldots Q_i \tag{9.5}$$

When the unreliabilities of all elements are equal, this expression reduces to:

$$Q_s = Q^i \tag{9.6}$$

where Q is the unreliability of each element.
Therefore the reliability is

$$R = 1 - Q_s$$
$$= 1 - Q^i$$
$$= 1 - (1 - R)^i \tag{9.7a}$$
$$= 1 - (1 - e^{-\lambda t})i \tag{9.7b}$$

Note that the reliability of a redundant system has increased because $Q^i < Q$ (see 9.7a).

The underlying assumption in developing equation (9.7) is that standby units have the same failure rate as the operational unit. In practice, such an assumption may not always be applicable. The failure rates of the standby units are likely to be lower, as they are under lower stress during normal operation, and therefore equation (9.7) gives a reliability bound (see Morgan and Gordon, 1989).

Failure analysis of spacecraft shows that failures are often correlated, i.e. failures of similar units occur simultaneously. This could be due to an identical manufacturing defect in the production of a given batch of components or an inherent design defect (e.g. in an electronic sub-system). The effect of such correlated failure is to lower the overall reliability to a value less than that predicted by the above analysis. One method employed to reduce the risk of this type of correlated failure is to procure units from different sources. However, design defects are generic to all satellites produced in a series. The first one or two satellites of the series are the most susceptible. Usually, such defects are rectified, or their impact minimized if a complete design change cannot be implemented, in the following satellites of the series.

To predict the overall reliability of a spacecraft and optimize the redundancy configuration, a reliability model is developed. Figure 9.2 illustrates a typical reliability model of a geostationary satellite. All the major sub-systems are shown in series. The reliability of the model is then obtained by series addition of reliabilities:

$$R_s = R_1 R_2 \ldots R_7 \qquad (9.8)$$

where R_1, R_2, etc. are shown in figure 9.2. The reliability of each sub-system is obtained by developing their individual reliability models. Figure 9.3 illustrates a simplified reliability model of a communication system. Applying the equation for series and parallel combination developed above, the reliability of the communication system is obtained as

$$R_s = R_{RX} R_{TX} \{1 - (1 - R_T)^2\} \qquad (9.9)$$

where reliabilities are specified in the figure. Substituting $R_T = 0.9$, notice that the reliability of the transponder system (the last term in (9.9)) increases to 0.99.

It should be noted that although reliability can be improved by the use of redundant units, the mass of the satellite increases. A useful figure of merit in quantifying the increase in reliability per unit mass can be defined as

$$F_\gamma = \frac{r}{M} \qquad (9.10)$$

where $\quad r = \dfrac{R'}{R}$

$R' =$ reliability with redundancy employed
$R \ =$ reliability without redundancy
$M =$ increase in mass due to added redundancy.

Sensitivity analysis performed for various alternative configurations provides the minimal mass system for a given reliability.

The addition of redundant equipment also increases the cost of the tran-

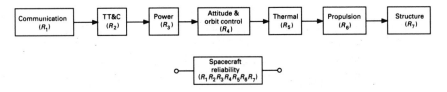

Figure 9.2   Reliability model of a communication satellite.

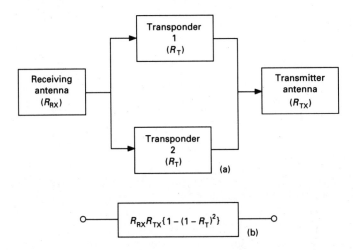

Figure 9.3   A simplified reliability model of the communication sub-system: (a) top-level block diagram of sub-system; (b) overall reliability.

sponder. There are two major cost components – the cost of the equipment together with the switching and failure sensing mechanisms used; and the associated increase in the weight of the spacecraft resulting in an increased launch cost (launch costs are usually based on the weight of the spacecraft). An optimization can again be performed as a function of cost, when cost minimization becomes critical.

9.4   Spacecraft sub-systems

A communication satellite essentially consists of two main functional units – *payload* and *bus*. The primary function of the payload is to provide communication. The bus provides all the necessary electrical and mechanical support to the payload. Figure 9.4 shows the main sub-systems of a typical communications satellite.

The payload is made up of a *repeater* and an *antenna* sub-system. The

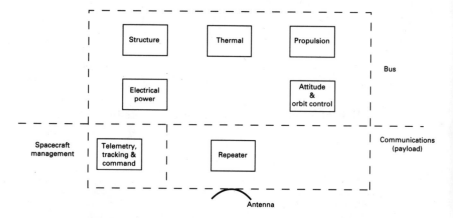

Figure 9.4   The main sub-systems of a geostationary satellite, shown partitioned according to their function.

repeater performs the required processing to the received signals, and the antenna system is used to receive signals from and transmit signals to the ground stations in the coverage area.

The bus (sometimes called a platform) consists of several sub-systems (see figure 9.4). The *attitude and orbit control* system (AOCS) stabilizes the spacecraft and controls its orbit. The *propulsion* system provides the necessary velocity increments and torques to the AOCS. The *telemetry, tracking and command* (TT&C) sub-system transmits the status of various sub-systems of the satellite (its 'health') to the satellite control centre and accepts commands from the control centre for performing essential functions on the spacecraft. This system also provides tracking support to ground stations. The electrical power supply system provides the necessary DC power. The *thermal* control system maintains the temperature of various sub-systems within tolerable limits. The *structure* provides the necessary mechanical support during all the phases of the mission.

A. Payload

As mentioned above, the payload comprises repeater and antenna sub-systems and performs the primary function of communication.

Repeater

The function of a repeater is to receive the uplink RF signals, and convert these signals to the appropriate downlink frequency and power for transmission towards the service area. Two types of repeater architectures are possible: (i) transparent and (ii) regenerative.

A *transparent* repeater (sometimes called a 'bent-pipe' because of the obvious analogy) only *translates* the uplink frequency to a suitable downlink frequency and power *without in anyway processing the baseband signal*.

A *regenerative* repeater, in addition to translation and amplification, changes the form of the baseband signal by a process such as demodulation. For example, consider a large number of single channel per carrier (SCPC) carriers accessing a satellite in a frequency multiplexed scheme. A regenerative repeater can demodulate individual carriers and time multiplex the signals on to one (or a few) carriers, and remodulate the signal using a better optimized modulation and/or multiple access scheme for the downlink before re-transmission. Thus the uplink and the downlink can be optimized separately, resulting in a more efficient system. A regenerative repeater offers several advantages (Maral and Bousquet, 1984) summarized as follows:

(a) increased communication capacity;
(b) greater resistance to interference;
(c) smaller ground stations.

At present, hardware for such repeaters is in the process of becoming established and therefore most existing satellites use transparent repeaters. The following description of the repeater is therefore limited to the realization of a transparent transponder.

Transparent repeater
The gain of a repeater is typically of the order of 100 dB. For instance, consider an earth station transmitting an EIRP of 60 dBW in the C band. The signal at the input of a satellite repeater using a global beam is therefore about -122 dBW. The required output power from the repeater is typically of the order of 10 dBW, which requires a repeater gain of the order of 132 dB. With such high end-to-end gain requirements, a repeater tends to become unstable unless the isolation between the receive and the transmit section is made high (typically 40 dB more than the loop gain). Isolation is achieved by using filters with high isolation between the transmit and the receive bands and careful RF shielding between the paths to avoid radiative coupling. However, even with such measures the desired high isolation is only possible when the frequency separation between the transmit and the receive bands is large. Therefore the uplink and downlink frequencies are usually assigned in separate bands – such as 6 GHz uplink and 4 GHz downlink. Generally the uplink frequency is larger than the downlink frequencies. One reason for such a choice of frequency was that power was at a premium on earlier satellites. Assuming that global coverage is required, then the required beamwidth of the satellite antenna is fixed at about 17°. With beamwidth fixed, the antenna gain is constant with frequency, but note

that the path loss reduces with frequency. Thus for a specified receive flux density on the ground, the satellite transmitter power requirement reduces as the frequency is reduced.

The main elements of a typical transparent repeater are shown in figure 9.5. Signals from the antenna and the feed systems are fed into a low-noise amplifier through a bandpass filter (BPF). The BPF attenuates all out-of-band signals such as transmission from the ground stations of adjacent satellite systems. The low-noise amplifier provides amplification to the weak received signals. The spacecraft antenna is pointed towards a relatively warm Earth having a noise temperature of about 300 K. Therefore there is no advantage in reducing the noise temperature of the LNA much below this value. Tunnel diodes were used in earlier satellites. At present, bi-polar transistors are used below 2 GHz and FETs above 2 GHz. Low-noise stages typically provide gains of around 20 dB, followed by a further amplification of 15–20 dB.

The conversion to the downlink frequency can be performed either in a single stage using one intermediate frequency or in multiple stages using more than one intermediate frequency. A single-stage conversion is simple to implement. Multiple-stage down-conversions are preferred when there are special requirements. For example, it may be necessary to interconnect earth stations operating in different frequency bands (e.g. 14 GHz, uplink – 4 GHz, downlink). Such an interconnection is performed readily at a common IF, made possible by the use of a multiple-stage conversion scheme. Sometimes the required transponder gain or filtering cannot be achieved if direct conversion is used because of technological limitations. In such cases a multiple-stage down-conversion scheme is useful, because the necessary amplification and filtering can be conveniently performed at an intermediate frequency.

Down-conversion is achieved by mixing the amplified uplink signal with a local oscillator. The local oscillator consists of a stable crystal oscillator operating at a relatively low frequency (for example, 100 MHz) followed by a multiplier chain, bandpass filters and amplifiers.

A bandpass filter is used at the output of the mixer to remove the unwanted out-of-band frequencies generated in the mixing process. It is essential that the amplification in the LNA and the down-conversion be performed linearly to avoid inter-modulation noise being generated by the signals arriving at the satellite from various earth stations of the network.

When a single conversion down-conversion is used, the down-converted signal is at the desired downlink frequency. The signal is fed into the input multiplexer, which divides the channels into sub-bands. We will refer to each sub-band as a *transponder*.

Channelization reduces the number of carriers passing through the high-power amplifier simultaneously, thereby minimizing the inter-modulation noise. Also, because satellite high-power amplifiers have a limited power-handling capacity, it is desirable to split the signals into sub-bands. In a satellite using spot beams, channelization is also used for interconnecting spot beams.

273

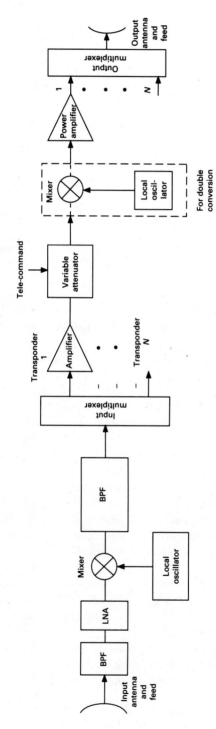

Figure 9.5 The main elements of a transparent repeater.

The input multiplexer essentially consists of a filter bank, each filter permitting a specific band of signal to pass through. The implementation scheme of the filters depends on the radio frequency in use. The filters need to provide the required rejection of the adjacent frequency bands with minimal amplitude variations and group delay in the passband. The signal in each transponder is amplified to boost up the relatively low-power level of signals at this stage. Bi-polar transistor and FET amplifiers are commonly used for amplification. Electronically switchable attenuators are introduced in the signal path to control the gain of the transponder. The attenuator is controlled through tele-commands.

The gain control facility is necessary to compensate for gradual gain degradation of a transponder over the lifetime of the satellite. The gain control function can also be used to optimize the overall link quality by apportioning noise contributions from the ground and space segments in an optimal manner. In a *dual-conversion* repeater, the composite uplink signal is downconverted into an intermediate frequency (IF) before demultiplexing. Hence the signals in each sub-band are at IF. The IF is amplified and up-converted to the downlink RF frequency after suitable amplification (see figure 9.5). RF signals are then fed into the high-power amplifier (HPA) stage, consisting of one or more stages of driver amplifiers, followed by a power stage. As discussed in chapter 4, the final stage of the HPA inherently possesses a relatively high non-linear characteristic. The non-linearity increases with increased drive levels. Therefore, in a multi-carrier environment, the final stage of the amplifier contributes significantly to the inter-modulation noise. To minimize the inter-modulation noise, the operating point is chosen to be in the linear portion of the transfer curve. Devices known as linearizers are sometimes used to improve the linearity. Another technique is to use an automatic level control (ALC) circuit at the driver stage, which maintains a constant drive level at the input of the final-stage amplifier irrespective of the input drive. This ensures that the same operating point, chosen to meet the specified inter-modulation criteria, is maintained for all traffic conditions. However, consequently the EIRP of individual carriers is a function of traffic load on the satellite. To avoid this sensitivity to traffic, in some satellites the ALC is designed to operate only above a certain threshold, below which the satellite HPA behaves as a linear device.

For high-power transmissions (e.g. direct broadcast satellites), at present travelling wave tube (TWT) amplifiers are commonly used, in particular at the higher end of the currently used radio frequencies – the K_u band. For low-power requirements and at the lower end of the frequency, solid-state amplifiers are increasingly being used because of their better linearity and lower power consumption. When power requirements exceed the levels available from a single solid-state amplifier unit, the outputs from several amplifiers are combined to provide the desired power. Solid-state amplifiers use bi-polar transistors up to a frequency of about 2 GHz and FET amplifiers above this frequency up to 20 GHz. At present, TWTs can provide higher efficiencies and their technology is well established.

The outputs from the transponders are combined into a composite signal in the output multiplexer. One possible configuration of the multiplexer consists of a bank of bandpass filters. The signals combine at the output of the filters, each filter providing the necessary isolation from the adjacent channels. In addition, the filters reject the harmonics generated by the non-linearity of the high-power amplifiers. The insertion loss of the filters must be minimized since power loss at this stage necessitates an increase in the size of the power amplifiers, resulting in an increase in the solar array in order to provide the higher DC power needed, together with an increase in the weight of the spacecraft. This drives up the cost of the spacecraft.

Finally, the output of the multiplexer is delivered to the antenna for transmission.

Antenna

System considerations

The function of the antenna is to transmit signals to and receive signals from ground stations located within the coverage area of the satellite. The choice of the antenna system is therefore governed by the size and the shape of the coverage area. At present, the practical maximal size of antenna which can be launched is about 4 m. Consequently there is a limit to the minimum size of antenna footprint on the ground. However, it is possible to deploy larger antenna in space by folding them to fit the launcher diameters during the launch, using complex mechanical arrangements, and unfurling them in space. At present, such arrangements are not favoured for operational satellite systems because of the cost and risk involved. Experimental satellites such as ATS-6 have demonstrated the concept by unfurling antennas of up to 10 m in space. It is possible that such antennas will become increasingly common in future.

The coverage area of a satellite is best represented by contours of constant received power on the ground, usually shown relative to the power at the beam centre. Thus, for example, if we consider a simple satellite antenna such as a horn, the -3 dB contour represents the area within which all the locations receive a minimum of half the maximal power.

When the complete disk of the earth, as seen from the geostationary orbit, is illuminated, the coverage is known as *global*. A global coverage requires an antenna beamwidth of about $17.4°$. Global coverage is mainly used for international communication, where communication needs to be established between widely separated earth stations.

If the service area is confined to a country or a region, it is wasteful to transmit power outside this area. The coverage in such systems is best provided by *spot* beams. From equations (4.1) and (4.5) we can deduce that a reduction in the beamwidth of the antenna also increases the antenna gain. Therefore for the same size of HPA, a spot beam gives a higher EIRP than a global beam. Thus spot beams are well suited for applications when satellite EIRP requirements are high, as in direct broadcast or mobile systems.

Further refinements in coverage are achieved through the use of *shaped* beams. In such a beam, the antenna pattern is shaped to follow the contour of the coverage region as closely as possible. Antenna patterns may also be shaped so that the power received on the ground is nearly constant over the coverage region instead of tapering towards the edge, as when using a conventional antenna.

Multiple spot beams are used when service is provided to and between zones of high traffic concentration separated by a large distance, such as between Europe and America.

Dual-polarized systems use antennas which permit orthogonally polarized transmissions. Cross-polar properties of such antennas have an important role in the link design.

All types of coverage are illustrated in figure 9.6 (Nabi Abdel *et al.*, 1990), which shows the coverage area of INTELSAT VII satellites developed to serve the international fixed satellite service through the 1990s. INTELSAT VII operates in the C (6/4 GHz) and K_u (14/11 or 12 GHz) bands. At the C band, there are two shaped hemispherical coverage beams H_1 and H_2; four shaped zone coverage beams (Z_{11}, Z_{12}, Z_{21}, and Z_{22}) and two circular dual polarized spot beams C_L and C_R steerable fully over the Earth. In addition, there are two C-band global beams (not illustrated). At the K_u band there are three fully steerable spot beams – elliptically shaped linearly polarized east (S_1) and west spot beams (S_2), and a circularly polarized beam S_3. The linearly polarized spot beams S_1 and S_2 are orthogonally polarized to each other.

The earliest satellites used a single global beam. There has been a rapid growth in the complexity of satellite antennas since. The driving force has been the growth in traffic and the need to accommodate the growth within a limited RF spectrum. For example, INTELSAT VII, mentioned above, provides a *four-fold* re-use of C-band frequency spectrum through the use of orthogonal polarization and spatial beam isolation, and a *two-fold* increase in K_u band spectrum through dual polarization.

Optimizing coverage areas is a complex trade-off analysis involving the advantages gained in terms of increase in EIRP, frequency re-use, minimizing interference and the resulting complexity in the antenna system. Considering the constraints on spacecraft antennas, elliptical beam shapes have been recommended by the ITU for its worldwide direct broadcast plan. A minimal-size elliptical shape for a given service area can be obtained by techniques such as those suggested in the literature (Chouinard, 1981). Computer programs provide a convenient method of optimizing such coverage areas (e.g. Richharia, 1984).

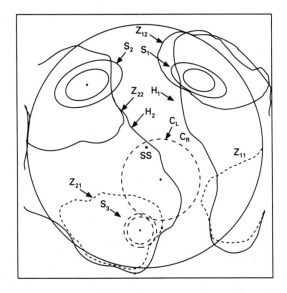

Figure 9.6   INTELSAT coverage showing various types of beams. H_1 and H_2 are C-band hemispheric-shaped coverage beams; Z_{11}, Z_{12}, Z_{21} Z_{22} are C-band zone coverage shaped beams; C_L and C_R are dual polarized fully steerable C-band circular spot beam; S_1 and S_2 are K_u band elliptical spot beams fully steerable over the earth; SS is the sub-satellite point (Nabi Abdel *et al.*, 1990).

A number of techniques are used to synthesize shaped beams. For example, the desired beam shape can be synthesized by a combination of small elemental circular beams. Phased array techniques are well suited for synthesizing complex coverage shapes. In one commonly used technique, a multi-element feed array is used. The desired shape can be obtained by controlling the amplitude and phase (the 'excitation coefficients') of each element of the feed. A 'beam-forming' network is used to provide the desired excitation coefficients.

Implementation issues
Fundamental-mode conical horn antennas are light and low cost, and therefore commonly used for providing global coverage. These types of horn, however, do not exhibit good cross-polar isolation and are therefore unsuitable for use in dual polarized systems. Hybrid-mode corrugated horns which provide high cross-polar discrimination are useful for such applications.

Reflector antennas are well suited for obtaining spot beams. Generally, the offset reflector configuration is favoured as this arrangement eliminates the inefficiency introduced by feed blockage. Further, its geometry permits the feed horns to be located close to the spacecraft body. However, the cross-polar discrimination property of the offset antenna system is poor. Gridded reflectors are used to improve the cross-polar performance. Such reflectors are made up of two sets of wire grids at right angles to each other, each set of grid permitting transmission of a single polarization.

Multiple beams are formed by using an array of feeds illuminating a single reflector. Each beam is formed by a cluster of feeds.

B. Bus

The function of a satellite platform is to support the payload operation reliably throughout the mission. Accordingly, the main demands on a satellite platform are as follows:

1. Maintain the position and orientation of a satellite at any specified orbital location and keep the antennas correctly pointed towards the service area; this function is performed by the *attitude and orbit control* system.
2. (a) Provide data to the ground control centre for monitoring the performance of the various spacecraft sub-systems.
   (b) Accept commands from the ground control centre for altering spacecraft configurations and performing vital manoeuvres.
   (c) Support ground stations tracking requirements.
   These functions are performed by the *telemetry, tracking and command* sub-system.
3. Provide DC power to all active components of a spacecraft; this function is performed by the *power sub-system.*
4. Maintain the temperature of the various spacecraft sub-systems within specified limits; this function is performed by the *thermal* sub-systems.
5. Provide the required mechanical and structural support.

Attitude and control system

A satellite maintains the desired orientation and orbital position through its attitude control sub-system. The attitude and control system must continue to perform all functions reliably throughout its lifetime because the loss of satellite attitude renders a spacecraft useless. To improve reliability, provision is made for the system to switch automatically to redundant equipment in case of failure of the operational equipment.

The demands on the attitude and orbit control system (AOCS) differ during the two main phases of the mission – the orbit-raising phase and the operational phase.

Two types of attitude control systems are in common use – spin stabilization and three-axis stabilization.

The attitude of the satellite is maintained during the orbit-raising phase so that essential communication with the satellite can be maintained and it can be orientated in the correct direction to impart the necessary thrust vector required by the orbit-control system. During the orbit-raising phase, the function of the orbit-control system depends on the launcher used. Most existing expendable launchers place a geostationary satellite initially in the transfer orbit. The orbit-control system is used to circularize the orbit to the geostationary orbit and move the satellite to the desired orbital location.

When on-station, a geostationary satellite is subjected to various short-term pertubations arising internally and externally. Further, the motion of the satellite in the orbit causes the satellite to rotate 360° around its own vertical axis every 24 hours. The attitude-control system maintains the antenna correctly orientated to the centre-of-coverage area vector.

The satellite is also perturbed by several external forces, causing it to drift away from its allocated location. This drift is nullified by the orbit-control system (see section 2.5).

Attitude control

The specifications of the attitude-control system depend on the desired spacecraft pointing accuracy which is a function of the satellite antenna beamwidth. The required attitude-control accuracies for spot, multiple and shaped beams are stringent.

The attitude of a satellite is maintained along three axes – pitch, roll and yaw, which are shown in figure 9.7. The attitude control may be either active or passive. A *passive* attitude-control system maintains the attitude by obtaining an equilibrium at the desired orientation without the use of active attitude devices. An *active* control system, shown in figure 9.8, maintains the attitude by the use of active devices in the control loop. As shown in the figure, the orientation in the pitch, roll and yaw axes are sensed and corrections are derived through a control and applied through actuators whenever there is a deviation from the desired reference. Actuators may also be activated by command from the ground.

Disturbances on a geostationary satellite

The main external perturbation on a geostationary satellite is caused by solar radiation pressure. Solar radiation pressure depends on the surface area of a satellite on which the sun-rays are incident. Therefore, satellites with large surface area suffer large torques which need compensation. Weak disturbing torques are, however, also generated by the Earth's magnetic fields and gravity. (Digressing to the low Earth orbits (LEO) briefly, the Earth's gravitational field can exert a significant torque to satellites in low orbital

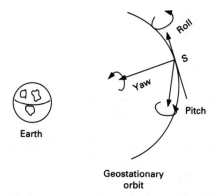

Figure 9.7   Pitch, roll and yaw axes. Roll axis is along the orbit, yaw axis is along the satellite–Earth vector and pitch axis is at right angles to the roll and yaw axis.

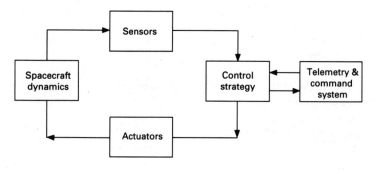

Figure 9.8   An active control system.

altitude. This torque is sometimes utilized for attitude control in low-cost LEO satellites. The spacecraft dimension in such satellites is designed such that the stable position due to gravitational acceleration occurs in the desired orientation.)

Additional disturbing torques are generated by sources internal to the spacecraft. One possible source of this type of torque is mis-alignment of thrusters. Whenever a mis-aligned thruster is fired, torques are generated in unwanted directions. Switching operation of mechanial relays or fuel sloshing can also give rise to such disturbing torque.

Sensors for attitude control

Sensors are used to estimate the position of the axis with respect to the specified reference directions. The choice of sensor depends on the desired attitude accuracy and the axis to be controlled. The reference directions commonly used are, the Earth, the Sun or a star such as *Polaris. Earth sensors* are commonly used for maintaining the roll and pitch axes. Orientation of the Earth with respect to the yaw axis is unsuitable for providing yaw errors, and therefore a Sun or star sensor is used for controlling this axis. There are several types of Earth sensor but all use the same basic principle. Infra-red emissions from the Earth, as observed on the geostationary orbit, are much higher than the background noise from cold space behind the Earth. Therefore the edges of Earth can be detected by sensors sensitive to infra-red emissions. The static sensing technique is shown in figure 9.9. A fixed focusing lens projects the image of the Earth on infra-red detectors. The detectors provide equal outputs along each axis when the Earth is centred, otherwise a voltage proportional to the offset is generated and fed back to the control system for generating the necessary correction. In a scanning sensor scheme, a mirror is made to scan the Earth continuously and the edges of the Earth are detected on an infra-red detector. A feedback control system is used to maintain the Earth centred correctly.

The infra-red Earth sensors have limited accuracy – about ±0.1°, because the edges of the Earth are not defined clearly.

The control accuracy can be improved by using the *RF sensing method.* In this method the satellite attitude is determined by tracking a RF signal transmitted from the Earth, much in the same way as earth stations track a satellite. This technique has been demonstrated on experimental satellites but has not yet been sufficiently tested for use with communication satellites.

As noted earlier, an Earth sensor cannot measure error in the yaw axis effectively. The Sun or a star such as *Polaris*, are commonly used. The advantage in using the Sun is the high flux density available, which reduces the sensitivity requirements of the sensor. The Sun, however, cannot be used in situations where it is in the same straight line with the Earth and the satellite. A star sensor overcomes this limitation but has the disadvantage that the received signal is of low intensity.

A *gyroscope* can also be used as an attitude sensor. The attitude information available from a gyroscope is relative to its spin axis. Under equilibrium conditions, the gyroscope gives zero error voltage. If the spacecraft orientation changes relative to this spin axis, a relative motion is developed in the gyroscope. This change can be detected and used for the attitude control. The achievable accuracy is high but a gyroscope requires the use of extra power for its operation and is less reliable.

Control system

The torque for attitude control can be obtained by the use of rotating wheels driven by motors. Such wheels are known as *reaction wheels.* Torque is transferred to the spacecraft body whenever the speed of the motor is changed.

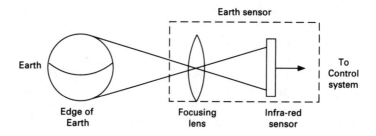

Figure 9.9   A static Earth-sensing technique.

Reaction wheels can be *zero momentum* systems or *momentum wheels*. Zero momentum systems use wheels which can turn in either direction whereas momentum wheels store momentum and rotate only in a single direction. It may happen that correction torques need be applied only in one direction. This would eventually cause the momentum wheel to reach maximal speed. The wheel must be slowed down to permit its use further. This is achieved by firing small independent thrusters so as to impart a disturbing torque to the satellite in the opposite direction. To apply corrections to this torque forces the control system to slow the wheel. This operation is often called 'momentum dumping', implying that the momentum is dumped into space.

The function of the control system is to generate a correcting torque from the sensed error. To obtain the correction torque the raw data obtained from the error sensors must be processed using a suitable control law. The control algorithm used for the purpose is of the form

$$N = I \, \ddot{\theta} \qquad\qquad (9.11)$$

where N = correcting torque
      I = moment of inertia of the spacecraft for the axis under
         consideration;
      $\ddot{\theta}$ = angular acceleration.

The algorithm takes into account factors such as natural resonances occurring on the satellite structure and may include the use of past attitude information. The complexity of the control law is a trade-off between the control accuracy and the on-board processing complexity.

The attitude and control requirements differ during the orbit-raising and on-station phases of the mission. The main functions of attitude control system during these phases are discussed below.

(a) *Orbit-raising phase*
Attitude control must be provided from the moment of separation of a spacecraft from the launcher. Most communication satellites are spin stabilized (discussed in a later section) during the orbit-raising phase. During this phase,

the attitude control must be able to dampen rotation of spin around the desired spin axis (known as *nutation*) which is caused during the separation of a spacecraft from the launch vehicle. An efficient lightweight technique for a basically stable spacecraft configuration (i.e. a spacecraft with its spin axis along the axis of maximal moment of inertia) is the use of an oscillating system such as a pendulum or a ball in a curved tube which produces a torque in a direction which dampens the nutation. For a configuration which can be unstable, an active control system with feedback from nutation sensors is used.

The attitude-control system also has the function of accurately determining the orientation of the spin axis. Sensors provide the spin axis orientation angles with respect to the Earth and the Sun. These data are transmitted to the ground control system and are used for planning the execution of the orbit-correction manoeuvres. The significance of accurate data can be realized by noting that the thrust requirement is high for orbit correction and an error in the orientation data leads to loss of fuel which could shorten the useful lifetime of a satellite.

(b) *On-station attitude and orbit control*
When a satellite has been positioned on-station, it is subjected to short-term disturbances. In addition, as mentioned previously, the satellite must be rotated by 360° per day around the pitch axis to maintain the on-board antenna pointing in the direction of the Earth. In chapter 2 we saw that satellites in geostationary orbits are perturbed by the gravitational effects of the Sun and the Moon and the Earth's asperity. These forces cause the satellite to drift from its assigned orbital location, which the AOCS needs to compensate.

Techniques of attitude control

There are two main techniques of satellite attitude control:

(i)   spin stabilization;
(ii) three-axis or body stabilization.

(i) *Spin stabilization*
Spin stabilization provides the necessary gyroscopic stiffness to the satellite by spinning either a part or the whole of the spacecraft. In a simple spinning satellite the entire body of the spacecraft is rotated. Figure 9.10 shows a simple spin-stabilized satellite. In such a satellite, to retain continuous Earth pointing, the antenna must possess a toroidal beam shape or employ electronic steering. Dual spin satellites have an antenna platform which is spun in the opposite direction with respect to the main spinning body, in such a way that the antenna system maintains continuous visibility of the

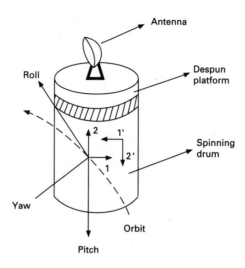

Figure 9.10   A spin-stabilized satellite.

Earth. The satellite may be spun either around the axis of maximal moment of inertia or minimal moment of inertia.

The solar arrays in a spin-stabilized configuration are mounted around the spinning drum. The despun platform uses a closed control loop to ensure that the antennas are continuously pointed to the Earth. It is vital that the earth lock is never lost and therefore on-board electronics are designed to re-acquire lock in case of failure. A spinning satellite can utilize a single thruster to control the attitude in two axes. This is achieved by firing a thruster at the instant when it is aligned along the desired axis during spin (see figure 9.10). The thruster is mounted well away from the spin axis to achieve maximum thrust. Disturbing torques cause the spin axis of a satellite to deviate from the nominal. The main cause is solar radiation pressure, which increases with increase in the offset between the centre of solar pressure and the centre of mass of the satellite. This deviation in the spin axis needs periodic corrections. Finally, the spin rate of the satellite decays with time and has to be constantly re-adjusted – achieved by firing thrusters tangentially to the circumference of the spinning drum.

Station-keeping in the north–south direction is maintained by firing the thrusters parallel to the spin axis in a continuous mode. The east–west station-keeping is obtained by firing thrusters mounted perpendicular to the spin axis.

(ii) *Body stabilized design*

In a three-axis stabilized satellite the body of the spacecraft remains fixed in space. In practice, it is difficult to achieve three-axis stabilization solely by the use of thrusters because of the lack of adequate gyroscopic stiffness.

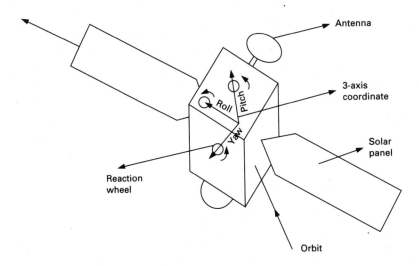

Figure 9.11   A three-axis stabilized satellite.

The required gyroscopic stiffness is achieved by using momentum wheels which rotate within the body of the spacecraft. Control is provided by accelerating or de-accelerating the momentum wheel. Figure 9.11 shows a schematic view of a three-axis stabilized satellite.

Unfortunately, a three-axis stabilized satellite cannot use those sensors that were in operation during the transfer orbit phase because spin stabilization, applied during this phase, utilizes a different type of sensor. Therefore additional sensors are incorporated for on-station stabilization.

Static Earth sensors are generally used for the pitch and the roll axes. The attitude of the yaw axis can be obtained either by a separate yaw-axis sensor or the roll and yaw axes can share the same sensor. Note that the roll axis becomes the yaw axis every quarter of an orbit but, because the change in the attitude of the yaw axis is generally insignificant during this interval, it is possible to share a sensor.

Station-keeping is achieved by firing the thrusters in the east–west or the north–south direction in a continuous mode.

An associated function of the attitude- and orbit-control system for a body-stabilized spacecraft is to provide solar-array Sun tracking. This is achieved by using a Sun sensor and another control loop.

Propulsion system

The function of the propulsion system is to generate the *thrust* required for the attitude and orbit corrections. The thrust requirement for orbit control is generally large. Mono- or bi-propellant fuels are commonly used for this purpose. For attitude control, thrusters are positioned away from the centre of mass to achieve the maximal thrust, the thrust being applied perpendicular to the direction of a spacecraft's centre of mass. For orbit control, the thrusters are mounted so that the thrust vector passes through the centre of mass because thrust vectors are either in the north–south or east–west directions. The force applied by a thruster depends on the flow rate of the fuel and its specific impulse:

$$F = \dot{W} I_{sp} \qquad (9.12)$$

where F = force
        $\dot{W}$ = weight flow rate
        I_{sp} = specific impulse.

Specific impulse is a commonly used figure of merit for thrusters and its unit is seconds.

Therefore the total impulse applied by a thruster in time t is

$$I = \int_0^t I_{sp} \dot{\omega} \, dt \qquad (9.13)$$

Figure 9.12 illustrates the principle of a mono-propellant propulsion thruster. Hydrazine is a commonly used fuel as it is a stable chemical with a large storage time and provides a relatively large specific impulse of 230 seconds. In the absence of gravity, it is necessary to apply external pressure to force out the hydrazine. This is accomplished by storing nitrogen under pressure (see figure 9.12). When the valve is released, the pressure of nitrogen forces the fuel to pass through the line. The fuel is next filtered to remove impurities and passed into the combustion chamber via a valve which permits only one-way fuel transfer. The hydrazine is finally passed through a catalyst which results in the decomposition of hydrazine in the combustion chamber and release of energy through a nozzle to provide the required thrust. The complete propulsion system uses several thrusters and includes redundant propulsion tanks and fuel lines.

A bi-propellant system consists of a fuel and an oxidizer, each being stored in separate tanks. These are combined whenever thrust needs to be imparted. Bi-propellant systems can provide large thrusts and are therefore used in apogee-kick motors and for north–south station-keeping manoeuvres.

Telemetry, tracking and command

Telemetry, tracking and command (TT&C) systems support the function of spacecraft management. These functions are vital for successful operation of all satellites and are treated separately from communication management.

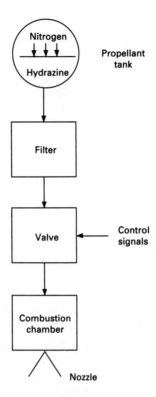

Figure 9.12   The principle of a mono-propellant propulsion thruster.

The main functions of a TT&C system are to:

(a) monitor the performance of all satellite sub-systems and transmit the monitored data to the satellite control centre;
(b) support the determination of orbital parameters;
(c) provide a source to earth stations for tracking;
(d) receive commands from the control centre for performing various functions of the satellite.

In effect these functions are performed by different sub-systems on a satellite but since all are related to the management of the satellite, the sub-systems are categorized together.

The function of the *telemetry* sub-system is to monitor various spacecraft parameters such as voltage, current, temperature and equipment status, and to transmit the measured values to the satellite control centre. The telemetred data are analysed at the control centre and used for routine operational and failure diagnostic purposes. For example, the data can be used to provide information about the amount of fuel remaining on the satellite, a need to

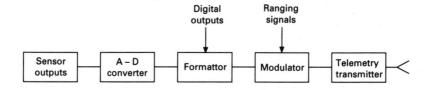

Figure 9.13   The main elements of a telemetry sub-system.

switch to a redundant chain or an HPA overload. The parameters most commonly monitored are:

(a) voltage, current and temperature of all major sub-systems;
(b) switch status of communication transponders;
(c) pressure of propulsion tanks;
(d) outputs from attitude sensors;
(e) reaction wheel speed.

Figure 9.13 shows the main elements of a telemetry sub-system. The monitored signals are all multiplexed and transmitted as a continuous digital stream. Several sensors provide analog signals whereas others give digital signals. Analog signals are digitally encoded and multiplexed with other digital signals. Typical telemetry data rates are in the range 150–100 bps. For low-bit rate telemetry a sub-carrier modulated with PSK or FSK is used before RF modulation. PSK is the most commonly used at RF. It is vital that telemetry information always be available at the satellite control centre, therefore a redundant chain is always available to improve reliability. There is also a provision to cross-strap the transmitters.

The telemetry signal is commonly used as a *beacon* by ground stations for tracking.

During orbit-raising when the main communication antennas are stowed, or under conditions when attitude control is lost, an antenna with near omnidirectional pattern is used. During the operational phase, the TT&C system uses the antenna system deployed for communications.

Distributed telemetry systems are increasingly being favoured. In this configuration, digital encoders are located in each sub-system of the satellite and data from each encoder are sent to a central encoder via a common, time-shared bus. This scheme reduces the number of wire connections considerably. This type of modular design also permits easy expansion of the initial design and facilitates testing during assembly of the satellite.

Command sub-system

The *command* system receives commands transmitted from the satellite control centre, verifies reception and executes these commands. Example of common commands are:

- transponder switching
- switch matrix reconfiguration
- antenna pointing control
- controlling direction speed of solar array drive
- battery reconditioning
- beacon switching
- thruster firing
- switching heaters of the various sub-systems.

Typically, over 300 different commands could be used on a communication satellite. From the example listed above, it can be noted that it is vital that commands be decoded and executed correctly. Consider the situation where a command for switching off an active thruster is mis-interpreted and the thruster remains activated – the consequence would be depletion of station-keeping fuel and possibly loss of the satellite as the satellite drifts away from its nominal position.

A fail-safe operation has to be achieved under low carrier-to-noise conditions (typically 7–8 dB). A commonly used safety feature demands verification of each command by the satellite control centre before execution. To reduce the impact of high-bit error rate, coding and repetition of data are employed. Further improvements can be obtained by combining the outputs of two receive chains. The message is accepted only when both outputs are identical.

Figure 9.14 shows the block diagram of a typical command system. The antennas used during the orbit-raising phase are near omni-directional to maintain contact for all possible orientations of the satellite during critical manoeuvres. The receiver converts RF signals to the baseband. Typical bit rates are around 100 bps. A command decoder decodes commands. This is followed by a verification process which usually involves the transmission of the decoded commands back to the satellite control centre via the telemetry carrier. The command is stored in a memory and is executed only after verification. The command system hardware is duplicated to improve the reliability.

The tele-command receiver also provides the baseband output of the ranging tone (see next section). This baseband is modulated on the telemetry beacon and transmitted back to the satellite control centre.

Tracking satellite position

To maintain a satellite in its assigned orbital slot and provide look angle information to earth stations in the network it is necessary to estimate the orbital parameters of a satellite regularly. The orbital parameters can be obtained by tracking the communication satellite from the ground and measuring the angular position and range of the satellite. Most satellite control centres employ angular and range or range-rate tracking. During orbit-raising

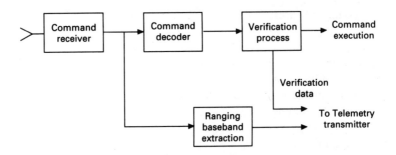

Figure 9.14   The main blocks of a command system.

when the satellite is in a non-geostationary orbit, a network of ground stations distributed throughout the globe is used for obtaining the orbital parameters.

The most commonly used method for angular tracking is the monopulse technique. Angular positions measured through a single station taken over a day are adequate for the determination of orbital parameters. The range of a satellite can be obtained by measuring the round-trip time delay of a signal. This is achieved by transmitting a signal modulated with a tone. The signal is received at the spacecraft and demodulated in the command receiver, the tone is then remodulated and transmitted back to the ground on the telemetry carrier. The time delay is obtained by measuring the phase difference between the transmitted and the received tones. Figure 9.15 shows the main blocks of a multitone ranging system.

In practice, the phase difference between the transmitted and the received tones can be more than 360°, leading to errors in multiples of tone time period. To resolve the ambiguity, multiple tones (e.g. 30 Hz to 30 kHz) are transmitted. Lower frequencies resolve the ambiguity and the high tone frequencies provide the desired accuracy. Consider a total phase shift in degrees ($\phi > 360°$):

$$\phi = 360n + \Delta\phi \tag{9.14}$$

where n = unknown integer
$\Delta\phi$ = measured phase shift.
The range of R is then given by

$$R = \lambda n + \frac{\Delta\phi}{360} \cdot \lambda \tag{9.15}$$

where λ = wavelength.

The magnitude of n can be determined by increasing the value of the wavelength. For instance, if $\lambda = 10\,000$ km (tone frequency = 30 Hz), we know that n can be either 3 for the range 36 000–40 000 km or 4 for the range 40 000–42 000 km because the range of a geostationary satellite lies

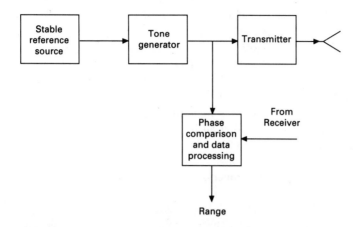

Figure 9.15   The main blocks of a multitone ranging system.

between 36 000 and 42 000 km. The measured accuracy with λ = 10 000 km will, however, be low. The error in the estimate is given by

$$\Delta R_e = \frac{\Delta\phi_e}{360} \lambda \tag{9.16}$$

where $\Delta\phi_e$ is the error in estimating the phase shift.

For example, when $\Delta\phi_e$ = 5°, the accuracy is 139 km, with a 30 Hz tone. The accuracy can be improved by using additional higher tone frequencies. A 300 kHz tone, for instance, would reduce the range error to 13.9 m.

Another technique employed for the measurement of range uses transmissions of pseudo-random digital data. In this method, the received signal sequence is correlated with a replica of the transmitted sequence. The time difference between the correlation peaks provides an estimate of the range.

Power sub-system

The function of the power sub-system is to provide DC power to all sub-systems throughout the life of a spacecraft. To meet this requirement, the power sub-system must generate DC power, regulate the power and provide an alternative energy source for periods when power cannot be generated by the spacecraft.

Power generation

Solar cells are used for power generation. The choice of this mode of electricity generation is on the basis that solar power is available for over 99% of a satellite's on-station lifetime.

The amount of power generated by a cell depends on the conversion efficiency of the cell and the intensity of the incident solar radiation. The average power of the Sun's radiation above the Earth's atmosphere is 137 mW/cm^2. However, because of a small eccentricity ($\sim$0.0167) in the Earth's orbit around the Sun, the distance between the two varies throughout the year. This variation in the distance causes a power variation of about $\pm$3.34% around the average intensity of solar radiation on a geostationary satellite. Further reduction in solar radiation is caused when solar arrays do not track the variation in Sun-position.

A solar cell consists of a p–n junction. An intrinsic (electrically neutral) layer is formed at the p–n junction of such a diode. When sunlight is incident on the intrinsic layer, solar energy is absorbed releasing electrons from the intrinsic layer and causing electric current to flow in an external circuit connected to the cell. The magnitude of the voltage developed depends on the incident solar power and the type of material used in the solar cell.

Silicon solar cells are widely used in communication satellites. The efficiencies of solar cells typically vary between 10% and 15% and are inversely related to the temperature. For example, an increase in temperature from 10°C to 70°C can reduce the power output of a non-reflective solar cell by $\sim$25%.

In space, solar cells are continuously bombarded by particles such as electrons and protons. The cells are therefore covered with protective layers of a material such as fused silica. Even then, the efficiency of a cell reduces over the years owing to the damage caused to the semiconductor structure. Figure 9.16 shows a typical variation in voltage of an array over the lifetime of a satellite. The solar arrays are designed to provide power up to the designed end-of-life of the satellite. A later section explains an approach to determine the solar array power and weight.

Several cells are connected in series to provide the desired bus voltage. A solar array typically consists of many thousands of cells. Hence it is necessary to configure the interconnections of cells to provide maximal reliability. If all the cells are connected in series, a failure in a single cell would result in a discontinuity and cause total failure. One possible configuration is to connect only some cells in series to form a 'stack', and many such stacks of cells can then be connected in parallel. If necessary, many such units can be connected in series. Such a configuration is more resilient to single point failure.

The solar array size for a specified power depends on the type of stabilization system used on the spacecraft. *Body-stabilized* spacecrafts employ solar panels which are folded during the launch and deployed in space. Many body-stabilized satellites use Sun-tracking solar arrays to maintain maximal solar flux density throughout. Thus all the solar cells in the array are active in the presence of the incident solar flux. In a *spin-stabilized*

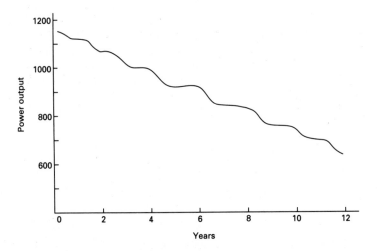

Figure 9.16 Reduction in efficiency of a solar array as a function of time.

satellite the solar cells are mounted on a rotating drum. Therefore, only a fraction of the cells is illuminated at any given time. The array size of a spin-stabilized satellite is therefore larger than that of a body-stabilized spacecraft for the same output DC power. However, the average temperature of the solar array in a spin-stabilized satellite is lower, which gives this configuration some efficiency advantage. The required surface area of a spin-stabilized satellite is ~2.7 times larger than that of a body-stabilized spacecraft:

$$F_A = \frac{A_c}{A_s} \eta \tag{9.17}$$

where A_c = area of the curved surface
A_s = projected area of the cylinder towards the Sun
η = advantage of spin-stabilized satellite due to lower operating temperature.

A geostationary satellite undergoes around 84 eclipses in a year, with a maximum eclipse duration of about 70 minutes. The eclipses occur twice a year for consecutive 42 days each time. The spacecraft DC power is provided by rechargeable batteries during eclipses. Batteries are allowed to discharge to a depth of 0.7–0.9 of the full charge, depending on the type of cell and taking into consideration that depth of discharge impacts the lifetime of a battery. Temperature control of the batteries is also essential to maintain performance and reliability.

Batteries are recharged when a satellite moves out of an eclipse. Charge is applied via the main power bus or a small section of the solar cell set aside for charging. Batteries are sometimes reconditioned by intentionally discharging them to a low charge level and recharging again. Reconditioning a battery usually prolongs its life. The operational status of the battery (i.e. recharge, in-service or recondition) is controlled by ground commands.

The mass of a battery constitutes a significant portion of the total satellite mass. Therefore a useful figure of merit to evaluate the performance of a spacecraft battery is the capacity (in watts-hour) per unit weight taken at the end of its life. To date, Ni–Cd batteries have been most commonly used because of their high reliability and long lifetime. Ni–H cells have a higher capacity per unit mass and tolerance to higher depths of discharge and are therefore beginning to be introduced.

Voltage regulation

The voltages generated from solar cells undergo both short- and long-term variations. Short-term variations are caused by rapid changes in temperature on the spacecraft. The voltage decreases with an increase in the ambient temperature. When a satellite emerges from an eclipse, the temperature can change from −180°C to 60°C in a few minutes, leading to array voltage change by a factor of about 2.5.

Long-term variations occur because of the periodic yearly variations in the intensity of solar radiation. The average DC power of a solar array varies over the year because of changes in the incident solar power caused by variations in the distance between the Sun and the Earth. Further long-term reductions in voltage occur owing to gradual deterioration in the efficiency of the solar cells over time.

For reliable and accurate functioning of spacecraft equipment it is essential to regulate such voltage variations. The two regulation approaches commonly used are:

(a) centralized;
(b) decentralized.

(a) *Centralized regulation*
In a centralized regulation, shown in figure 9.17, the solar array voltage is regulated centrally and the regulated voltage is available to all the loads on the bus. A fraction of the generated voltage is used for charging the battery. The amount of charge can be controlled through ground commands. The output of the battery can be switched to a reconditioning mode or to the main bus. The regulator circuit can take the form of a series or shunt regulator. The monitor and control sub-system provides various control signals, sends monitored data to ground via the telemetry sub-system and receives commands from the command sub-system. The advantages of this scheme are:

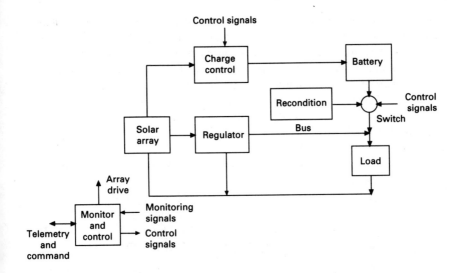

Figure 9.17　A centralized regulation scheme.

(i)　availability of a well-defined voltage;
(ii)　simplified power conditioning for loads.

The main disadvantages of the scheme are:

(i)　possibility of a single point failure;
(ii)　large demands on the regulator circuit because of high variations in load requirements.

(b)　*Decentralized regulation*
In a decentralized power regulation scheme, each load regulates its own voltage. To avoid dangerously high voltage when a satellite emerges from an eclipse, a voltage limiter is used with each load. The main advantages of this approach are:

(i)　simplicity in the overall power regulation requirements;
(ii)　lower overall mass.

The disadvantages are:

(i)　complex power conditioning requirements on individual loads;
(ii)　need to provide over-voltage protection to each load.

In a *multiple bus* concept, both regulated and unregulated bus structures are used. This approach avoids the complete satellite failure should one section of the power sub-system fail. As an example, the TT&C system, which is vital for the success of a mission, could be grouped on a separate bus with a more reliable configuration.

Thermal control

The thermal conditions on a geostationary satellite undergo a wide range of variations. Thermal variations occur during the launch of a satellite and throughout its operational life. All the vital components of a satellite must be maintained within a specific temperature limit throughout the mission. In this section we shall discuss the main factors influencing the thermal environment of a spacecraft and the principles of thermal control.

Factors influencing thermal conditions (Collette and Herdan, 1977)

A satellite undergoes different thermal conditions during the launch and operational phases. Table 9.1 summarizes the main sources of heat and their effects during the *launch phase* of the mission. It should, however, be noted that the communication systems which generate the maximal heat within a satellite (in particular, the high power amplifier stage) are switched off during this phase. The main sources of heat affecting a geostationary satellite during its operational phase are:

(i) solar radiation;
(ii) heat generated within a satellite due to dissipation.

There are also other minor heat sources such as albedo (solar radiation reflected off the Earth).

As already noted, an eclipse has a significant impact on the design of the thermal system, causing temperature variations from around $-180°C$ to $+60°C$. When a satellite is under eclipse, there is no incident thermal energy and therefore the ambient temperature falls well below $0°C$. The ambient temperature rises rapidly from the moment the satellite emerges from the eclipse.

Principles of thermal control system design (see Morgan and Gordon, 1989)

A geostationary satellite operates in vacuum and therefore conduction and radiation are the only mechanisms for dissipation of heat. A thermal equilibrium is achieved when the sum of incident thermal energy and the heat lost from the spacecraft are equal. The heat lost by a body through radiation is governed by the Stefan–Boltzmann law of radiation as follows:

Table 9.1 Summary of thermal environment and its effect on equipment during the launch phase of a communication satellite.

Phase	Source	Effect	Equipment affected
Low Earth orbit			
(a) Shroud not jettisoned	Aerodynamic friction	Shroud is heated but interior can also become heated	Large-area, small-mass external equipment, e.g. antenna dishes, solar array panels
(b) Shroud jettisoned	Aerodynamic friction, albedo, solar radiation	Satellite exposed	All sub-systems unless special attitude manoeuvre and launch window used
Transfer orbit			
(a) Before apogee-kick motor firing	Aerodynamic friction near perigee and some albado	Effect not severe	
(b) During apogee-kick	Heating due to motor firing	Heat absorbed by equipment close to motor	Equipment adjacent to apogee-kick motor

$$H = \varepsilon A \sigma T^4 \tag{9.18}$$

where H = heat lost by a body
 ε = emissivity (discussed later)
 A = surface area
 σ = $5.6703 \pm 0.007 \times 10^{-8}$ W/m^2 K^4 (Stefan–Boltzmann constant)
 T = absolute temperature.

As an example, consider a surface of 1 m^2 with an emissivity of 1 (i.e. a black body) at a temperature of 25°C. Using equation (9.18) it can be calculated that the surface radiates heat of 450 W/m^2. Now if a thermal energy of 1450 W/m^2 is incident on the body, the net thermal energy into the body is 1000 W/m^2. A thermal equilibrium is achieved when the temperature of the surface increases such that the surface radiates 1450 W/m^2.

Emissivity, ε, is a measure of a material's capacity to radiate heat. The value lies between 0 and 1, a perfect emitter having an emissivity of 1. *Absorptivity*, α, of a material is its capability to absorb heat. Its value also lies between 0 and 1, a perfect absorber having an absorptivity of 1. The ratio α/ε determines whether the material is a net emitter or absorber of heat. If α/ε is less than unity, the material is a net emitter of heat. The thermal design of a spacecraft is based on the selection of material with emissive and absorptive properties as appropriate for each part of the satel-

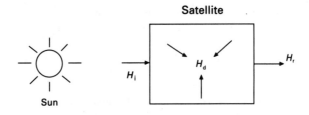

H_i = Heat input from external source
H_d = Heat dissipated inside the spacecraft
H_r = Radiated heat

Figure 9.18　A basic model for determining the average spacecraft temperature.

lite such that the equilibrium temperature is within acceptable bounds.

Figure 9.18 shows the basic principle used in determining the average spacecraft temperature. An average heat H_i/unit area is incident over a surface area A, facing the Sun and another component of heat H_d is added by dissipation within the satellite. The absorbed heat is balanced by the heat radiated by the surface facing away from the Sun. The average temperature of the spacecraft is obtained as

$$H_i + H_d = H_r \tag{9.19}$$

where H_i, H_d and H_r are the average incident, dissipated and radiated heat respectively. The equation for thermal equilibrium is

$$\Delta H_i A_i \alpha + H_d = \varepsilon A_r \sigma T^4 \tag{9.20a}$$

$$T = \left[\frac{1}{\varepsilon A_r \sigma} (\Delta H_i A_i \alpha + H_d) \right]^{-4} \tag{9.20b}$$

where　ΔH_i is the incident thermal energy
　　　　A_i is the effective absorption area
　　　　A_r is the effective radiative area.

A satellite is made up of different materials and therefore its average temperature is estimated by integrating the average temperatures of the individual surfaces. A detailed thermal model is prepared involving parameters such as emissitivity, absorptivity and surface areas. Such models are complex and hence are best optimized using computer-based analyses. An initial hardware model is developed on the basis of these analyses and thermal–vacuum simulation tests are performed on the hardware model.

Re-adjustments to the model are made based on the test results, and the process is repeated until an acceptable solution is developed.

Thermal control techniques

Thermal control can be achieved either by passive or active techniques. Passive techniques are simple and reliable and therefore preferred wherever feasible. Such techniques involve the use of materials with the desired emissitivity/absorptivity to achieve thermal equilibrium according to requirements. For example, the power amplifier stages of a payload dissipate large amounts of heat and therefore these stages are mounted close to radiators having good emissivity.

Insulation blankets are used around the apogee motor to provide protection against damaging temperatures caused during apogee motor firing. Fillers are used to provide high conductivity paths from the hot spots (e.g. near high-power amplifier stages). A combination of such measures, judiciously applied, provides the desired thermal control.

Active thermal control techniques are necessary to supplement passive techniques when the latter are ineffective – such as during an eclipse when the low ambient temperature could cause temperatures to fall below the tolerance. Commonly used techniques include:

(i) Electric heaters
Electric heaters are generally used during eclipses. They are controlled either by heat sensors on-board or activated by ground commands.
(ii) Use of hinged pipes
Hinged pipes are used to expose or cover specific areas to/from cold space for heat dissipation, as required.
(iii) Use of heat pipes
Heat pipes are used to transfer heat from a heat source to a radiator. In a heat pipe, fluid is vaporized at the hot end. The vapour travels towards the cold end and condenses, thus giving off heat.

The thermal design of spin- and body-stabilized satellites differ. In a spin-stabilized spacecraft, equipment is mounted within a spinning drum. The rotation of this drum keeps the temperature inside it to within 20–25°C for most of the time, except during eclipses when the ambient temperature falls considerably. To minimize the impact of low temperatures, the drum is isolated from the main equipment by heat blankets. The main heat dissipation has to occur at the north face because the apogee-kick motor covers the south face.

The fluctuation of heat over a day is larger in a body-stabilized spacecraft because the spacecraft orientation remains fixed in space. Therefore an insulation blanket is placed around the spacecraft to protect the equipment inside. Heat dissipation in this configuration can take place from both the north and the south surfaces. Thus the equipment can be mounted more evenly within the spacecraft.

Structure (Morgan and Gordon, 1989)

A spacecraft undergoes severe shock and vibrations during the launch phase, which must be absorbed by the structure of the spacecraft. On-station, the structure must provide accurate alignment of antennas and sensors. All the requirements must be satisfied within the limited volume and shape of the launch vehicle. The diameters of existing expendable launch vehicles are typically around 2.5–3.5 m. This constraint limits the maximum size of non-furlable-type antennas which can be used. Note that this type of antenna configuration is very common in existing satellites. The length of the fairing is of the order of 8–10 m and therefore does not impose notable constraints. A spacecraft must include a mechanical adapter to interface with the launch vehicle. Figure 9.19 (Morgan and Gordon, 1989) shows various parts of the structure interface of a centaur launcher with an INTELSAT satellite.

The structure design is also affected by the needs of the attitude-control system. For instance, a spin-stabilized satellite requires the centre of gravity to be in certain preferred directions.

The structure design usually employs a mathematical technique known as the finite element method. The satellite structure is modelled as a combination of a large number of small elements each with the required mechanical properties. A computer simulation is then used to study the performance of the structure under various stressed conditions to optimize the structure design.

The materials chosen for the spacecraft structure must be such that no deformity can occur under the worst loading conditions – typically experienced during the launch phase. Aluminium and magnesium alloys, carbon-fibre reinforced plastics and beryllium products are the most commonly used materials.

At this stage readers are familiar with all the basic systems used in a satellite and, to develop further insight of the real world, they are encouraged to study the description of satellites which are published regularly in the literature, as new satellites are developed (e.g. Mancuso and Ciceker, 1990; Neyret *et al.*, 1990).

9.5 Spacecraft mass and power estimations

During the early planning stages of a satellite communication system, simplified spacecraft mass and power estimation techniques are useful in assessing the sensitivity of these parameters to changes in system parameters such as the satellite EIRP, eclipse operation and the redundancy configurations.

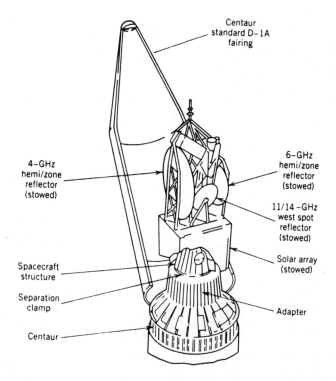

Figure 9.19   Interface of a Centaur launcher with an INTERSAT satellite.

Several simplified models have been used for this purpose (e.g. Kiesling *et al.*, 1972; CCIR, 1982; Pritchard, 1984). Here a model developed by Pritchard (Pritchard, 1984) is described to illustrate one approach to the modelling technique. The estimates from the model have been shown to provide adequate accuracy for early system planning. The model uses broad operational requirements together with the prevailing technological factors to estimate the mass and power of a communication satellite.

The required RF transmitter power, number of transponders, eclipse service capability and satellite lifetime are used as the initial input to the model. These parameters are a function of the desired traffic capacity, channel quality objectives, coverage area and space segment cost.

The RF transmitter power level together with the number of transponders are used to estimate the payload DC power through simple relationships. Other parameters, such as satellite lifetime, eclipse operation conditions and house-keeping requirements, are then introduced to estimate the total primary power and array size.

The mass esimates of the various sub-systems are based on the state of the prevailing technology and the use of empirical relationships obtained from available databases on communication satellites. Simple orbital mechanics is used to determine the fuel requirements for different phases of the mission. The individual masses obtained thus are summed to provide the total in-orbit mass of the satellite.

It must be emphasized here that the accuracy in mass or power estimates obtained by any model using such statistical modelling techniques depends on the accuracy of assumptions used in respect of the state of the technology and the reliability of the databases. Changes in technology or additions to the databases can be accommodated by applying the relevant revisions to the model.

Primary power sub-system

The total primary DC power, P_t, for a satellite transmitter is given by

$$P_t = \sum_{i=1}^{n} \frac{RF_i}{\eta_i} \qquad (9.21)$$

where RF_i = RF power of the ith transmitter
       n = total number of transmitters
       η_i = efficiency of the ith transmitter.
The total power P_T for the transponder is

$$P_T = P_t + P_r + P_h \qquad (9.22)$$

where P_r = DC power of receiver
       $P_t + P_r = aP_t$
       a is a constant between 1.03 (large satellite) and 1.1 (smaller or
          more complicated transponder)
       P_h = power required by TT&C, attitude-control, propulsion and
          thermal sub-systems.
The total primary power to be supplied by a solar array during equinox includes a battery-charging component. The charging power, P_c, is given by

$$P_c = U/\eta_c t_c \qquad (9.23)$$

where U = energy required to charge the battery
           = $P_e t_e/d$
       P_e = required power during eclipse
       t_e = eclipse duration
       d = depth of discharge
       η_c = charging efficiency
       t_c = charging time.

The primary power to be supplied by the array during the equinox is

$$P_{EQ} = P_t + P_r + P_h + P_c \tag{9.24}$$

At the solstice:

$$P_{so} = P_t + P_r + P_h \tag{9.25}$$

Note that for a given array size the available power is smaller at the summer solstice (the solar flux available at the summer solstice is 0.89 times lower than that at the equinox). Therefore if the heater power requirement is low, the array size is determined by the summer solstice power since the battery charging power at equinox is more than compensated by the extra solar power available.

In addition, the design must take account of long-term solar array degradation, expressed as

$$P = P_i e^{-kN} \tag{9.26}$$

where P_i is the beginning-of-life power

N is the design lifetime

k is a constant to allow for the deterioration of cells at the end of life (for example, with 20% deterioration of cells in 7 years, $k = 0.025$).

An additional margin of 5% is assigned to account for miscellaneous losses. The array size A can then be calculated from this expression

$$A = (P_A/G\eta_s\eta_a S) \tag{9.27}$$

where P_A = array power at the beginning of life

G = solar constant (1353 W/m^2)

η_s = solar cell efficiency

η_a = a factor to allow for miscellaneous losses (typically 0.85)

S = shadowing factor from spacecraft structures such as antenna (typically 0.9).

The array size derived from equation (9.27) represents the effective array area. For three-axis stabilized satellites this effective area is close to the array panel area. However, as noted earlier, when the satellite is spin stabilized the required surface area is $\sim 2.7\ A$. In practice, the effective temperature of solar cells for a spin-stabilized satellite is lower because they are shadowed for part of the revolution. The lower temperature of the solar cells permits a higher efficiency and a compensating effect on the available power is introduced. Therefore the area of the spinning drum is greater than a Sun-oriented solar array by a factor (Collette and Herdan, 1977)

$$F = \frac{\pi}{(1 + \Delta Tn_t)} \sim 2.7 \qquad (9.28)$$

where ΔT = difference in operating temperature between a Sun-oriented panel array and a drum array (typically 30°C)

η_t = solar cell efficiency factor (typically 0.5% improvement per ° temperature reduction).

The efficiency of solar cells is being improved continually and it is essential that the most recent values for cell efficiency, degradation factor and temperature dependence be used.

Mass estimate

The mass of the primary power system is given by

$$M_P = M_a + M_b + M_c \qquad (9.29)$$

where M_a = array mass
M_b = battery mass
M_c = power control mass.

The following quasi-empirical formulas have been derived from communication satellite data of the 1970 and 1980 periods:

$$M_a = \frac{SP_A}{r} + 10 \qquad \text{kg} \qquad (9.30)$$

$$M_c = 0.01P_A + 10 \qquad \text{kg} \qquad (9.31)$$

where S = spin factor – 2.7 for a spin-stabilized and 1 for a three-axis stabilized satellite

P_A = array power
r = array factor in W/kg ($\sim$30 W/kg but improving steadily).

The power control mass can be determined from the relationship

$$M_b = \frac{U}{\beta} = \frac{P_e T_e}{d\beta} \qquad \text{kg} \qquad (9.32)$$

where U = energy required by battery
β = battery-specific mass in Wh/kg (about 30 Wh/kg for a Ni–H or Ni–Cd battery)
d = 0.7–0.9 (depending on battery type – Ni–H batteries have a higher discharge capability than do Ni–Cd batteries).

Payload mass

The payload mass, M_{pl}, consists of the sum of the transponder and antenna masses:

$$M_{pl} = bR_{ti} \sum_i n_i M_{ti} + \sum_j M_{aj} + X \qquad (9.33)$$

where R_{ti} = redundancy in the ith transmitter
  b = factor to account for mass of receivers, switches, up/down converters, filters and other components (b = 1.1–1.5, depending on complexity)
  M_{ti} = mass of ith type transmitter
  M_{aj} = mass of jth antenna
  n_i = number of ith type transmitters
  X = mass of other payload components.
The following relationship is applicable for an average reflector/feed conbination:

$$M = 12 + 3.5D^{2.3} \text{ at the C band} \qquad (9.34)$$

$$M = 0.9 + 3.0D^{2.3} \text{ at the K band} \qquad (9.35)$$

It should be noted that sometimes an antenna configuration can be realized by more than one technique. For example, several techniques are available for shaping a beam, and a designer may prefer to select a configuration based on past experience. The mass estimate at this stage of planning is therefore only an approximation.

Mass of platform

The mass of the platform is related to the dry mass (see next section) of a satellite by the linear relationship shown in figure 9.20, obtained by applying regression to the data obtained from 12 spinning and 18 body-stabilized satellites (1972–82 data set). The resultant relationship is

$$M_f \text{ (3-axis stabilized)} = -10 + 0.5M_D \qquad (9.36)$$

$$M_f \text{ (spin stabilized)} = 30 + 0.45M_D \qquad (9.37)$$

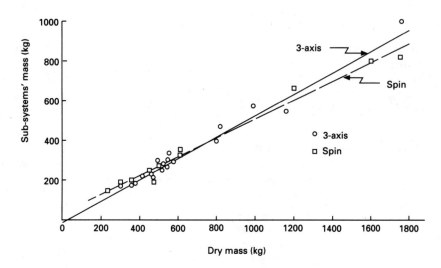

Figure 9.20   Sub-systems mass without payload and primary power as a function of a satellite's dry mass (Pritchard, 1984).

Dry mass

Dry mass is the mass of the satellite in orbit without propellant. Thus

$$M_D = M_{pl} + M_p + M_f \qquad (9.38)$$

where M_{pl} = payload mass
        M_p = primary power mass
        M_f = platform.
Substituting (9.36) and (9.37) in (9.38):

$$M_D \text{ (3-axis)} = 2.0(M_{pl} + M_p - 10) \qquad (9.39)$$

$$M_D \text{ (spin)} = 1.81(M_{pl} + M_p + 30) \qquad (9.40)$$

Wet mass

The wet mass of a spacecraft is the sum of the dry mass and the propellants. A general equation for the beginning-of-life mass is

$$M_0 = M_D e^{(\Delta v_0 + N\Delta V_i)/gI} \qquad (9.41)$$

where Δv_0 = velocity increment for correction of initial orbit errors
     I = specific impulse of propellants (seconds)

g = acceleration due to gravity (9.807 m/s^2)

N = design lifetime

ΔV_i = fuel repaired for regular station-keeping.

Typical specific impulses of commonly used fuels are: 225 seconds for hydrazine and 280–290 seconds for bi-propellant fuel. The value of Δv_0 is approximately 100 m/s. This allows 50 m/s to correct initial launch errors of about 400 km in altitude and 0.5° in inclination. An accurate launch permits the fuel allocated for launch error correction to be utilized for station-keeping, thereby effectively increasing the lifetime of the satellite. The other 50 m/s is allocated to contingencies such as relocation of satellites to other orbital positions. The station-keeping fuel requirement comprises two components – a term for east–west station-keeping and another for north–south station-keeping. The fuel requirement for the east–west station-keeping manoeuvre depends on the longitude of the orbital location, and this requirement varies from year to year. The north–south station-keeping fuel requirement depends on the date of launch and can amount to a maximum of about 50 m/s for a year (see appendix B). A further 10% is allocated for attitude control. The required beginning-of-life mass is then

$$M_0 = M_D e^{(100 + 50N)/gI} \tag{9.42}$$

with north–south station-keeping and

$$M_0 = M_D e^{(100 + 5N)/gI} \tag{9.43}$$

without north–south station-keeping.

Further refinements in these initial estimates are possible by using detailed orbital mechanics equations.

Mass in transfer orbit

The mass M_L in transfer orbit can be obtained from the relationship

$$M_L = M_0 e^{\Delta v/gI} \tag{9.44}$$

where I = value of specific impulse of fuel used by the apogee- or perigee-kick motors

Δv = velocity required for injection into transfer orbit, to circularize the orbit and to correct the inclination ($\Delta v \sim 1515$ m/s for 8.5° inclination transfer orbit to geostationary orbit – this is Ariane's transfer orbit). Transfer orbit weights generally range between 1.7 and 2.3 times the in-orbit weight of a communication satellite.

9.6 Space segment cost estimates

In the planning stages of a satellite communications system it is often necessary to compare alternative system configurations as a function of cost in order to obtain a minimal cost solution or to estimate the funds required for the project. A cost model of the space segment is therefore a useful planning tool.

It is extremely difficult to develop an accurate general cost model because costs are influenced by commercial factors, the state of the technology or in some cases regional and political factors. Consequently, an accurate cost estimate can only be obtained by communicating with the vendors and issuing a request for proposals. To obtain realistic costs it is best to use data from the prevailing market prices, once a configuration has been chosen and optimized. However, at the initial planning stages such detailed cost estimates are generally not necessary.

Here we shall discuss a general cost model, bearing in mind the limitations mentioned above. A spacecraft system's costs consist of two elements – non-recurring and recurring. Non-recurring cost is the cost associated with research and development of various spacecraft hardware/software. The recurring cost element consists of the spacecraft production cost together with the launch cost. Several cost models have been proposed (e.g. Kiesling *et al.*, 1972; Hadfield, 1974; CCIR, 1982). Here the space segment recurring cost model proposed by Hadfield (1974) is briefly stated *as an illustration.* Interested readers may wish to develop their own version of such a model, using a similar or alternative approach. The model is applicable to a single or multiple satellite system using a circular or elliptical orbit at altitudes above 180 km, and to super-synchronous either in a single or multiple orbital plane. The model is based on empirically derived relationships between cost and basic system parameter, and is therefore well suited for studying simple trade-offs between various alternatives. The launch cost element has been developed by a regression analysis of the available launch cost data (1970s database). The payload cost component is based on a heuristic approach and its validity confirmed against real data.

The total space segment cost (launch plus payload) for a single satellite in a circular orbit up to geostationary orbit is given by

$$\text{Cost (million \$)} = 0.026(W_s)^{2/3} \left(1 + K + \frac{H}{8000} \right) \qquad (9.45)$$

where W_s = weight of satellite in pounds

K = a constant proportional to the payload complexity (≈ 3 for the satellite technology of the 1970s) whose value increases with complexity

H = orbit altitude in statute miles.

The cost relationship was compared with data and was found to be gener-

ally valid for geostationary satellites.
For elliptical orbit the relationship is

$$\text{Cost (million \$)} = 0.026(W_s)^{2/3} \left(1 + K + \frac{H}{21\,000} \right) \qquad (9.46)$$

The equation is valid for apogee equal to or less than about 70 000 miles. Equations (9.45) and (9.46) do not include a probability factor for successful launch.

9.7 Spacecraft development programme

The development of a communication satellite is a complex project, requiring careful planning and good management techniques. The engineering management generally involves the following stages.

At the *conceptual design phase* several alternative configurations are proposed and evaluated. On completion of this phase the *definition phase* begins. During this phase, a single configuration is selected after an extensive review of the proposed solutions by experts, keeping in mind the main objectives and constraints of the mission (e.g. commercial, scientific, technological development).

In the next phase the detailed specification of the spacecraft is developed and a preliminary design review (PDR) conducted to examine the proposed solution critically.

In the *development/manufacturing phase* a critical design review (CDR) is performed. At the end of this phase the design is frozen and hardware development commences.

Spacecraft development stages

Several models of a spacecraft are produced during its development. A *breadboard* model is developed for the purpose of studying the viability of the chosen spacecraft design. An *engineering* model is developed to define the mechanical and electrical performance of the satellite. Reliability and environment conditions are not tested on the engineering model and therefore space-qualified components are not used on this model. Structural and thermal models of the satellite are also developed at this stage. The structural model is used to check the capability of the design to withstand the environmental conditions, and the thermal model validates the thermal design of the satellite.

A *prototype* model is manufactured next. The prototype model has the capability to withstand the same conditions as the operational satellite. Environmental tests at levels of up to 150% of the nominal are conducted. After the successful completion of qualification tests in a prototype model, the *flight* model is manufactured using identical parts and techniques. The tests conducted on the flight model are known as the acceptance tests, performed at similar stress levels as the real environment. The spacecraft is finally sent to the launch site where it is mounted on the adapter of the launch vehicle; fuel is injected here and the final tests are conducted before the launch.

Problems

1. (a) Develop a reliability model of a transponder in a transparent repeater. The TWT has a limited life and is considered less reliable than most other sub-systems.
   What is the reliability of the transponder assuming a reliability of 0.7 for TWT and 0.95 for all other systems?
   Now consider a parallel redundancy introduced into the TWT. Determine the improvement in reliability achieved.
   (b) Extending the number of transponders to 6, develop a reliability model for the repeater without introducing any redundancy and calculate the reliability of the system using the reliability values given above.
   Now introduce the following redundancies and develop the reliability model for each. Calculate the reliability in each case:
   (a) one redundant TWT for all 6 transponder chains;
   (b) two redundant TWTs for all transponder chains;
   (c) two redundant TWTs and one redundant receiver chain.
   Suggest the preferred configuration, giving reasons for your choice.
2. This problem requires a certain amount of literature survey and could be taken on as a small project.
   The aim of the investigation is to develop a reliability model of geostationary communication satellites, based on data obtained from the literature on satellite failures from the 1960s to the present. The model can be refined according to the ingenuity of the student. For example, the data can be partitioned according to the phase of the mission, such as failures in launch, in operation, etc. Other categorizations could be based on the type of stabilization; complexity of satellite, based on the coverage requirement and number of transponders; year of launch; operational frequency band, etc. Further investigations could involve a detailed analysis of each failure, if such data are available.
   A model of this type is very useful for the purpose of planning space segment deployment configurations, in particular, when the reliability of the space segment is of paramount importance, such as when satellites are used for supporting distress-related messages.
3. An earth station used for obtaining the range of a satellite uses a multitone ranging system. Suggest the tone frequencies which might be used if a

range accuracy of 100 m is required.

4. Determine the maximum number of television channels that a satellite using a spot beam of gain 32 dB can provide given the following parameters:

EIRP/channel 55 dBW
Maximum power at end of the satellite lifetime limited to 4 kW because of technical constraints
DC-to-RF conversion efficiency 30%
85% of the power is used by the RF power amplifier.

What are the technical constraints which limit the maximum available DC power from a satellite?

5. A satellite is required to provide 2 kW of power at the end of 15 years of lifetime. Determine the beginning-of-life power of the solar array, stating the assumptions made. Assuming a conversion efficiency of 20% and an average solar power of 137 mW/cm², what should be the size of the solar array for (a) Sun-tracking body-stabilized and (b) spin-stabilized satellites? State clearly and justify the assumptions made.

6. (a) Consider a regional geostationary satellite system to provide voice/data communications to mobile terminals. The total EIRP of the satellite is 50 dBW. The satellite is three-axis stabilized and has a single transponder. The satellite system must have full eclipse operation and a standby in-orbit satellite. Using the model illustrated in the text, estimate the in-orbit and transfer orbit weight of the satellite, stating and justifying all the assumptions made.

(b) Using the cost model described in the text, estimate the total space segment charges. The non-recurring costs may be taken as 50% of the recurring costs obtained by the cost model.

What should be the space segment charges, assuming that each duplex connection via the satellite requires 14.5 dBW on average? The traffic over a day may be assumed to follow a sinusoid with 100% satellite utilization during the busy hour and a minimum of 20% utilization during the non-busy hour trough. An average utilization of 25% of capacity can be taken for the two-day weekend. It is necessary to break even within four years after the start of service. An interest of 20% per year is to be paid on the invested capital.

State all the assumptions.

References

CCIR (1982). *Report of Interim Working Party*, PLEN/3, XVth Plenary Assembly, Geneva.

Chóuinard, G. (1981). 'Satellite beam optimization for the broadcasting satellite service', *IEEE Trans. Broadcasting*, Vol. BC-27, No. 1, March, pp 7–20.

Collette, C.R. and Herdan, B.L. (1977). 'Design problems of spacecraft for communication missions', *Proc. IEEE*, Vol. 65, March, pp 342–356.

Hadfield, B.H. (1974). 'Satellite cost estimation', *IEEE Trans. Commun.*, Vol. COM-22, No. 10, October, pp 1540–1547.

Kiesling, J.D., Gilbert, B.R., Garner, W.B. and Morgan, W.L. (1972). 'A technique for modelling communication satellites', *COMSAT Technical Review*, Vol. 2, No. 1, Spring, pp 73–103.

Mancuso, P. and Ciceker, M. (1990). 'Design and performance characteristics of the Astra spacecraft', *AIAA 13th International Communication Satellite Conference*, March, Paper AIAA-90-0822-CP, pp 360–374.

Maral, G. and Bousquet, M. (1984). 'Performance of regenerative/conventional satellite systems', *International Journal of Satellite Communications*, Vol. 2, pp 199–207.

Morgan, W.L. and Gordon, G.D. (1989). *Communications Satellite Handbook*, Wiley, New York.

Nabi Abdel, T., Koh, E. and Kennedy, D. (1990). 'INTELSAT VII communications capabilities and performance', *AIAA 13th International Communication Satellite Conference*, March, Paper AIAA-90-0787-CP, pp 84–94.

Neyret, P., Dest, L., Betaharon, K., Hunter, E. and Templeton, L. (1990). 'The INTELSAT VII spacecraft', *AIAA 13th International Communication Satellite Conference*, March, Paper AIAA-90-0788-CP.

Pritchard, W.L. (1984). 'Estimating the mass and power of communication satellites', *International Journal of Satellite Communications*, Vol. 2, pp 107–112.

Richharia, M. (1984). 'An interactive computer program for elliptical beam optimization', *International Journal of Satellite Communications*, Vol. 2, pp 263–268.

10 Earth Stations

10.1 Introduction

Earth stations are a vital element in any satellite communication network. The function of an earth station is to receive information from, or transmit information to, the satellite network in the most cost-effective and reliable manner while retaining the desired signal quality. Depending on the application, an earth station may have both transmit and receive capabilities or may only be capable of either transmission or reception. Further categorization can be based on the type of service. Usually, the design criteria are different for the Fixed Satellite Service (FSS), the Broadcast Satellite Service (BSS) and the Mobile Satellite Service (MSS).

A fundamental parameter in describing an earth station is the G/T (antenna gain to system noise temperature ratio – see section 4.5). This figure of merit represents the sensitivity of an earth station. A higher value implies a more sensitive station. Thus, depending on the value of G/T and service provided, various categories of earth stations are possible. Table 10.1 shows a possible categorization of earth stations based on these criteria.

In the following sections we shall discuss basic design considerations of earth stations, followed by descriptions of the main sub-systems of a typical earth station and important features of various types of earth station.

10.2 Design considerations

The design of an earth station depends on a number of factors. Some of these are:

- Type of service: Fixed Satellite Service, Mobile Satellite Service or Broadcast Satellite Service
- Type of communication requirements: telephony, data, television, etc.
- Required baseband signal quality at the destination
- Traffic requirements: number of channels, type of traffic – continuous or bursty
- Cost, reliability.

Two broad stages may be identified in the design process. The first stage is based on the overall system requirements from which the required earth station parameters such as G/T, transmit power, access scheme, etc. emerge. An earth station designer then engineers the most cost-effective configuration to achieve these specifications.

Table 10.1   The various categories of earth station.

Service	Earth station type	Approximate G/T (dB/K)	Comments
FSS	Large	40	
	Medium	30	
	Small	25	
	Very small	20	Transmit/receive
	Very small	12	Receive only
MSS	Mobile earth station		
	Large	−4	Tracking required
	Medium	−12	Tracking required
	Small	−24	Without tracking
BSS	User earth station		
	Large	15	Used for community reception
	Small	8	Used for individual reception

It is necessary to minimize the overall system costs which include costs (development and recurring costs) of space and ground segments. Several trade-offs are applied in the design optimization. During the earlier days of development, the available effective isotropic radiated power (EIRP) from satellites was low and hence earth stations tended to be complex and expensive. Earth stations used large antennas (≈ 30 m) and were very expensive ($5–10 million). The current trend is to minimize earth station complexity at the expense of a complex space segment, especially in applications which are directed towards a large user population (e.g. direct broadcasts, mobile and business use). The availability of low-cost earth stations is vital to the economic viability of the overall system in such applications.

Some of the trade-offs applied can be understood by examining the equation developed in an earlier chapter. The link equation, in a rearranged form, can be written as

$$G/T = C/N_o - \text{EIRP} + (L_p + L_m) + k \qquad (10.1)$$

where C/N_o = carrier to total noise power spectral density
EIRP = satellite effective isotropic radiated power
L_p = path loss
L_m = link margin
k = Boltzmann constant (in dB)
A useful optimization criterion is to minimize the earth station sensitivity (quantified as G/T), as this results in a minimal-cost earth station. It is

therefore necessary to minimize the parameters on the right-hand side for a specified baseband signal quality.

In general, this is achieved by using high satellite EIRP and modulation schemes which are robust to noise (in other words, the desired baseband quality at the receiver can be achieved with lower carrier-to-noise ratio). When digital baseband is used, further reduction in earth station G/T is possible by using coding. In fact, many small earth stations are only economically viable through the use of coding.

Other factors which have an impact on earth station cost are earth station EIRP, satellite tracking requirements, traffic handling capacity, interface to user or terrestrial network and network architecture. For a given application, the optimization is constrained by international regulations (e.g. antenna patterns have to meet regulatory requirements) and the state of the technology.

International regulations

Most of the fixed satellite service frequency bands are shared with terrestrial systems. For systems to coexist the International Telecommunication Union (ITU) has specified certain constraints in the transmitted EIRP of satellites.

Such constraints have an impact on the design of earth stations. For example, consider the use of earth stations in applications where a small size of terminal is essential (e.g. an earth station located in a user's premises). To be able to provide good signal quality with a limited available flux density on the ground, the G/T cannot be reduced below a certain value. Moreover even if G/T were reduced by the use of a smaller antenna, a reduction in the antenna size also increases the antenna side lobe levels, resulting in more interference to and from adjacent satellite systems.

It should be noted that limiting the satellite EIRP for applications such as direct broadcast and mobile communication would preclude the use of small antenna diameters, essential for such applications. This fact is recognized by the ITU and hence the frequency band allocations for these services are exclusive (i.e. not shared with other radio services), permitting much higher satellite EIRP. The limitations (if any) in these applications are mainly due to technological constraints of the space and ground segments.

Technical constraints

The transmitted power from a satellite is limited by the maximum DC power which a satellite can generate (at present ≈ 5 kW) and the lack of availability of reliable power amplifiers. The maximal spacecraft antenna gain is limited by the practical constraint imposed on the satellite antenna diameter. Note that for a given antenna size the gain falls with a decrease in frequency and therefore the EIRP limitation is more acute at lower frequencies (e.g. for the L band, commonly used for mobile satellite communication).

Technical constraints apply to earth station hardware and software. Generally, the cost and size of earth stations have been steadily falling. The size reduction has been dramatic in certain applications. For example, in mobile applications portable telephones are commonplace, pocket-sized pagers are becoming available and pocket-sized telephones are being planned. Similarly, direct-to-home receivers now (1994) typically use a 50 cm dish and cost under £200.

From the above discussion it is evident that several trade-offs are necessary in the optimization process. The use of computer-aided techniques simplifies the design process considerably (e.g. Richharia, 1985). It should be noted that design optimization criteria for services vary.

To ensure compatibility and maintain acceptable signal quality, service-providers standardize earth station size for use in their network.

With the basic specifications laid out, an earth station designer's next task is to develop an optimum configuration. The specified *G/T* may be achieved by a number of combinations of antenna and low-noise amplifier (LNA). It is possible to choose a combination of either a small-diameter (low cost) antenna and a relatively low-noise (expensive) LNA, or a large antenna and a LNA with a higher noise figure. It should however be noted that the diameter of the antenna also has an impact on the EIRP of the station. A small value of gain may give rise to an unduly large high-power amplifier (HPA) requirement. Therefore this variable must also be taken into account in the optimization.

Additional factors which must be considered include the cost of other equipment, the floor area, environmental factors (e.g. temperatures to which equipment may be subjected, wind load on antenna, etc.), interference considerations (sites must be well away from potential interfering sources, such as microwave radio relays) and recurring costs. For a mobile earth station, antenna and receiver mounting space imposes a severe constraint when considering direct broadcast receivers. Very low cost is essential for commercial viability of the system. Aesthetic antenna design is becoming increasingly important.

10.3 General configuration

Figure 10.1 shows a general configuration of an earth station. Signals from the terrestrial network are fed to an earth station via a suitable interface. The baseband signals are then processed, modulated and up-converted to the desired frequency, amplified to the required level, combined with other carriers (if necessary) and transmitted via the antenna. The feed system pro-

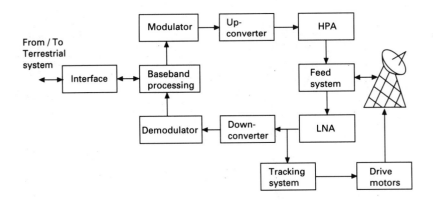

Figure 10.1   A general configuration of an earth station.

vides the necessary aperture illumination, introduces the required polarization and provides isolation between the transmitted and received signals.
Signals received by the antenna are amplified in a low-noise amplifier, down-
converted to an intermediate frequency (IF), demodulated and transferred to the
terrestrial network via an interface (or directly to a user in some applications).

Other sub-systems such as tracking, control, monitoring and power supply
provide the necessary support. The exact configuration of an earth station
depends on the application. For example, a TV receive-only station has a
very much simplified block schematic diagram. On the other hand, earth
stations for interconnecting large traffic nodes are considerably more complex. Major features of various types of earth station are described in a
subsequent section. The following RF sub-systems, common to most earth
stations, are described next:

(i)   Antenna;
(ii)   Feed;
(iii) Tracking;
(iv)   Low-noise amplifier;
(v)   High-power amplifier.

Antenna system

Most earth stations use reflector antennas since such antennas can readily
provide high gain and the desirable side lobe characteristics. A reflector
antenna consists of a parabolic reflector which is illuminated by a primary
radiator – usually a horn. The reflector diameter can vary from around
30 m for an INTELSAT Standard-A earth station to less than 60 cm for a
direct broadcast satellite receiver. A high efficiency is essential because an

antenna's cost is sensitive to the size of its diameter. Additionally, the radiation pattern of antennas must have low side lobes to minimize interference from and to other radio systems. The efficient utilization of two natural resources – the radio spectrum and the geostationary orbit – are affected by the side lobe characteristics of an earth station antenna. Hence the CCIR has laid down guidelines in the form of a reference radiation pattern. These guidelines are followed by all system designers. For example, it has been recommended that the gain, G, of 90% of the side lobe peaks of an FSS earth station antenna of more than 100 wavelengths (λ) diameter and for a frequency range of 2–10 GHz must not exceed

$$G = 32 - 25 \log(\theta) \quad \text{dBi} \qquad (10.2)$$

where θ = off-axis angle with respect to the antenna boresight (off-axis
        angle is limited to within 3° of the geostationary arc)
   $1° \leqslant \theta \leqslant 20°$

The reference pattern given by equation (10.2) was based on the available data of earth station antennas of the 1960s. This pattern has recently been modified to (CCIR, 1986)˙

$$G = 29 - 25 \log(\theta) \quad \text{dBi} \qquad (10.3)$$

The effect of this change is a reduction in interference caused or received by an earth station. The main purpose of the change was to increase the effective utilization of the geostationary orbit, effected by a reduction in the interference caused/received to/from adjacent satellite systems. Thus the improved specification has permitted adjacent satellite spacings to be reduced from about 3° to $\approx 2°$.

For practical reasons the CCIR reference antenna pattern for antenna diameters, D, less than 100λ is modified to

$$G = 52 - 10 \log(D/\lambda) - 25 \log(\theta) \quad \text{dBi} \qquad (10.4)$$

Similar reference patterns have been suggested for other services such as BSS.

Earth stations can have either axi-symmetric or asymmetric (also called 'offset') antenna configurations based on their geometry.

Axi-symmetric configuration

In an axi-symmetric configuration the antenna axes are symmetrical with respect to the reflector, which results in a relatively simple mechanical structure and antenna mount. The axi-symmetric antenna configuration has been used very widely until recently. Depending on the feed arrangement, several types of configurations are possible. Two of the most commonly used arrangements are:

(i)   prime-focus feed;
(ii)   the Cassegrain and Gregarian systems.

(i) *Prime-focus feed*
This arrangement, shown in figure 10.2, consists of a parabolic reflector antenna which is fed from a primary feed source located at the focus of the parabolic reflector. Owing to the geometry of the arrangement, the signal reflected from the parabolic reflector possesses a planar wavefront in the aperture plane, essential in producing the desired radiation pattern.

Although simple to implement, such a feed arrangement results in a larger antenna noise temperature because the feed horn is pointed towards a relatively hot earth and therefore picks up a significant amount of noise. Additional thermal noise is added by the dissipative loss in the cable/waveguide located between the feed and the low-noise amplifier (LNA) – unless the LNA is mounted close to the feed. In the transmit mode, some power is lost in the cable/waveguide used to connect the high-power amplifier (HPA) to the antenna. For these reasons the prime-focus feed is mainly used in small earth stations ($<\approx 3$–5 m) where the inherent simplicity of the scheme provides an economic solution.

(ii) *Cassegrain and Gregarian systems*
A *Cassegrain* antenna system consists of a parabolic reflector and a hyperbolic sub-reflector sharing the same focal point F_1 as shown in figure 10.3. The primary feed is located at the second focal point F_2 of the sub-reflector. The electromagnetic waves from the primary radiator are reflected off the sub-reflector to the main reflector. The geometry of the arrangement ensures the desired planar wavefront in the aperture plane. Cassegrain antenna configuration is a low-noise system because of the low magnitude of its main noise components – the antenna noise and the thermal noise from the transmission line. The low antenna noise contribution is attributed to the fact that the feed system is pointed towards a relatively cool sky. Further, because of the easy accessibility of the feed, LNAs can be mounted close to the feed, reducing noise contributions from the feeder line loss. Additional advantages include a reduction in power requirements of the earth station HPA and easy access to the electronics unit which can be mounted at the base of the reflector. Most large earth stations use the Cassegrain system.

The hyperbolic sub-reflector of the Cassegrain system can be replaced by an ellipsoid. This configuration is known as the *Gregorian* configuration. Here the focal point of the main parabolic reflector and the ellipsoid are at the common point F_1. This configuration is less commonly used than the Cassegrain system.

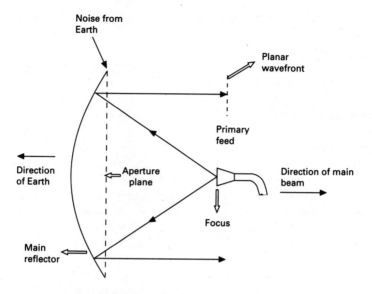

Figure 10.2   The prime-focus feed arrangement.

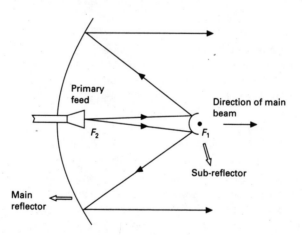

Figure 10.3   A Cassegrain feed system.

Asymmetric configuration

The performance of an axi-symmetric configuration is affected by the blockage of the aperture by the feed and the sub-reflector assembly. The result is a reduction in the antenna efficiency and an increase in the side lobe levels. The *asymmetric* configuration can remove this limitation. This is achieved by offsetting the mounting arrangement of the feed so that it does not ob-struct the main beam (see figure 10.4). As a result, the efficiency and side lobe level performance are improved. The latter improvement is desirable because of the more stringent performance requirements recently set by the CCIR. Hence the trend is to use this configuration where possible – especially for lower antenna sizes. The geometry of the configuration is, however, more complicated and difficult to implement in large earth stations.

Antenna mounts

Several types of antenna mount are in common use. Some of the most com-monly used mounts in medium and large earth stations are:

(i)   the azimuth–elevation mount;
(ii) the *X–Y* mount.

The *azimuth–elevation mount* (figure 10.5a) consists of a primary ver-tical axis. Rotation around this axis controls the azimuth angle. The hori-zontal axis is mounted over the primary axis, providing the elevation angle control.

The *X–Y mount* (figure 10.5b) consists of a horizontal primary axis (*X*-axis) and a secondary axis (*Y*-axis) mounted on the *X*-axis and at right-angles to it. Movement around these axes provides the necessary steering.

In recent years *polar mounts* are beginning to be used increasingly in small earth stations. A polar mount permits scanning of the complete geostationary arc by rotation around a single axis which is made parallel to the Earth's polar axis. The mount is well suited for applications where it is necessary to steer to new geostationary satellites frequently (e.g. reception of different DBS broadcasts). Various types of fixed antenna mount are used in small stations. The choice depends on factors such as the available mounting space and the cost.

It is necessary to stabilize the antenna platform for mobile earth stations (e.g. on board a ship). The stabilization system costs constitute a major proportion of the overall earth station antenna costs (see section 10.4).

Feed system

The primary feed system used in existing earth stations performs a number of functions. Depending on the type of earth station, these functions may be:

(i)   to illuminate the main reflector;
(ii) to separate the transmit and receive bands;

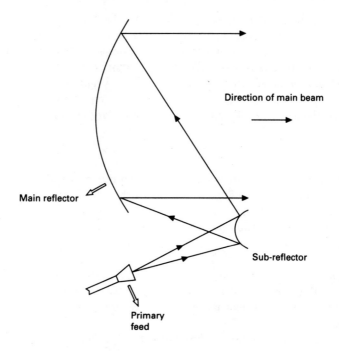

Figure 10.4　An offset feed arrangement.

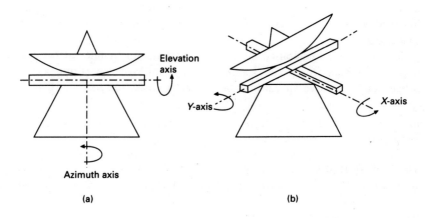

Figure 10.5　(a) An azimuth–elevation mount. (b) An *X–Y* mount.

(iii) to separate and combine polarizations in a dual polarized system;

(iv) to provide error signals for some types of satellite tracking system.

A horn antenna is commonly used as the primary feed at microwave frequencies. A horn antenna consists of an open waveguide which is flared at the transmitting end so that the impedance of the free space matches the impedance of the waveguide. This ensures an efficient transfer of power. The shape of the aperture can be rectangular or circular. Circular aperture horns, known as conical horns, are widely used as primary feeds in earth stations. Depending on the excitation mode, conical horns may be categorized as single-mode, multi-mode or hybrid-mode types. The mode of a waveguide defines the electromagnetic fields distribution across the waveguide (or horn) aperture. For example, in a rectangular waveguide TE_{mn}, the subscript m denotes that there are m half-cycle variations along the X-axis of the waveguide and n half-cycle variations along the Y-axis. (The X-axis and the Y-axis are the axes in the plane of the paper, if the wave propagates into or out of the plane.) The single-mode horns are excited by the TE_{11} mode and are used as primary radiators in many earth stations. This type of horn, however, does not have symmetrical radiation patterns in the E and H planes. Moreover, its cross-polarization properties are often inadequate for dual polarised systems. These limitations can be overcome, to a large extent, by using horns with multi-mode excitation capability such as the TE_{11} and TM_{11} modes. The use of hybrid-mode EH_{11} excitation can provide even larger bandwidth, lower cross-polarization and side lobes. Hence horns with a capability of hybrid mode excitation are often used in earth stations where such performance characteristics are essential. The most commonly used hybrid mode horn is the corrugated horn where the mode is excited by using annular rings within the flare of the horn.

Figure 10.6 shows a block diagram of an orthogonal polarization feed assembly. A higher-mode coupler (mode extractor) provides the error signals to the monopulse tracking system, if such a method is useful (see next section). The orthogonal mode junction (OMJ) assembly is used to separate the dually polarized transmit and receive signals. The orthogonal mode transducer (OMT) separates the two linear orthogonally polarized signals into separate ports on the receive side and combines two linearly polarized signals into a composite linear orthogonally polarized signal on the transmit side. Because OMT operates on linearly polarized signals, polarizers are used to convert a circular polarization to a linear (and vice versa). Polarizers are therefore not required for linearly polarized systems. Some earth stations have the capability to compensate polarization variations introduced by atmospheric effects (see section 3.3) by means of a feedback control system. The polarization properties of an antenna are mainly affected by the characteristics of the primary radiator and the polarizer.

Other arrangements of the feed system are possible, depending on the

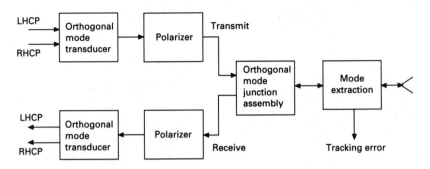

LHCP = Left-Hand Circular Polarization
RHCP = Right-Hand Circular Polarization

Figure 10.6   Various elements of a feed system.

application. For example, in a 6/4 GHz linear single polarized system it is only necessary to isolate the received carriers from the transmitted carriers. This is achieved by using a diplexer. Moreover, polarizers and OMT are not required.

In recent years, beam waveguide feeds have been increasingly introduced into large earth stations. Figure 10.7 shows a commonly used configuration of a beam waveguide feed Cassegrain antenna (Mizusawa and Kitsuregawa, 1973). A beam waveguide consists of a system of reflectors through which a radio beam is passed before transmission or after reception. This arrangement permits the high-power and low-noise amplifiers to be installed at the ground surface, offering a great deal of convenience in maintenance and operation. An added advantage is that a rotary joint is not required because the beams are guided in free space rather than via a waveguide. A rotary joint is necessary in all other types of mechanically steered antenna systems to permit antenna movement and has the disadvantages of increasing the transmission line loss and adding to costs.

Tracking system

Tracking is essential when the satellite drift, as seen by an earth station antenna, is a significant fraction of an earth station's antenna beamwidth, i.e.

$$\delta\theta_s > \theta_{hp}/N \tag{10.5}$$

where $\delta\theta_s$ = maximum satellite drift over a day as seen from earth station
        θ_{hp} = antenna half-power beamwidth
        N = application-dependent constant (typically 5–10).
Existing satellites typically move in the range 0.5–3°/day. Therefore, anten-

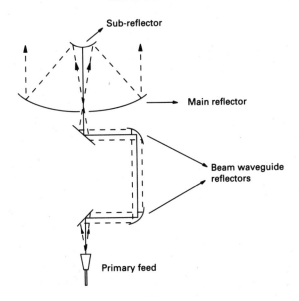

Figure 10.7   A beam waveguide.

nas with large beamwidths (e.g. DBS receivers) do not require satellite tracking. However, larger earth stations do require some form of tracking, the tracking accuracy being determined by the given application.

An earth station's tracking system is required to perform some or all of the functions discussed below.

(a) *Satellite acquisition*
Before communication can be established it is necessary to 'acquire' a satellite. One method is to program the antenna to perform a scan around the predicted position of the satellite. The automatic tracking system is switched on when the received signal strength is sufficient to lock the tracking receiver to the beacon. In its simplest form, a satellite can be acquired by moving the antenna manually around the expected satellite position.

(b) *Automatic tracking*
After acquisition a satellite needs to be tracked continuously. This function is performed by the automatic tracking system (often called an auto-track system). Auto-track systems are closed-loop control systems and are therefore highly accurate. This tracking mode is the preferred configuration when accuracy is the dominant criterion. The most commonly used techniques are discussed in a subsequent section.

(c) *Manual track*
To avoid a total loss of communication due to a failure in the tracking

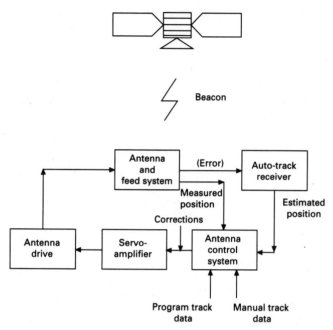

Figure 10.8   The main elements of a satellite tracking system.

system, earth stations generally also have a manual mode. In this mode an antenna is moved through manual commands.

(d) *Program track*
In this tracking mode the antenna is driven to the predicted satellite position by a computer. The satellite position predictions are usually supplied by the satellite operators. In the earlier days of satellite communications this method was used commonly, but now this facility is confined to expensive earth stations and satellite control stations. It may be noted that since a program track system is an open-loop control system, its accuracy is mainly governed by the accuracy of the prediction data.

Main functional elements

Figure 10.8 shows the main functional elements of a satellite tracking system. Communication satellites transmit a beacon which is used by earth stations for tracking. The received beacon signal is fed into the auto-track receiver where tracking corrections or, in some auto-track systems estimated positions of the satellite, are derived. In other auto-track techniques the feed system provides the required components of error signals. The outputs of the auto-track receivers are processed and used to drive each axis of the antenna to the estimated satellite position.

In the manual mode, an operator sets the desired angles for each axis on a control console. This position is compared with the actual antenna position, obtained through shaft encoders, and the difference signal is used to drive the antenna.

In the program track mode the desired antenna position is obtained from a computer. The difference in the actual and the desired antenna positions constitutes the error and is used to drive the antenna.

Auto-track system

There are three main types of auto-track system which have been commonly used for satellite tracking:

(i)   conical scan;
(ii) monopulse;
(iii) step-track.

The basic principles of these techniques are described below. Several improvements in tracking techniques have been introduced in recent years. Some of these techniques are also discussed briefly.

(i) *Conical scan*
The conical scan technique has evolved from the lobing technique used in Radars (Radio Detection and Ranging). In this technique an antenna beam is switched between two positions. When an approaching target is at the centre of these beams the echoes from each beam are equal in magnitude, but at other positions unequal. The antenna position is adjusted such that the amplitudes of echoes are equalized.

This concept was extended to a continuous rotation of a beam around a target, giving rise to the conical scan technique. Figure 10.9(a) (adapted from Skolnik, 1980) shows the principle of this technique. An antenna beam is rotated around an axis (rotation axis) which is offset from the beam axis by a small 'squint' angle. Whenever the satellite is off the target axis, the envelope of the received beacon is modulated at the rate of beam rotation. The amplitude of the resulting waveform, $b \cos(\theta + \phi)t$ [waveform B in figure 10.9b] provides the magnitude of the angular error and the phase delay, ϕ, with respect to the envelope of the rotating mechanism, $a \cos(\theta)t$ [waveform A in figure 10.9b], the direction of the satellite.

The tracking receiver in the conical scan technique uses the full received power for extracting tracking errors and hence the sensitivity requirement of the auto-track receiver is reduced. Additionally the phase stability of the receiver is not critical. However, since the error signals are derived from beacon amplitude variations, the accuracy of the system is affected by amplitude disturbances having spectral components within the error signal band-

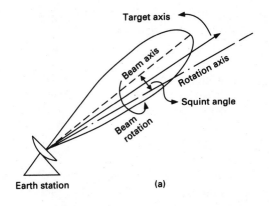

Earth station   (a)

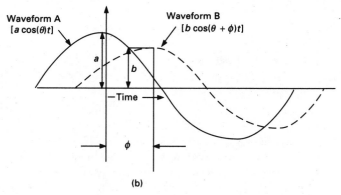

(b)

Figure 10.9   (a) Conical scan configuration. (b) Associated waveforms
(adapted from Skolnik, 1980).

width. Moreover, the maximum gain of the antenna is not realized because
of the squint introduced into the main beam. Further, the mechanical moving parts are usually mounted at quite inaccessible positions on the antenna
structure which leads to difficulties in maintenance. As a result of these
limitations, conical systems are being surpassed by other tracking techniques
in most new earth stations.

(ii) *Monopulse technique*
In the monopulse technique the errors for driving the antenna system are
derived by *simultaneous* lobing of the received beacon – hence the name
static-split or monopulse. The inherent susceptibility of the conical scan technique to amplitude fluctuations is eliminated since errors are derived from
simultaneous measurements. Several monopulse schemes such as amplitude
comparison, phase comparison or amplitude and phase comparison are possible. The amplitude comparison technique is the simplest and is commonly

used for satellite tracking. The basic principle of its operation can be understood from figures 10.10(a) to (c) (adapted from, Skolnik, 1980) – shown for a single axis. Two horns are offset and mounted in a plane (figure 10.10a). The radiation patterns of individual horns are shown in figure 10.10(b) and the superposed pattern in figure 10.10(c). Two types of pattern can be distinguished – a sum pattern Σ and a difference pattern Δ. The difference pattern output with respect to the sum pattern is zero when the satellite is centred, otherwise the output is proportional to the tracking error. It should be emphasized that the difference pattern must be detected with respect to the sum pattern to obtain the error. This can be achieved by a coherent detection process. It can be seen from figure 10.10(c) that the difference pattern changes phase around the beam maximum and therefore it is important that the phase stability of the receiver be good.

Several techniques can be used to determine the sum and difference signals. One commonly used technique is to use microwave circuits known as hybrids. Referring to figure 10.10(a), power from two horns A and B is fed into a microwave hybrid. A hybrid consists of two input arms A and B and two output arms the 'sum' arm and the 'difference' arm. The property of the hybrid is that the input powers appear as a sum in the 'sum' arm and as a difference in the 'difference' arm. In another technique, known as the mode extraction technique, modal characteristics of waveguides are used to obtain these signals. This technique is based on the principle that in some types of primary feed horns, higher-order modes are generated when a satellite is displaced from the antenna boresight. The generated mode can be processed to provide an estimate of the tracking error. Several types of mode extraction implementation are possible. The principle used in one of the earliest implementations is described here (Cook and Lowell, 1963).

The feed system consists of a circular horn permitting the propagation of the dominant TE_{11} mode and a higher-order TM_{01} mode. The higher-order mode is generated whenever the satellite is offset from the boresight. For small offsets, the amplitude of the TM_{01} mode is proportional to the pointing error and the relative phase difference between the TE_{11} and TM_{01} modes provides the angular information. The errors are given by

$$\delta A = K_a A_{01} \cos(\phi)$$
$$\delta E = K_e A_{01} \sin(\phi)$$

(10.6)

where δA = azimuth error
δE = elevation error
K_a and K_e are constants
A_{01} = magnitude of TM_{01} mode signal
ϕ = relative phase difference between the modes.

The feed system of a monopulse system is large in size and complex but

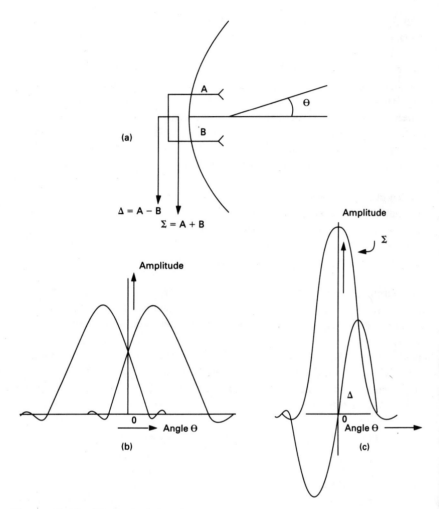

Figure 10.10   The principle of a monopulse tracking system: (a)
mounting arrangement; (b) patterns of each horn; (c)
superposed patterns (adapted from Skolnik, 1980).

requires very little maintenance. The auto-track receiver requires several
channels of coherent receiver with good phase stability. Although bulky and
expensive, the system provides a high degree of accuracy and is preferred
in large earth stations where tracking accuracy is the dominant criterion.

(iii) *Step-track system*
In the step-track technique, error signals are derived from amplitude sensing.
The operation is based on maximization of the received signal by moving
the axes in small steps (hence the name 'step-track') until a maximization is

effected. The control logic and receiver requirements are very simple, permitting a simple low-cost solution well suited for medium and small earth stations (Richharia, 1986). However, since the tracking information is derived by amplitude sensing, the system is susceptible to beacon amplitude perturbations caused by factors such as scintillation and signal fades. The system also has a lower accuracy and slower response than other types of tracking system. The tracking accuracy of the technique depends on the step size and the signal-to-noise ratio. For high signal-to-noise ratios, the standard deviation of tracking error approaches the step size. The accuracy of the technique can be improved by using improved maximization algorithms (see next section).

The merits and demerits of these conventional techniques are summarized in table 10.2. At present, the step-track technique is used where low cost and simplicity are essential (for example, medium-sized FSS earth stations) and the monopulse technique is used where tracking accuracy is the main criterion (for example, large FSS earth stations).

Recent tracking techniques

There have been some interesting recent developments in auto-tracking techniques which can potentially provide high accuracies at a low cost (Hawkins *et al.*, 1988).

In one proposed technique the sequential lobing technique has been implemented by using rapid electronic switching of a single beam which effectively approximates simultaneous lobing. The high rate of switching is achieved by the use of an electronically controlled feed. This technique, sometimes referred to as electronic beam squinting, requires a simple single channel receiver and has been reported to achieve a tracking accuracy approaching that of the monopulse technique (Dang *et al.*, 1985).

In another approach, the satellite position is computed by optimal control techniques. The relatively complex computations are readily performed by an inexpensive microcomputer. The satellite position is obtained by optimally combining the antenna position estimate obtained from an accurate gradient tracking algorithm with predictions obtained from a simple, self-learning satellite position model. The algorithm can switch between either of the two schemes or optimally combine the satellite position estimates depending on the system state. For example, in the presence of signal fluctuations the antenna is updated by using only the data from the prediction mode. The technique retains the simplicity offered by the step-track technique but overcomes its limitations at a marginal extra cost (Richharia, 1987). Accuracies approaching monopulse technique are possible at a fraction of the latter's cost, but the system has a slower response (see table 10.2) and achieves specified accuracy at the end of a learning phase lasting several hours when the satellite motion model is being developed.

Table 10.2 A comparison of auto-track systems.

Auto-track technique	Advantages	Disadvantages	Applications
Conical scan	• One RF channel • RF phase stability not important, hence simpler receivers possible • Tracking accuracy good • Medium response time	• Mechanical moving parts with difficult maintenance requirements • Accuracy sensitive to amplitude interference	• Medium/large earth stations Note: Not in common use now
Monopulse	• No mechanical moving parts, hence very little maintenance required in the feed system • Very high tracking accuracy • Fast response	• At least two channel coherent receivers required • Good RF phase stability required • Expensive • Feed system large and complex	• Large earth stations (e.g. INTELSAT Standard-A)
Step-track	• Simple design • Low cost • One RF channel • RF phase stability not important • Communication channel signal can be used, hence no extra requirements for feed	• Tracking accuracy low • Slow response time • Accuracy sensitive to amplitude interference	• Low-cost and simple earth stations (e.g. INTELSAT Standard-B) • Ship earth stations
Intelligent track	• All advantages of step track systems • Highly accurate • Resistant to amplitude fluctuations of beacon after the 'learning' phase	• Slow response time • Susceptible to amplitude fluctuations during initial acquisition • Full accuracy is achieved several hours after acquisition	• Large, medium or small earth stations

There has also been some interest in employing the phased array technique for satellite tracking especially in applications where the important design criteria are agility, low-profile and aesthetics. In this technique an antenna beam can be steered by exciting elements of an array antenna electronically. If a phase-shift is introduced between successive elements of an array, the beam formed by the array is tilted in a direction determined by the sign of the phase-shift and the amount of tilt by its magnitude. The proposed areas of applications include land mobile terminals and DBS receivers.

Low-noise amplifier

In the earliest earth stations, MASERs were used as the front-end amplifier. These devices are relatively narrow band (40–120 MHz), require liquid helium temperatures and hence are expensive with difficult maintenance requirements. As a result, these were replaced by parametric amplifiers which could provide wide bandwidths with the required low-noise temperatures at lower cost and complexity. Several improvements have been made to parametric amplifiers over the years. These have been made possible by the availability of improved devices (e.g. varactors, Gunn oscillators) and the use of thermoelectric cooling by the Peltier effect. In recent years the advent of gallium arsenide field-effect transistors has greatly simplified the front-end amplifier design of earth stations. These devices provide similar orders of noise temperature and bandwidths as parametric amplifiers but at a lower cost. Some of the important characteristics of various low-noise amplifiers are summarized in table 10.3.

High-power amplifier

Earliest stations required the use of very high transmitted power and hence invariably used microwave tubes such as a klystron. At present, the smallest stations use solid-state power amplifiers (≈ 10 W) and the largest use travelling wave tube (TWT) amplifiers or klystrons (10–15 kW). An important consideration is the inter-modulation noise which arises within a high-power amplifier during multi-carrier operation. Other considerations include the impact of amplifiers on earth station prime power requirements, the method of combining used by the carriers and the scope for future expansion.

Two most commonly used high-power amplifiers in large earth stations are the TWT amplifier and the klystron. TWT amplifiers can offer bandwidths of the order of 500 MHz and are capable of providing powers of up to 10 kW. Sometimes a linearizer is used to improve the linearity of the amplifier, especially if a multi-carrier operation is desired.

Table 10.3   Typical noise temperatures of low-noise amplifiers.

Low-noise amplifier	Frequency range (GHz)	Typical noise temperature (K)	Comments
Parametric amplifier	3.7– 4.2	30	Thermo-electric cooling
– cooled	11 –12	90	
Parametric amplifier	3.7– 4.2	40	Temperature of enclosure controlled thermo-electrically
– uncooled	11 –12	100	
GaAs FET	3.7– 4.2	50	Typically, first-stage
– cooled	11 –12	125	GaAs FET cooled thermo-electrically
GaAs FET	3.7– 4.2	75	
– uncooled	11 –12	170	

Klystrons are narrow-band devices, typically offering bandwidths of the order of 40 MHz, tunable over the entire 500 MHz bandwidth. Maximal powers are of the order of 3 kW. These amplifiers are therefore mainly used for relatively narrow-band carriers (e.g. television). A multiple amplifier configuration becomes essential (see below) when more than one such carrier is required. However, klystrons have a higher efficiency, longer tube life, lower cost and are simpler to maintain and operate than TWTs.

The configuration of high-power amplifiers depends on the type of application. For a multi-carrier operation, two types of configurations are used depending on the stage where the carriers are combined. In a *single amplifier configuration* all the carriers are combined before the amplifier and therefore only one HPA is used (see figure 10.11a). In this configuration it is desirable to operate the HPA on a linear portion of the HPA characteristic to minimize the inter-modulation noise. In a *multiple amplifier configuration* each HPA amplifies one or a few of the total carriers. All the amplified signals are then combined at the output of HPAs (see figure 10.11b). It is therefore possible to operate HPAs nearer to their full power rating which improves the overall efficiency of the earth station. However, this is achieved at the expense of a larger number of HPAs. In most earth stations HPA redundancy is provided to improve the reliability because many earth sta-

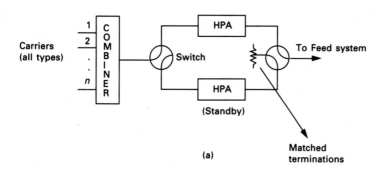

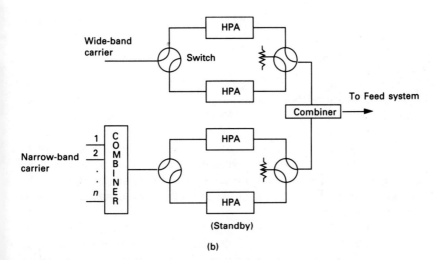

Figure 10.11   High-power amplifier configurations: (a) single amplifier; (b) multi-amplifiers.

tion failures are caused by HPAs. Switches at the input and output can be controlled either manually or automatically. The standby HPA is terminated into a matched load.

10.4   Characteristics

In this section, the main characteristics of earth stations used in various types of service are discussed. The section begins with a description of typical FSS earth stations. Two types of earth stations are considered – a large earth station and a very small terminal. This is followed by a description of earth stations used in the MSS. Again, two classes of earth station are discussed. The section concludes with a description of a receive-only earth station, used in the BSS.

Fixed satellite service earth stations

The size of an earth station is dictated by its traffic requirements. Thus the antenna size varies from 30 m for an INTELSAT-A station to about 1.5 m for a typical K_u band very small aperture terminal (VSAT). The earliest FSS stations were large. With an evolution in technology, the size of earth station hardware has reduced considerably and their operation has been simplified.

In this section, two types of FSS earth station are described – a large earth station of INTELSAT standard-A type and a VSAT.

(a) Large earth stations

Figure 10.12 shows a block schematic diagram of a large earth station – typical of an earth station used in the INTELSAT network.

The antennas used in large earth stations are of the reflector type. Various types of Cassegrain antenna systems have been used. Because of the advantages offered by the Beam waveguide feed Cassegrain system, this type of configuration is being used increasingly. At present, the use of offset antenna is rather limited owing to the difficulty of implementing large antennas of this variety.

A typical feed system employed in a large earth station is shown in figure 10.6 (discussed in section 10.3). Polarization purity is of special importance in large dual polarized earth stations. This is achieved by such techniques as the use of corrugated horns and high-performance polarizers.

Most large earth stations need to track the satellite accurately because of their use of narrow beamwidth antennas. Typically, a tracking accuracy of less than one-tenth of half-power beamwidth is essential. The monopulse technique is, therefore, commonly used. The step-track technique is commonly used in medium-sized earth stations.

Several types of HPA configurations are possible (see previous section). The configuration and the HPA rating depend on the number, type and size of transmitted carrier. At present, both the single and multiple HPA configurations are used. Depending on the required bandwidth, either TWTs or klystrons are used. The choice of HPA rating is dictated by an earth station's current requirements, together with possible future expansion.

An attempt is made to choose a low-noise amplifier having a noise temperature of at least equal to or less than the antenna noise temperature. It is

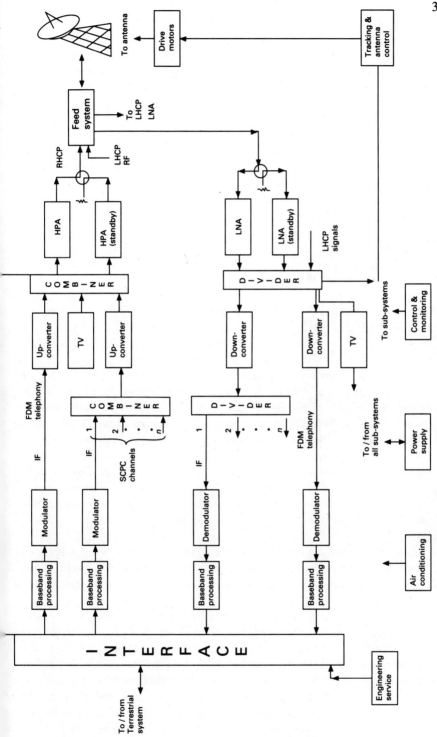

Figure 10.12 A general configuration of a large FSS earth station.

of interest to note that sky noise temperatures above 10 GHz are higher than at 4 GHz because of atmospheric effects. Several types of low noise amplifier are available now (see table 10.3). The choice depends on the required G/T, cost, ease of installation/maintenance and reliability.

The function of the transmit communication equipment is to process and modulate the baseband and convert the IF signal to an RF level suitable for amplification and subsequent transmission. In the receive path, the output from the LNA is amplified, down-converted and demodulated to the baseband. Depending on the transmission requirements, several types of communication equipment are possible:

(i)   FM/FDM telephony and television;
(ii) SCPC –
      (a) fixed assigned
      (b) demand assigned;
(iii) TDMA/DSI or DNI.

The salient features of each type of communication equipment are described below.

(i) *FM/FDM telephony and television (see chapters 4 and 8 for details)*
The FM/FDM telephony and television signals considered here are of the analog type. The baseband signals are filtered and pre-emphasis is applied as necessary (telephony or television). The audio in television transmission is introduced by frequency modulating a sub-carrier above the television baseband. SCPC channels may be included, if necessary (see figure 10.12).

Two types of frequency translators are commonly used – single conversion or dual conversion type. In a single conversion type translator, the IF is converted to RF directly, whereas in a dual conversion technique, the IF is up-converted to a second IF, followed by a translation to the required RF. Similar (but reverse) techniques are applied on the receive chain to down-convert RF to IF.

Two types of FM demodulators are used – the conventional and the threshold extension type. The threshold extension type demodulator is useful when the available carrier-to-noise ratio is low (see chapter 5 for details).

(ii) *Single channel per carrier terminal configuration*
The INTELSAT single channel per carrier (SCPC) is representative of this type of transmission and is therefore discussed here. Table 10.4 shows the main transmission parameters of the INTELSAT SCPC system. There are two main categories of operation – fixed assigned and demand assigned. Both types of operation permit voice and voice–band data transmission. The main difference between the earth stations of these two types is the requirement of an additional unit for channel management and signalling re-

Table 10.4 Main transmission characteristics of INTELSAT SCPC transmission.

Parameter	Description	Comment
Bandwidth	45 khz	
Transmission rate	64 kbps	
Voice coding	7-bits PCM/A-law	Voice activation
Modulation	4-phase CPSK	
Earth station EIRP	60.5–69.8 dBW	Depends on link configuration and spacecraft used
Operating C/N	15.5 dB	$1-3 \times 10^{-9}$ with FEC
Nominal BER	10^{-6}	

lated to demand assignment for the demand-assigned scheme (see chapter 8 for details).

(iia) *SCPC terminals – fixed assigned*
The main blocks of a fixed assigned SCPC terminal are shown in figure 10.13 (Nosaka, 1984). The following three functional units can be identified:

(i)   channel units;
(ii) common equipment;
(iii) frequency converter.

Each voice channel is coded into a PCM frame in the codec. The channel synchronizer performs the function of timing. A similar but reverse operation is performed by this unit on the receive signals. The 4-phase PSK modem performs the function of modulation and coherent demodulation. The voice detection circuit detects the presence of voice and activates the transmit output only during voice spurts. This technique reduces the average satellite EIRP and inter-modulation noise in the satellite transponder, thereby increasing the capacity. The frequency synthesizer is capable of tuning to any desired frequency within the band. Data are fed to the data codec via a data interface. The data codec performs the function of FEC coding and synchronization. All channels are summed in the IF summer. The IF system is shared by all channel units and is therefore referred to as the common equipment. The sub-system consists of a 70 ± 18 MHz filter/equalizer, an IF amplifier and a pilot combiner (required only in the reference stations). The output of the common equipment is then fed into the HPA for transmission.
The receive system consists of an LNA, a down-converter and various

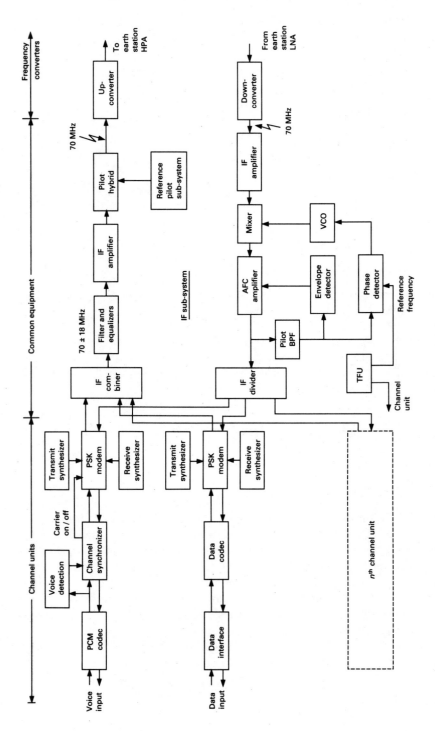

Figure 10.13 The main blocks of an SCPC terminal (Nosaka, 1984).

Table 10.5 Main transmission characteristics of INTELSAT TDMA.

Parameter	Description
Bit rate	120.832 Mbps
Frame length	2 ms
Voice coding	8 bits PCM/A-law
Modulation	4-phase CPSK
Earth station EIRP	89 dBW
Bit error rate	As per CCIR recommendations

IF stages followed by a divider, as shown in the figure. An automatic frequency correction is an integral part of any SCPC system and is necessary for synchronization of the earth station local oscillator to the network reference frequency (pilot). The timing and frequency unit (TFU) provides a reference frequency to the channel units and the automatic frequency correction (AFC) loop. IF signals are divided and fed to SCPC demodulators. The output of demodulators provide the received baseband.

(iib) *SPADE terminal – demand assigned*
The demand assigned terminal in the INTELSAT network is referred to as single channel per carrier PCM multiple access demand assigned equipment (SPADE). The configuration of a spade terminal is similar to fixed assigned SCPC terminal. However, there is a need for a demand-assigned signalling and switching unit (DASS) here. All SPADE network signalling is performed on a common signalling channel. The associated signalling and switching functions are handled by DASS (see chapter 8).

(iii) *TDMA terminal*
Various versions of TDMA systems have been implemented throughout the world. The INTELSAT TDMA terminals discussed here are widely used and representative of this class of transmission. The transmission parameters used in the network are summarized in table 10.5. Two types of transmission are used – TDMA/DSI (digital speech interpolation) and TDMA/DNI (digital data without interpolation). The digital speech interpolation technique is used to increase the capacity of the satellite network by utilizing the pauses in a telephonic conversation (see section 8.4).

Figure 10.14 (Nosaka, 1984) shows a typical configuration of a TDMA traffic terminal. The DSI (or DNI) modules process the incoming signals and provide traffic sub-bursts to the common TDMA terminal equipment (CTTE). Mapping of terrestrial channels containing a voice to satellite channel is performed on this block by a transmit channel assignment processor. A

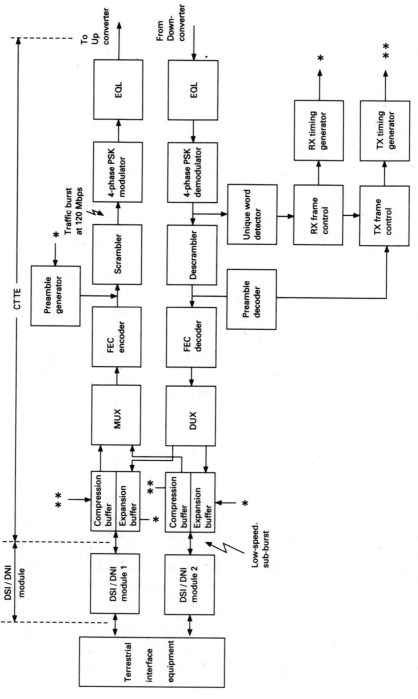

Figure 10.14 A typical configuration of a TDMA traffic terminal (Nosaka, 1984).

similar but inverse operation is performed by the modules on the receive path. The functions of the CTTE are:

(i)   traffic burst generation,
(ii) timing control;
(iii) modulation/demodulation.

The sub-bursts from the DSI/DNI modules are processed in the compression buffer. These sub-bursts are combined in the multiplexer to form the traffic part of the TDMA burst. These data are then coded in the FEC encoder, combined with a preamble and then scrambled. Scrambling is used to minimize spectral spikes (which may arise in lightly loaded conditions) within specified limits. The resulting bursts at 120 Mbps are 4-phase PSK modulated through an equalizer circuit and transferred to the RF section of the terminal which is used to equalize the transmit chain amplitude and group data response. The transmit burst timing is derived by applying the necessary delay to the received unique word of the reference burst.

The receive section consists of a down-converter, an IF section which provides the necessary amplification, filtering and equalization, and a 4-phase PSK demodulator which coherently demodulates the signal. The demodulator must recover carrier from each burst, since transmissions of earth stations in the network are independent of each other. This is achieved by recovering the carrier-and-bit timing from the relatively short carrier-and-bit time recovery symbols of the preamble portion of each burst. Careful optimization of the demodulator is necessary to achieve this. The preamble portion of the reference burst is also used for obtaining the transmit burst timing and receive timing of the terminal required in the compression and expansion buffer respectively. The descrambler, FEC decoder and demultiplexer perform the reverse of the transmit functions.

Interface with terrestrial network
The traffic originating in the terrestrial network often has to be reformatted for the satellite network. This function is performed by an interface at the earth station, the configuration of which depends on the type of traffic. Traffic signals may be available as frequency division multiplexed (FDM) analog telephony channels or time division multiplexed (TDM) streams with data or digitized telephony channels. Until a few years ago, most of the signals were analog but digital networks are being employed increasingly.

Such signals are demultiplexed at the earth station and rearranged on the basis of destination. A similar but inverse operation is performed on the receive side. The process of rearrangement depends on the type of access – FDMA or TDMA. In a FDMA scheme, channel allocations are done on the basis of destination carrier frequency.

When a TDMA transmission is used, the FDM signals from the terrestrial network need to be converted to a TDMA format. Several approaches are possible. In one approach, the FDM signals from the terrestrial networks are demultiplexed into individual voice channels, digitized in a PCM codec and time division multiplexed, before being transferred to the TDMA terminal. In another more economical approach, the FDM signals are directly converted into a TDM stream (and vice versa for the receive signal) by using transmultiplexers. A transmultiplexer can rearrange channels and additionally incorporate the facility for monitoring the channel quality. When a TDM stream is available a direct digital interface (DDI) equipment is used as an interface. Such an interface is more economical than the FDM interface since the requirements of sub-systems such as frequency division multiplexer/demultiplexers are no longer necessary. The configuration of the DDI depends on the synchronization strategy employed between the terrestrial network and the TDMA system.

Sub-systems for support
The power supply unit provides power to all sub-systems. Special care is essential to ensure that transmission outages are minimized in case of power supply failure, by using standby generators. Air conditioning is essential for reliable operation of all large earth stations.

The engineering service circuits are used by earth station staff for the purpose of maintenance, etc. The operation of the entire earth station is controlled from a control console. However many earth stations now have a capability of unattended operation.

(b) Very small aperture terminal (VSAT)

The recent revolution in information technology together with advances in satellite technology have resulted in the proliferation of very small aperture terminals (commonly called VSAT) (Chakraborty, 1988). VSATs are increasingly being used for information exchange from a single point to multipoint in broadcast applications and from multi-point to a single point in data collection systems. Satellite technology, because of its inherent capability of large coverage, offers a cost-effective solution for such applications. The network consists of a large fixed earth station, known as the hub, for communicating with one or several VSAT networks. The total user population in a single network may be of the order of several hundreds. The system is controlled by a network control centre which is often co-located with the hub. VSATs dispersed throughout a coverage region access the satellite through a random access protocol. Several types of protocol have been studied for such applications (see section 8.6).

The earliest VSATs were operated in the C band and employed the spread spectrum technique to counter interference in this heavily congested band (Parker, 1985). The current trend is to utilize the less congested K_u band which also permits the use of small antennas. The cost and reliability are the major issues in the design of a VSAT. The main cost-sensitive subsystems are the antenna (typically 1.2–1.8 m in the K_u band), solid-state power amplifiers (1–2 W) and the low-noise amplifier. A cost-effective solution is to use a low-noise converter, in which the low-noise amplification and down-conversion are performed in the same unit. Coding is usually employed because of the power-limited downlink. Block codes are simple to implement but offer lower coding gain for such applications, whereas convolution coding, although complex, can provide higher gain. At present, convolution coding with Viterbi soft-decision decoding is cost-effective. Although several modulation schemes are possible, some form of BPSK, such as differential BPSK, is favoured owing to cost, reliability and implementation considerations (see worked example of the end of chapter 4).

Mobile satellite service earth stations

Satellites are well suited for a large area mobile communication. However the practical constraints imposed on the design of mobile earth stations meant that the introduction of this service had to wait until the technology matured to a stage where these constraints could be overcome in a cost-effective manner.

The main features in design optimization of an MSS earth station are:

(i) limited mounting space implies that the antenna size on mobiles is severely restricted;
(ii) minimization of earth station cost is important for service uptake since the terminal cost is shared by relatively few users – especially in land mobiles;
(iii) traffic flow through the earth station is low (typically, a single channel is adequate).

Typical examples of mobile earth stations are Standard-A and Standard-C terminals for use in the INMARSAT network. Standard-A terminals can be interfaced with voice, telex or data networks, whereas Standard-C terminals can only be interfaced with telex or data networks. Standard-A terminals are relatively large in size with tracking antennas and are used as shipborne terminals. Standard-C terminals are small and inexpensive with non-tracking antennas. These terminals, initially developed for small vessels, are now also being used as land mobile terminals. Other standards are evolving for various mobile applications. Here we shall discuss these two standards as they are representative of this class of service.

Table 10.6   Primary characteristics of INMARSAT Standard-A ship earth
station.

Parameter	Value	Comments
Transmit band	1636.5–1645.0 MHz	25 kHz steps
Receive band	1535.0–1543.5 MHz	
Receive G/T	≥ -4 dB/K	Clear sky conditions 5° elevation
EIRP (nominal)	36 dBW	
Telephony	1 channel (multichannel versions available)	FM/SCPC
Telegraphy (receive)	1200 bit/s BPSK	TDM stream
Telegraphy (transmit)	1 channel 4800 bit/s BPSK	TDMA
Request	2 channels	Either one selected via random access

(a) Large mobile earth station (Standard-A)

A Standard-A ship earth station consists of a stabilized antenna system, an
outdoor electronics unit and an indoor electronics unit connected to peripherals. The outdoor section (sometimes called above deck unit) consists of a
stabilized antenna with an auto-track facility and an outdoor electronics unit
which amplifies and down-converts the signal to an intermediate frequency.
The main electronics are mounted below deck (e.g. in a ship's radio room)
and are sometimes referred to as the below deck unit. The peripherals typically include a telephone, a telex terminal and a display/control unit for
operation. These peripherals may be integrated for convenience and cost
reasons. Table 10.6 shows the main specifications of a typical Standard-A
ship earth station.

Above deck unit
An important consideration in the design of a ship earth station is the antenna system. To meet the specified G/T a parabolic dish of around 90 cm
is used – this corresponds to an antenna gain of ≈ 21 dB and a beamwidth
of $\approx 15°$ in the L band. A ship's movement can be relatively large and

hence some means of platform stabilization and satellite tracking are essential. Several stabilization schemes such as flywheel stabilization, and 4- and 3-axis servo mount have been considered (Johnson, 1978; Nouri and Braine, 1983). In addition to stabilization, the antenna system also includes a suitable means for satellite acquisition and an auto-track system. The step-track technique with large integration periods (≈ 30 seconds) is commonly used because of its low cost and simplicity.

The signals from the below deck unit are up-converted to a frequency in the unlink band (≈ 1640 MHz) amplified to 20–40 W (depending on the antenna size) and transmitted at an EIRP of 36 dBW. A diplexer isolates the high-power signals from the sensitive receiver circuits.

Receive frequencies are 101.5 MHz lower than the transmitted frequencies. The low-noise amplifier typically has a noise figure in the range 1–2 dB. Amplified signals are then down-converted and transferred to the below deck unit. In some receiver architectures, down-conversion is performed in the below deck unit.

Several configurations of the below deck unit are possible. A receiver with BPSK demodulator is used for receiving digital data from the broadcast signals whereas an FM modem is used for SCPC voice communications. Further, a TDMA modulator is essential for the transmission of data traffic. A microprocessor-based system is used for most control functions. These functions include:

- Selecting the transmit/receive frequencies according to network control messages
- Decoding the received broadcast signal – a TDM stream
- Sending/receiving signals to and from the antenna system
- Interface with the peripherals

(b) Small mobile earth station

Standard-C (INMARSAT, 1988) terminals can be used on small vessels and in land mobile applications. One of the main objectives in the specification of this standard has been to minimize the cost. Therefore a significant effort has been spent on minimizing the antenna complexity and earth station EIRP requirements. Figure 10.15 (INMARSAT, 1988) shows a block diagram of the Standard-C terminal and table 10.7 lists the primary characteristics of the terminal. Low G/T requirement permits the use of simple non-tracking antennas. Such antennas can be easily mounted on vehicles. The penalty paid for this reduced antenna complexity is the very low-bit rate capability of the service.

Standard-C system can provide either store-and-forward messages or end-to-end services. In the store-and-forward service, the complete messages are formatted in the SES (or CES) and transmitted on a simplex basis when a

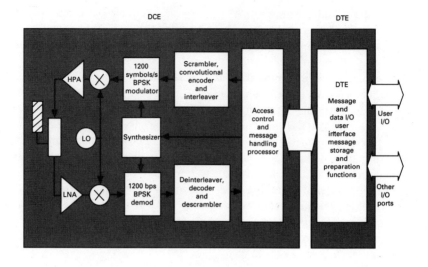

Figure 10.15   A block diagram of an INMARSAT Standard-C terminal (INMARSAT, 1988).

Table 10.7   Primary characteristics of INMARSAT Standard-C mobile earth station.

Parameter	Description	Comments
Transmit band	1626.5–1646.5 MHz	5 kHz steps
Receive band	1530–1545 MHz	
G/T	−23 dB/K	5° elevation clear sky conditions
Minimum EIRP	12 dBW	5° elevation
Information data rate	600 bps	
Modulation	BPSK – 1200 symbols/s	
Request	1 channel	Random access channel

channel is available. In the end-to-end service, permanent or semi-permanent connections are established for the duration of a call. Additional options include full duplex circuits, polling, data reporting and group call reception.

The mobile terminal consists of two main functional parts – Data Terminal Equipment (DTE) and Data Communication Equipment (DCE). A message is formatted in the DTE for transmission and transferred to the DCE. A convolution encoder and interleaver together with an ARQ scheme are used for encoding the transmit bit stream to protect against difficult propagation conditions of the channel. A synthesizer permits the SES to tune to any required frequency in increments of 5 kHz. The message remains stored in the DCE and is transmitted whenever a channel is available. On the receive side the DCE transfers received messages to the DTE (see INMARSAT, 1988 for details).

Satellite television receivers

Prior to the introduction of high-power direct broadcast satellites, low-power television transmissions were mainly intended for programme distribution and community viewing. These FSS transmissions were not authorized for direct viewing by the public but individuals began to receive such programmes. A thriving industry grew around the application, especially in the USA. These terminals (commonly called Television Receive Only or TVRO terminals) were relatively expensive because of the low-power flux density of satellite transmissions available on the ground.

Recognizing the potential of broadcasts from satellites directly to home, a worldwide plan was formulated by the ITU for direct broadcast satellite systems (DBS). Several countries are now beginning to implement this plan and consequently a considerable development effort has gone into the production of satellite receivers at a price affordable by individuals. The users of such sets are mainly interested in receiving good-quality pictures reliably rather than its technology, unlike the TVRO terminal users who were also keenly interested in the technological aspects of the receiver. As a result several improvements have been introduced into DBS receivers.

The design optimization criteria for direct broadcast receivers (the individual reception type service) are based on the following BSS features:

- The present ITU plan permits high-power transmissions from satellites, thus permitting small antennas to be used in homes
- The terminals are receive-only
- Factors such as low cost, reliability and aesthetics are vital.

The antenna sizes for a DBS receiver are generally between 60 and 90 cm. Polar or simple fixed mounts are commonly used. When DBS transmissions are circularly polarized the feed system uses a circular-to-linear polarization converter. Parabolic antennas with offset feed provide the most reliable and cost-effective solution at present. However, flat antennas, which merge well with surroundings, are also used (Ito *et al.*, 1988).

A low-noise block down-converter (LNB) is attached close to the antenna to minimize the degradation in system noise. This block converts the incoming K_u band frequencies to an IF of around 1 GHz. The unit typically consists of an LNA, followed by a bandpass filter and a down-conversion stage which consists of a mixer, a local oscillator followed by a filter.

The receiver performs further down-conversion to an IF of 70 or 140 MHz. The local oscillator usually employs a phase lock loop (PLL) synthesizer system. The IF filter bandwidth determines the predetection bandwidth. An AGC provides a constant input signal to the FM demodulator – such demodulators generally use a threshold extension principle. Several techniques such as FM negative feedback tracking, tracking filter and PLL threshold are used (Konishi and Fukuoka, 1988) (see chapter 5).

To obtain the audio and video components from received signals, it is necessary to have a knowledge of the baseband format. Different transmission techniques are used in various parts of the world. The most commonly used systems are the PAL, NTSC and SEACOM systems. However, there are certain inherent limitations in these systems, especially when considering satellite transmissions. These include the relatively large use of time spent on picture synchronization, sensitivity to noise and undesirable effects such as cross-chrominance, cross-colour and cross-luminance effects. Moreover, the use of sub-carriers for sound transmission is wasteful of power. These characteristics are not attractive for a high-definition television system (HDTV). Consequently new transmission techniques are being investigated. A technique called Multiplexed Analog Components (MAC) is now being implemented in Europe. In the MAC system the audio, chrominance and luminance components are digitized, multiplexed and transmitted. This technique overcomes most of the limitations of the conventional systems mentioned above.

The base-band processing may also involve the use of descrambling since many operating companies scramble the transmitted picture to avoid unauthorized viewing. At present, several analog techniques are used for video scrambling, and digital techniques are used for the scrambling of audio signals if high security is considered necessary.

Problems

1. What are the main considerations in the design of an earth station? With the help of a block diagram, discuss the operation of a typical large earth station. How is this configuration different from that of a Very Small Aperture Terminal (VSAT)? State reasons for this difference.
2. An earth station operating at 4 GHz is part of a domestic satellite system. The antenna has a diameter of 3.22 m and an efficiency of 55%. The sky noise at the operating frequency and elevation is 50 K. The feed and waveguide losses are 0.35 dB. A directional coupler inserted for monitoring purposes has a loss of 0.09 dB. A waveguide switch inserted in the path for switching to a standby low-noise amplifier has an insertion loss of 0.09 dB. The low-noise amplifier used has a temperature of 105 K and a gain of 30 dB. The following receiver stages have a total noise temperature of 600 K.
   Calculate the G/T of the earth station at the input of the LNA.
   If there is a further loss of 0.5 dB due to antenna mis-point, do you think the system noise will increase? How will the G/T be affected?
3. Compare the performance of the various types of tracking system used in earth stations.
   Suggest a suitable tracking system for the following earth stations, stating the reason for your choice:
   (a) a large earth station in tropical regions of Earth;
   (b) a medium-sized ship earth station;
   (c) a medium-sized FSS earth station to be located in a mid-latitude location.

References

CCIR (1986). *Recommendations and Reports*, Vol. IV.1, ITU, Geneva.

Chakraborty, D. (1988). 'VSAT communications networks – an overview', *IEEE Communications Magazine*, Vol. 26, No. 5, pp 10–23.

Cook, J.S. and Lowell, R. (1963). 'The autotrack system', *Bell Systems Technical Journal*, pp 1283–1307.

Dang, R., Watson, B.K., Davies, I. and Edwards, D.J. (1985). 'Electronic tracking systems for satellite ground stations', *15th European Microwave Conference*, Paris, September, pp 681–687.

Hawkins, G.J., Edwards, D.J. and McGeen, J.P. (1988). 'Tracking systems for satellite communications', *IEE Proc.*, Vol. 135, Pt. F, No. 5, October, pp 393–407.

INMARSAT (1988). *System Definition Manual for the Standard-C Communications System*, March, Inmarsat, London.

Ito, K., Ohmaru, K. and Konishi, Y. (1988). 'Planar antennas for satellite reception', *IEEE Trans. Broadcasting*, Vol. 34, No. 4, December.

Johnson, M.B. (1978). 'Antenna control system for a ship terminal for MARISAT', *Proc. IEE Conf. Maritime and Aeronautical Satellite Communication and Navigation*, IEE Conf. publ. no. 160.

Konishi, Y. and Fukuoka, Y. (1988). 'Satellite receiver technologies', *IEEE Trans. Broadcasting*, Vol. 34, No. 4, December, pp 449–456.

Mizusawa, M. and Kitsuregawa, T. (1973). 'A beam waveguide feed having a symmetric beam for cassegrain antennae', *IEEE Trans. Antenna and Propagation*, Vol. AP-21, November, pp 884–886.

Nosaka, K. (1984). 'The earth segment', in Alper, J. and Pelton, J.N. (eds), *The INTELSAT Global Satellite System, Progress in Astronautics and Aeronautics*, AIAA, New York, pp 135–189.

Nouri, M. and Braine, M.R. (1983). 'Design considerations for a larger INMARSAT standard ship earth station', *International Journal of Satellite Communications*, Vol. 1, pp 123–131.

Parker, E.B. (1985). 'Cost effective data communication for personal computer applications using micro earth stations', *IEEE Journal on Selected Areas in Communication*, May, pp 449–456.

Richharia, M. (1985). 'Computer aided system synthesis and costing of direct broadcast satellite systems', *IETE Technical Review (India)*, Vol. 2, No. 8, pp 277–283.

Richharia, M. (1986). 'Design considerations for an earth station step-track system', *Space Communication and Broadcasting*, Vol. 4, pp 215–228.

Richharia, M. (1987). 'Noise-resistant unit automatically tracks geosynch satellites', *Microwaves & RF*, November, pp 119–124.

Skolnik, M.I. (1980). *Introduction to Radar Systems*, McGraw-Hill, New York.

11 Future Trends

11.1 Introduction

The status of current technology and future directions in satellite communications were summarized in chapter 1. Subsequent chapters discussed fundamental concepts in satellite communication system design and described all its major system components. In this chapter we shall examine future trends, based on various factors which are likely to influence the growth of satellite communication systems. The chapter begins with a brief review of growth trend to date. Factors likely to influence the growth of satellite communication systems are outlined next. This knowledge, together with the current growth trend and prevailing interests are then used to project likely growth areas in each service. Possible areas of development in spacecraft and earth station technology are highlighted next. A section discussing various advanced concepts follows. The chapter concludes by citing examples of two experimental programmes which could be the forerunners of the next generation of geostationary satellites.

In the initial years of satellite communications, costs and risks were high because of limitations in technology. Most resources were therefore expended in improving the technology. Application areas were limited to the fixed satellite services where satellite communication systems provided distinct advantages despite high costs. Satellite service was primarily used for connecting high traffic density points, such as trans-oceanic international routes. Existing cable links on trans-oceanic routes were expensive, capacity limited and inflexible relative to satellites. The advent of satellites offering various inherent advantages released a pent-up demand for capacity. This resulted in an ever increasing demand for space segment capacity and a healthy growth in technology. The use of satellites extended from international systems to regional and domestic systems.

Significant research and development efforts were made in subsequent years to improve all the aspects of technology. The power generation capabilities of satellites have increased significantly; satellite and ground antenna sub-system performance have improved, making it possible to use shaped and spot beams and dual polarization operation. Launch and spacecraft reliabilities have improved and spacecraft lifetimes have increased from a few years to 12–15 years.

Similarly, earth station technology has improved considerably, benefiting from advances in solid-state technology and mass production. Earth station size and costs have decreased dramatically, giving satellite systems the capability to provide service directly at users' premises and to portable telephones.

Multiple access schemes, modulation and coding techniques have improved, leading to more efficient utilization of the space segment capacity.

Mobile satellite systems have become a reality as a result of the availability of high-power satellites, low-cost earth station hardware and improvements in multiple access, modulation and coding techniques. Broadcast satellite systems, transmitting television or sound programmes directly to homes, serve hundreds of thousands of households in several parts of the world. Overall, the net result has been a significant reduction in the operational costs of satellite systems.

Generally, the growth in satellite communications technology has followed a similar pattern throughout the world. In most instances the introduction of technology has been sponsored by a government and the first applications have been in the FSS for interconnecting international and national PSTNs. Following this phase the satellite service provision has either been transferred to the commercial world under a government licensing body (such as in the USA) or been retained by the government (such as in India). As the space segment and earth station costs have reduced, the service has been extended to provide communication facilities to companies via privatized networks. Finally, the technology has reached individuals, offering products such as satellite broadcast receivers, portable communication systems capable of messaging/telephony/paging services. A significant growth is expected in the area of personal communications during and beyond the 1990s, as increasing numbers of applications are offered. Realizing the benefits of satellite systems, an increasing number of countries have begun to exploit the potential of this medium. The developed status of technology has enabled late entrants to utilize the full range of services from the outset. The use of technology has benefited both the developed and developing countries.

At this stage, a question arising in the minds of a curious reader may be, 'where is this growth leading us?' Let us attempt to assess the possible areas of growth in the next decade (~ 2000 AD). Growth in satellite communication technology cannot occur in isolation but is bound to be influenced by technological improvements/innovations in other transmission media and also by the prevailing commercial and political environments. In some applications, such as point to multi-point broadcasting over large geographical areas, satellites offer distinct advantages and therefore a healthy growth is expected in such applications. For many applications, the service-provider has a choice of a number of transmission media – satellites, optical fibre systems, microwave radio-relays, cellular and other types of mobile communication systems, etc. Therefore satellite systems will have to become increasingly competitive commercially to survive and grow. In yet other types of applications, the availability of several alternative transmission media will give the service-provider a capability to combine the best features of each to achieve the most cost-effective system. An example already evident is the combination of satellites with optical fibres, each used where suited best.

11.2 Influencing factors

We noted above that the growth of a transmission medium cannot occur in isolation. Here the factors which could influence the development trends in satellite communications are summarized (Mahle *et al.*, 1987; Pelton and Wu, 1987).
The main factors to be considered are:

* Advances in fibre optic technology.
* Increasing geostationary orbit congestion.
* Gradual introduction of integrated switched data networks.
* Growing demand and awareness of mobile communications.
* Growing demand for low-cost earth stations and for their closer proximity to the traffic source.
* Regulatory environment which is influenced by conflicting demands arising from different countries (e.g. developing versus developed country); regional interests versus global interests; the needs of satellite communication systems versus those of terrestrial or other satellite systems (e.g. mobile satellite service versus radio determination service or cellular radio systems).
* Growing awareness of the capabilities of direct broadcast systems.
* Satellite communication industry's natural urge to compete and survive against competition from other transmission media.

11.3 Future applications

Based on the unique advantages offered by satellite communication systems and taking into account the influencing factors summarized above, some of the possible future applications and emerging growth areas are identified as follows:

* Direct communication to customers' premises bypassing the public switched network.
* Interconnecting large traffic sources such as city telephone exchanges in regions with little or insufficient infrastructure.
* Restoration of outages in optical fibre systems.
* Providing ISDN facility to remote areas.
* Teleconferencing.
* Video distribution between fixed points.
* Communications and distress alert facility for ships, aeroplanes, land vehicles and individuals.
* Personal communications such as newspaper delivery, portable radios receiving high-quality signals perhaps leading to 'wrist-watch' size radios

early in the next century, newspaper and mail delivery to remote areas, extension of medical consultancy to remote/inaccessible areas and others (see, for example, Ida *et al.*, 1984).

• Direct-to-home and community broadcasts.
• High-definition television system.

With this background, we shall now examine each service individually.

(i) Fixed satellite service

Impact of optical fibre links

Let us first consider the impact of fibre optic links on satellite communication systems, as this topic has been a subject of considerable interest in recent years. Fibre optic systems are expected to influence mainly the growth of the fixed satellite service. It is now recognized that a portion of satellite communications' traffic will migrate to the optical fibre medium where optical fibre systems are installed. To remain competitive, satellite systems must lower costs and improve quality. Quality improvement refers to improvement in factors such as the effect of echo and delay, and network reliability. The traffic environments most likely to be affected are international and inter-city routes of large developed countries such as the USA.

It is interesting to note that during the past 20 years satellites and optical fibres have both grown by about 20 times. Let us compare the main features of each.

At present, the main advantage of optical fibre vis-à-vis satellite communications are:

• less delay and echo;
• lower risk (e.g. no launch or satellite failure);
• less prone to noise;
• longer life (> 25 years).

Whereas the main advantages of satellite communication in this respect are:

• capability of point to multipoint broadcast;
• cost-effectiveness on thin routes;
• suitability for broadcast and mobile communications;
• minimal extra transportation cost associated with the 'last mile';
• short restoration time (typically < 1 hour) in case of a satellite failure;
• danger of physical damage to satellites non-existent (unlike optical fibres);
• cost and difficulty associated with cable laying non-existent.

It is worth noting that when comparing optical fibre systems with satellite systems on international routes, the 'last mile' advantage of satellite communications is often forgotten. On such routes optical fibre terminations are often far away from the traffic points. As an example, consider the trans-oceanic optical fibre systems connecting the USA and Europe. At the European end, cable links can terminate only in countries adjacent to the Atlantic Ocean. Thus a country located inland has to incur extra transportation cost by the use of a terrestrial 'tail'. Further, not all the nations may be willing to use terrestrial links through other countries. Satellite communication reduces this so-called 'last mile' problem by delivering the traffic much nearer to the destination.

Another interesting issue is related to the equitable development of regions (Casas and Fromm, 1988). Optical fibre systems tend to concentrate around the developed areas giving further impetus for growth to an already developed region, at the expense of less developed regions. However, satellites because of their inherent broadcasting capability give an equal weightage to all regions. Thus, for example, the advantages of ISDN could reach the remotest regions instantly using a satellite system whereas it is likely to be a long time before optical fibre links to the region can be economically justified.

We shall briefly illustrate the advantage of satellite systems over optical fibre systems on thin traffic routes. For point-to-point communication applications, a comparison between optical fibre and satellite systems can be made by estimating the cost versus transmission distance for each medium. The break-even point can be defined as the distance beyond which satellites begin to offer cost advantage. Satellite system costs are insensitive to distance but depend on the amount of traffic, whereas optical fibre system costs increase with distance but are relatively insensitive to traffic. A comparison of point-to-point links shows that satellites become increasingly cost-effective as traffic between the points reduces. Thus it can be concluded that satellites will be used increasingly on thin routes, but on heavy routes a portion of satellite traffic may migrate to optical fibre systems where these systems coexist.

We have noted above that in many applications service-providers could combine the advantages of each system to offer the most cost-effective solution. Some of the possible satellite–optical fibre synergistic applications are:

(i) A satellite system is used for concentrating data from various low-density traffic sources. The concentrated traffic is then transferred to a central location via an optical fibre link.

(ii) Satellites provide back-up to optical fibres in case of optical fibre outage (e.g. on a trans-Atlantic route).

(iii) Optical fibre links transport video programmes from the studio to an earth station for satellite broadcasts.

Impact of ISDN

Another area likely to influence satellite communications is the introduction of the ISDN. It is expected that signal quality and time delay requirements can be met by single hop satellite circuits. Thus satellite and optical fibre systems will be competing on an equal basis for the ISDN market on some routes. However, satellites have some notable advantages (Casas and Fromm, 1988), summarized below:

(i)   Possibility of rapid introduction of ISDN over large geographical areas.
(ii) Capability of broadband ISDN from the outset – whereas broadband capability using optical fibres can only be introduced gradually, giving satellite systems a clear lead.
(iii) Capability to shift traffic according to *demand* – on a long-term basis as terrestrial links are gradually introduced, or on a short-term basis as traffic demand changes seasonally, over a day or in response to an extraordinary event. This rapid switchover of capacity can be achieved in a number of ways. For example, in a TDMA system, time slots can be redistributed among earth stations according to demand.

New technology and applications

The proliferation of digital traffic implies that in the FSS, TDMA technology is likely to grow at both extremes – high-bit rate (e.g. 220 Mbps) and very-low-bit rate (e.g., 1–3 Mbps). The use of a low bit rate is consistent with the projected interest in the use of VSATs. In spite of a rapid growth in data traffic during the 1990s, voice traffic is expected to remain dominant throughout this decade. Therefore considerable research and development effort is being devoted to improving voice coding techniques. It may be recalled that reduction in coding bit rate permits a more efficient utilization of bandwidth and satellite power. For example, a reduction in voice coding bit rate/channel from 64 kbps to 8 kbps can theoretically increase the capacity of a transponder by approximately a factor of 8. Voice coding rates approaching 2.4 kbps are expected to become widely available in the 1990s.

To improve the utilization of satellite transponders, higher-order modulation schemes are being considered. For example, the use of 8-phase modulation could permit transmission rates of over 220 Mbps in a transponder bandwidth of 72 MHz, making the capacity of one such transponder equivalent to an optical fibre link, with all the added advantages of satellites.

A number of new applications of FSS may appear in the 1990s. For example, satellites could be utilized to communicate directly between small terminals via a large earth station (called a hub). When two small terminals are in direct contact via an FSS satellite, the received signal quality is inadequate for good quality communication owing to the limited satellite EIRP transmissions permitted by international regulations. One possible configuration by which communication between small terminals can be established is to interconnect them via a large station (the hub). Note that this type of system configuration requires double-hop links, introducing a one-way transmission delay of about half a second. This amount of delay can cause annoyance in a conversation unless the parties make an effort to 'learn' to communicate. In remote areas without any other communication facility, users may be quite willing to tolerate the inherent limitation, if they are made aware of the problem. The availability of low-cost terminals, existing infrastructure and the technical base (trained and experienced personnel, availability of transponder, etc.) could result in growing use of this mode of communication. Such systems could be rapidly deployed by including a voice communication facility to the infrastructure of an existing VSAT network.

Extensive work has been carried out by INTELSAT (Pelton and Wu, 1987) to forecast traffic growth trends in the FSS. Trade-off studies modelling the total network (i.e. ground and space segments) show that small FSS earth stations provide lower cost/circuit. Thus it is expected that generally large FSS stations will be phased out gradually and replaced by medium-to-small earth stations. To be able to serve smaller earth stations and at the same time make efficient use of RF spectrum and geostationary orbital arc, satellites may need to use improved techniques. One possible approach is to use on-board processing satellites. On-board processing refers to a number of functions such as demodulation/modulation, switching, etc. performed either at RF or baseband. The advantages include interference resistance and the capability to optimize uplinks and downlinks separately. When on-board switching is employed at the baseband, the satellite resembles a telephone switching centre with all the traffic points visible simultaneously. Hence traffic routing is simplified considerably. For example, in a domestic system the traffic destined for each telephone switching centre can be combined on-board and transmitted on individual carriers. The advantages are simple earth station design together with simple and low-cost terrestrial routing. Remote areas without any terrestrial links can be readily accommodated in the network. A satellite with such an on-board processing facility is often called a 'switchboard in the sky'.

Other techniques with a potential of providing more efficient use of space segment are discussed later.

(ii) Mobile satellite service

At present in populated areas such as cities, terrestrial cellular mobile systems are far more cost-effective than satellite mobile communication systems. However, satellite systems offer the only viable solution for providing mobile communications to remote areas. Satellite systems are expected to continue to play a complementary role in mobile communications – that of providing service to remote areas such as oceans, air corridors, unpopulated or scantily populated land, and to those areas of the world unserved by terrestrial mobile systems.

The mobile satellite service is the last entrant to the public communication domain. Its introduction had to be deferred until the 1980s because of inadequate technical development, perhaps because of lack of commercial and political support earlier. Awareness of the capabilities of MSS is now growing steadily, accompanied by a steady reduction in terminal and call charges. The MSS is therefore expected to be one of the largest growth areas in the 1990s. Recent announcements of plans by several organizations to provide communication to hand-held telephones by the late 1990s and INMARSAT global paging in the mid 1990s herald another milestone in the short history of satellite communications – bringing satellite communications into the area of personal communications (e.g. Lundberg, 1991).

At present, INMARSAT is the only major provider of mobile satellite service. Its network provides worldwide communication service to ships, aircrafts and land vehicles via a network of land earth stations. Several types of mobile terminal are used in the network, each matched to specific communication needs. The organization has recently enhanced its existing capabilities by including *portable telephones* (known as INMARSAT-M) capable of being mounted on yachts and land mobiles. As mentioned above, worldwide *paging* service providing service to pocket-size terminals are about to be introduced. Judging by the way INMARSAT's first-generation INMARSAT-A terminals have evolved over the past ten years, both in cost and size, it seems likely that in the next few years (~1995–98) terminal manufacturers could offer near hand-held versions of INMARSAT-M terminals at a price affordable by individuals.

Several regional mobile satellite systems (e.g. the USA and Canada) are planned to emerge by 1995 giving a wider choice to the users in the region. Several companies, many in the USA, have proposed a variety of mobile communication systems offering services varying in range from low-bit rate messaging service to medium-data rate and voice service. Several proposed systems plan the use of low earth orbit (LEO) or medium earth orbit (MEO) satellite constellations in the space segment. Network concepts vary in complexity. On one end are architectures deploying low earth orbit satellite constellations, using on-board processing satellites interconnected in space via inter-satellite links (ISL) and providing voice capability (e.g. Richharia *et al.*, 1989). At the other extreme are LEO networks using simple satellites to provide store-and-forward low bit rate data communications without the need for using ISLs (Maral *et al.*, 1991). The main problems with LEO/MEO constellations appear to be the extreme complexity of such networks, especially if real-time worldwide networking via inter-satellite links is required, and the financial risk involved in fostering unproven technology too quickly.

With the race for personal satellite telephones already on, new techniques, revolutionizing satellite communications, will emerge. However, simple LEO networks to provide store-and-forward communications are feasible even with current technologies – again commercial interests will decide their success, but such networks appear more viable in the current technical/commercial climate. This type of service is planned to be available by the mid 1990s.

One of the problems in the use of geostationary or low earth orbit for high bit rate land mobile satellite systems is the need for large propagation margins at low elevation angles, making low elevation land mobile links unreliable. When using a geostationary satellite this implies that at high latitudes (e.g. many parts of Europe) communication is unreliable especially for land mobiles. When using an LEO system for land mobile communications this implies that link connectivity will always be erratic in environments with a large number of obstructions (e.g. wooded areas or city environments). The problem can be minimized by increasing the number of satellites in the constellation so that several satellites are simultaneously visible from a terminal, enabling the terminal always to select the most robust communication link. However this solution is bound to increase the complexity and cost of the network. To overcome this problem, the use of highly elliptical orbits (HEO) has been proposed (IEE, 1989). Satellites in HEO always appear at high elevation angles in a specified high-latitude coverage area. The link margins therefore are reduced considerably and hence the link reliability increased. Such an orbit configuration also requires the use of several satellites and on its own can at best provide regional coverage. A global service therefore would need satellites in other types of orbit. It should be mentioned that, at present, land mobile satellites are receiving considerable attention. The premise is that land mobile systems are likely to offer the largest market.

Recently, both MEO and LEO constellations have been chosen as the space segment configuration to provide hand-held terminals. However, it seems reasonable to assume that the full potential of the existing GSO system will also be realized. In a recent interview, Arthur Clarke summarized the situation very aptly: "May the best orbit win." It may well be that other types of orbit only supplement the Clarke orbit!

A recent interesting development has been the combination of navigation and communications capabilities into a single package. Hybrid terminals combining these two functions are already commercially available. These terminals utilize a navigation satellite system such as the global positioning system (GPS) of the USA or a local terrestrial system to obtain their location, and a communication system such as INMARSAT network to transmit this information to a central location where these data are processed and used as necessary. An example of an emerging application in this area is fleet management by large trucking companies. Here a central location maintains and manages the movement of each vehicle in the fleet via a communication–navigation system, saving cost and time. It is not surprising that INMARSAT has added a navigation payload to its third-generation satellite to augument the existing GPS navigation satellite systems.

The aeronautical community has favoured satellites for transferring location and other safety-related data. A significant growth is also expected in aeronautical voice and data traffic during the 1990s. Once again, the INMARSAT system is currently at the forefront of this technology.

One of the most difficult issues in mobile satellite service is that of spectrum sharing. The frequency range between ~800 MHz and ~2.5 GHz is best suited for mobile satellite communications. There are a number of other services, including terrestrial land mobile systems, radio determination services and radio astronomy, which also prefer the use of this band. Thus the main problems as far as the MSS service-providers are concerned are to obtain adequate spectrum for the MSS system from the international frequency management body and then to agree a mutually agreeable sharing arrangement of the allocated spectrum between themselves. An international meeting known as the World Administrative Radio Conference (WARC) convened in 1992 and allocated additional spectrum to the MSS to meet rising demands. Despite these allocations, spectrum sharing between service-providers is expected to become progressively difficult. To maximize spectrum utilization the use of spot beam satellites is essential, even though generating spot beam at the L band is more difficult than at higher frequencies such as 11 GHz.

Advanced techniques such as on-board processing could be employed by the end of the 1990s. Spread spectrum modulation technique may begin to appear attractive for operation in an increasingly interference-limited and multipath environment.

(iii) Broadcast satellite service

In most developed countries, broadcast satellite systems are beginning to penetrate a market traditionally dominated by terrestrial systems. Broadcast satellite services have been introduced in several countries such as Japan, the UK and other European countries. BSS are also well suited to serve the developing world and communities isolated in remote areas of some developed countries. Satellite broadcast is sometimes the most cost-effective and quickest solution for such areas.

In Europe the customer base of broadcast channels transmitted by ASTRA satellites continues to grow steadily despite severe terrestrial competition. It is interesting to note that medium-power ASTRA satellites operating in the fixed satellite service have edged out a more powerful satellite designed specifically for direct-to-home broadcast. This happened as a consequence of a dramatic reduction in the cost of receivers brought about by the rapid development in terminal technology, together with the commercial advantage gained by the ASTRA system, through an early market entry. This is a good example of the increasing influence of commercial pressures on satellite communication technology.

Among developing countries, India's satellite broadcast system is notable. The Indian domestic broadcast satellite system (INSAT) is used to distribute television programmes of national interest to widely separated terrestrial television and radio transmitters (Rao *et al.*, 1987). Education and social awareness programmes are beamed daily to low-cost community receivers in remote towns and villages. The programmes are either received directly for community viewing or, where possible, received and re-transmitted through terrestrial transmitters. The recent introduction of commercial channels by another operator is proving to be very popular. The main reason for this success is the wide choice of channels (not available until now) together with the affordable costs at which cable television companies are providing the service. This example indicates that the potential growth areas of BSS are those regions of the world where there is generally a lack of good viewing material – provided, of course, that the viewing cost is made affordable to the local community. Other regions which could benefit from this technology are African countries.

Summarizing, it can be stated that during the 1990s the BSS can be expected to grow in various forms and continue to increase its customer base throughout the world. It is worth noting here that the worldwide WARC plan for direct broadcast formulated in 1977 and 1981 has so far not been implemented to the extent anticipated.

WARC 92 has allocated spectrum for direct broadcast of sound to portable receivers. Direct-to-home sound broadcasts are already in use in some countries in Europe. Transmissions are made on television channels as subcarriers and received by fixed receivers intended for receiving television broadcasts. However, WARC has allocated spectrum specifically for broadcasts direct to portable radio receivers. Several organizations have shown interest in developing this technology further, and some companies in the USA have already put forward commercial proposals. Severe opposition and competition from existing terrestrial broadcast system operators, especially in the developed world, are possible. However, there are vast areas of the developing world which could benefit from this technology.

11.4 Technology trends

In this section we shall highlight the main technological trends both in the space and ground segments.

(i) Spacecraft technology

In general, spacecraft technology has improved in all aspects but, to keep within the scope of the section, only the main developments are highlighted. Alleviating *spectrum shortage* and effective utilization of *spacecraft power* have been two major areas of research and development. Dual polarization systems, spot and shaped beam technology and the use of high-frequency bands have been some of the major achievements. At present, the use of the dual polarization technique is limited to systems operating below 6 GHz. The technique is expected to extend to higher-frequency bands as demand for spectrum increases. Judging by the trend in the growth of C and K_u band satellites, the traffic forecast and increasing interference environment, it is anticipated that the number of K_a band satellites should gradually increase in the next decade (Bargellini, 1984).

Techniques for enhancing spectrum usage continue to remain of interest. One possibility is to reduce the antenna beamwidth even further. At present, beamwidths of spot beams are of the order of several degrees but values approaching 0.5–1° are possible provided that losses associated with the beam-forming networks are reduced significantly. One beam-forming technique under consideration is the use of microwave integrated circuit modules containing all the elements of a beam-forming network – low-noise amplifiers, high-power amplifiers and phase shifters. Each module can be attached directly to an individual radiating element thereby minimizing all the losses. Amplitude and phase shift of each module are then controlled individually to provide the desired beam shape. This technique could also be utilized for the realization of *hopping beam system* (discussed later).

Technologies expected to bring competitive advantage for satellite communications are Gallium Arsenide technology and the use of microwave integrated circuits.

Low-noise amplifiers (LNA) for satellites have already reached a limit in sensitivity. A satellite's communication antennas are Earth-facing and therefore their minimum achievable noise is ~300 K (the Earth's temperature). Making LNA noise temperature much less gives diminishing returns. In future, advances in LNA technology may be limited to reducing the size and weight of the LNA module. The possibility of integrating this module directly to the radiating aperture has already been mentioned.

In transmitter technology, solid-state power amplifiers (SSPA) with DC to RF efficiency approaching 50% are possible in the foreseeable future. SSPAs are likely to replace TWTs for power applications below ~20 W in the C and K_u bands. The use of TWT is then expected to remain mainly confined to applications requiring high power such as broadcast satellites and at K_a band frequencies.

For *on-board processing*, digital techniques involving fast Fourier transforms and other advanced digital signal processing are under development.

The basic concepts of on-board processing (e.g. modulation/demodulation, coding, switching, etc.) are no different from those used on the ground. A major difference for on-board application is the need to use small mass and power devices having high reliability and resistance to radiation.

A basic form of on-board processing involving a programmable RF switch matrix is already in use in the INTELSAT VI satellite. Further developments in the areas of demodulation/modulation and baseband processing technology are in progress at a number of organizations. A number of experimental satellite systems such as ACTS (Holmes and Beck, 1984) and ACTS-E (Ida *et al.*, 1984) include on-board processing packages. The technology is expected to mature in the later half of the 1990s.

After the basic satellite technology was developed at the end of the 1970s, satellite circuit cost-reductions were mainly obtained by *lifetime extension* of satellites. Some of the techniques used for satellite lifetime extension are the use of Nickel–Hydrogen batteries which have a long lifespan; increasing the amount of on-board fuel; and permitting larger orbital inclination. Further extensions to the lifetime are possible by using ion thrusters. It is estimated that, by combining all these techniques, the lifetime of a satellite can be extended to over 20 years, making satellite lifetimes comparable with those of optical fibre systems. However, some studies have shown that extending the lifetime of conventional satellites beyond 16–18 years results in diminishing returns because of factors such as technical obsolescence. Therefore it appears that satellites designed to operate much beyond this 16–18 year period must have in-built future proofing (see section 11.5 on Advanced concepts).

It is likely that because of the complexity in the payload of future satellites, the spacecraft architecture will include self-monitoring functions to minimize human intervention.

(ii) Earth station technology

Earth station technology has also greatly benefited from the extensive research and development effort in the past decades. The emphasis has been on reduction of size and cost of earth stations with minimal impact on performance.

Studies show that overall the FSS system cost is best optimized by using medium-to-small earth stations. Therefore the number of small-(VSAT type)-to-medium earth stations is expected to increase significantly. Many earth stations are likely to have the capability of unattended automated operation with facilities to diagnose and report faults to a central site.

The costs of direct broadcast receivers have reduced dramatically throughout the world. For example, in the UK a package consisting of an antenna system, LNA/down-converter together with interface to a domestic television receiver costs less than the cost of a good television receiver. To encourage subscription, some companies have offered terminals at considerably lower prices, and this trend could continue. Further cost reductions are possible through technology improvements and mass production, as satellite broadcasts increase their customer base to serve millions across the world.

Indications are that the largest growth is likely to be in the mobile satellite service. Forecasts indicate that by 1996 there could be over 70 000 mobile terminals in the INMARSAT network alone. Assuming that hand-held terminals become available towards the end of the 1990s, the number of users is estimated to rise to over two million by the turn of the century.

Let us now focus on some technological aspects which continue to receive attention. Echo control technology is one such area. Further improvements are expected with the use of digital techniques and delay compensation protocols. Improved voice coding and video compression techniques are expected to bring bit rates down to 2–4 kbps and 750 kbps respectively at affordable costs. Prototypes of such coders are already available. Variable-rate modems and codecs are under development and could provide the flexibility of altering bit rate at short notice. Another area of development is the earth station antenna system. In mobile and direct broadcast applications the need for aesthetics together with improved performance are becoming increasingly important. Low profile antennas are currently beginning to appear in mobile terminals and development efforts continue for improving direct broadcast receiver antennas. For large earth station antennas the effort is focused on side-lobe level reduction techniques to permit closer spacing of satellites in the geostationary orbit. The use of VLSI and LSI is expected to reduce the size of sub-systems. VLSI techniques are expected to be invaluable in mobile terminals, such as paging receivers which are expected to be small enough to be carried in pockets!

11.5 Advanced concepts

A number of advanced concepts have been proposed during the past decade, some of which might be implemented during the 1990s.

The use of *high-frequency bands* such as the 20–30 GHz band is under active consideration. Many experimental satellites have been launched (e.g. the OLYMPUS satellite developed by the European Space Agency) and more are planned in the future (e.g. ACTS funded by the National Aeronautical and Space Administration). There are a number of advantages in using the K_a band:

(i) It is possible to use very small earth station antennas, enabling the location of terminals on customers' premises.
(ii) As this band is not in common use either by terrestrial or space sys-

tems, inter-system system coordination problems are minimal.

(iii) Large bandwidth is available.

(iv) Frequency re-use by geographical separation is more easily achieved because satellite antennas can be made more directive.

There are also certain shortcomings in the use of this band which have prevented its wide use to date. One of the main problems is signal attenuation caused by rain. The attenuation is much greater here than in either the C or K_u bands. Thus, to achieve link reliability comparable to the C or K_u band warrants a much larger system overhead as extra link margin or the need to use diversity. Moreover, because of the present limited technological development status of this band, the overall system cost is higher.

Higher-frequency bands such as 40/50 GHz have also been proposed (Ida *et al.*, 1984).

Satellite *on-board processing*, already mentioned in a preceding section, is another promising concept with a potential to bring substantial cost reductions. Some of the advantages of on-board processing are as follows:

1. Uplinks and downlinks can be optimized separately because the satellite's on-board demodulation/modulation process decouples the links. To clarify the concept, consider a TDMA system. With the transparent satellites in use now, all the earth stations in the network must operate at a common bit rate irrespective of individual traffic requirements. By decoupling uplinks and downlinks, regenerative repeaters permit optimization of links according to the traffic requirements of individual earth stations. Each earth station transmits/receives at a bit rate commensurate to its traffic. Consequently earth station design can be better optimized giving benefits, in particular, to small users.
2. It is possible to use much lower uplink EIRP than required by conventional transponders, making it possible to use a much smaller earth station. This could be of great benefit to mobile terminals.
3. The effects of transponder non-linearity and inter-system interference are minimized because of decoupling between the uplink and the satellite path.

Up to now, the number of spot beams has been limited typically to less than 10. Certain studies propose the use of *dozens of spot beams* to service small terminals.

Another novel concept is the *hopping beam system* (Campanella *et al.*, 1990). In such a system, coverage is provided by a single narrow beam which is rapidly scanned (or 'hopped') in non-overlapping zones across the coverage area. All the earth stations in the network are synchronized to the hopping rate of the beam and communicate only when illuminated by the hopped beam. Thus at any instant, a significant portion of satellite EIRP and bandwidth is available to a single zone. To implement this concept, a satellite must have on-board processing and storage capability to route information correctly. Note that in a *multiple spot beam system* each beam is dedicated to a specified coverage area and therefore the satellite resource is shared between all the beams.

The concept of *inter-satellite link* (*ISL*) has also been studied extensively (e.g. IJSC, 1988). The technology and benefits of ISLs have been demonstrated in missions such as the Tracking and Data Relay Satellite System (TDRSS). The TDRSS system is used to relay data from low earth orbiting satellites to a few conveniently located earth station in the USA via host geostationary satellites. ISLs are used to connect low earth satellites with the host satellites. However, the use of ISLs has not been demonstrated in commercial satellite communication systems, even though there are a number of advantages accruing to the use of ISLs. One advantage is the possibility of extending the coverage area of a geostationary satellite network by linking satellites directly in space. The effective coverage of such a network is the sum of the coverages available from each satellite separately. An earth station in the coverage area of one satellite obtains an extended visibility via the ISL and the adjacent satellite. The maximum separation between two satellites is decided by the permitted propagation delay and unobstructed visibility between satellites (note that, beyond a specific separation, the Earth begins to shadow an ISL). In another architectural concept, ISLs are used to form a cluster of co-located satellites, each satellite in the cluster having a different function. For example, the satellites of a mobile satellite service and a fixed satellite service network could be linked and clustered together, and calls from one service could be directly connected to the other via an ISL. Similarly, a domestic satellite system could be directly connected to an international network. Some of the recent low earth orbit satellite concepts propose the use of ISLs for interconnecting satellites in the LEO constellation to achieve global interconnectivity. Microwave frequencies at 20 and 60 GHz and laser inter-satellite links have been investigated.

It is interesting to note that a type of clustered satellite operation is already in use in the INMARSAT network. However in this application satellites remain independent entities without any ISL. The purpose of such a *dual-satellite operation* is to enhance the capacity of the network. This is made possible by locating the satellites within a few degrees of each other so that the mobile earth stations (all of which have large beamwidth) are unable to distinguish between them. It is then possible to utilize the power and frequencies of both satellites. To avoid interference the frequency plan of each satellite must not overlap. In a power-limited case the power from each satellite is summed, whereas in a bandwidth-limited case the spectrum of each satellite is summed.

Recently, a great deal of interest has been generated in using constellations of *Low* or *Medium Earth Orbit* satellites for mobile communications.

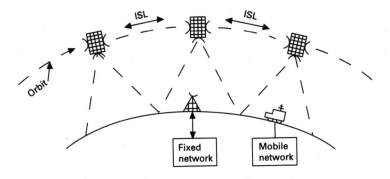

Figure 11.1   An MSS architecture using a low earth orbit satellite
constellation. Messages are routed from the source to the
destination via inter-satellite links.

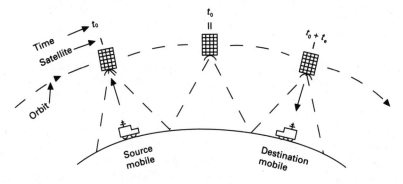

Figure 11.2   An MSS architecture for store-and-forward messaging, using
a constellation of low earth orbit satellites. Messages are
delivered by the source mobile to satellite I at t_0 and
delivered to the destination mobile after t_e seconds when the
mobile comes into view of satellite I.

In such a system, fixed earth stations are connected to mobile stations via a
constellation of low earth satellites. The space segment architecture depends
on an acceptable message delay and the coverage area. As already mentioned earlier, at one extreme are constellation designs consisting of satellites interconnected via ISLs, capable of providing worldwide voice ·
communications (e.g. Richharia *et al.*, 1989), and at the other are store-and-
forward message delivery system tolerating delays of several hours using a
constellation of simple unconnected satellites (e.g. Maral *et al.*, 1991). Figures
11.1 and 11.2 show the space segment architecture of these two networks.

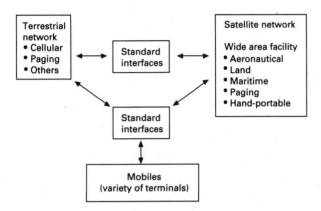

Figure 11.3   Architecture of future public land mobile
telecommunication system.

Another architecture of current interest is an integrated terrestrial–satellite mobile communication system (e.g. Richharia and Evans, 1990). It has been recognized that, at present, satellite systems cannot compete with cellular systems in populated area. However, satellites by virtue of large coverage areas, are well suited to serve the remaining areas of the Earth be it sea, an air corridor or a remote mountain. Thus there are obvious benefits in combining the advantages of each system such that the users obtain the benefits of each. Various levels of synergy are possible. In a fully integrated network, the network intelligence decides the best transmission medium at any given time (with a manual capability in each terminal to select any desired option). Thus, with a single number and terminal, the users get optimal service at all times, together with continent-wide or even global-roaming capability. At present, terrestrial and satellite systems are separate. To obtain a synergic advantage in such an architecture, users may at best use *dual mode terminals*. With such a terminal the user switches to the desired network according to needs, with a certain degree of local automation built in – for example, the terminal may indicate that the user has migrated outside the terrestrial coverage area.

The ideal solution for mobile users is the ability to use the same mobile phone throughout the world or at least over large parts of the world. The CCIR/CCITT are studying such a concept which is called *future public land mobile telecommunication systems* (FPLMTS). The goal of this network is to provide telecommunication service to mobile terminals through an integrated network. The network architecture consists of terrestrial and satellite system elements in which a variety of mobile terminals are connected to the network through standardized interfaces, as shown in figure 11.3. The ultimate goal is to have this all-pervading network operational throughout the globe.

It has been mentioned that the fundamental limitations of a satellite communication system are large transmission delay and path loss. Improvements to the existing echo control techniques and delay compensation protocols are underway for minimizing the problem of transmission delay. On a futuristic note, some radical solutions have been proposed (Pelton and Wu, 1987). The use of such concepts could virtually eliminate the inherent disadvantage of the geostationary orbit. The use of *tethered satellites* is one such concept. Here a satellite is made to appear stationary with respect to the Earth at a lower altitude by suspending it from a geostationary satellite with the help of a tether. To compensate the torque on the geostationary 'anchor' satellite, another tether is suspended upwards as a counter-weight. Some trials to tether satellites have been conducted by NASA with limited success. Delay is minimized because of the much shorter range of the satellite. Another concept is the use of *very light weight satellites* maintained in a 'geosynchronous' orbit at say 800 km, powered by energy beamed from the ground. The use of the *LEO satellite constellation* concept as proposed for mobile satellite service is another possible solution. Whether such 'zero delay' satellites do become reality in the twenty-first century is a matter of conjecture. If implemented, they could provide nearly all the capabilities of optical fibre coupled with the added advantages of satellite systems.

We have noted earlier that satellite circuit costs are reduced by extending the operational lifetime of the satellite. Several ideas have been proposed for *satellite lifetime extension*. These include:

1. The use of space platforms with the capability of being refurbished/retrofitted/repaired by robots and service vehicles. The lifetime is estimated to extend up to 40 years by such a system.
2. The use of satellite clusters with the capability of being activated and updated over time according to need.

11.6 Experimental programmes

Research into new satellite communications techniques generally requires considerable investment and involves risks such as launch failures and malfunction of components not yet space qualified. Moreover, some spacecraft missions may only be necessary for obtaining data, such as sensitivity of a component (e.g. digital memory) to radiation or to try new concepts, such as ion thrusters for attitude control. Consequently, commercial interest in funding such projects is rather limited even though many of these concepts, if proven, may be exploited commercially. To encourage commercial growth, most innovative research is usually sponsored by government bodies such as the National Aeronautical and Space Administration (NASA) in the USA, the European Space Agency (ESA) and similar agencies in countries such as Japan and India. Many universities and research laboratories also contribute to the growth of technology. Here, as an example, we shall very briefly consider the ACTS (Holmes and Beck, 1984; Graebner and Cashman, 1990) and ACTS-E (Ida *et al.*, 1984) which could both be the forerunners of emerging techniques.

The Advance Communication Technology Satellite (ACTS) program is being sponsored by NASA to foster advanced technology. The ACTS satellite will demonstrate a number of techniques such as the use of 20/30 GHz for communications, the beam scanning technique and the use of on-board processing with baseband switching. The on-board processing functions include demodulation of TDMA bursts, decoding of FEC coded channels and demultiplexing of individual channels. This is followed by multiplexing of channels according to destination, adaptively encoding each channel and transmitting bursts to the destination via the appropriate spot beam. This satellite was launched in 1993.

Similarly, Japan's Advanced Technology for Experiment (ACTS-E) is planned to gain experience in a number of forefront technologies. Plans include the use of 50/40 GHz for high-speed/high-capacity point-to-point links and personal mobile communication; a 27/22 GHz broadcast system including high definition television; a 14/12.5 GHz fixed and broadcast system using steerable beams; investigation of the on-board processing capability of switching, scanning-beam control and beam inter-connections; a VHF-millimetre wave propagation experiment; and uplink communication experiments using *lasers*. The use of lasers has not yet been considered favourably for Earth–satellite links because of the expected large degradation in the Earth's atmosphere. However, real propagation data are scant and such a laser experiment would provide invaluable data.

Thus, judging from the current trend, it is anticipated that satellite communication systems will continue to grow at a healthy pace having benefited from these exciting innovations.

Problems

1. (a) It has been mentioned that inter-satellite links (ISLs) can be used for interconnecting mobile and satellite service networks. What are the advantages and limitations (if any) in integrating an MSS and an FSS network in space for the following types of network:
   (i) domestic MSS and FSS;
   (ii) international MSS and FSS?
   (b) Develop an expression for determining the inter-satellite distance between two geostationary satellites. List all the assumptions in your derivation.
2. This problem is aimed at the advanced student since further study and some research are necessary. It has been mentioned in the text that inte-

grating satellite and terrestrial mobile systems provides several advantages. Your tasks are as follows:
(i) Outline the main advantages of such an integrated network.
(ii) Consider the integration of a mobile satellite system (e.g. INMARSAT) with a terrestrial system having a potential of wide usage (e.g. the Pan-European GSM system or any other). Highlight the main differences in their system architecture (e.g. bit rates, access scheme, etc.).
(iii) Suggest a possible architecture for an integrated network. Assuming that the integration between the two systems taken in task (ii) is evolutionary, discuss the stages in the integration process. If your study concludes that integration in certain aspects is not possible (e.g. supported bit rates differ), identify these parameters clearly, stating reasons as to why integration is not possible.
[To start off the project, consult for example Richharia and Evans (1988, 1990). These papers discuss various network issues and include several useful references. It should be mentioned that this type of system architecture is currently of significant interest.]
3. Compare the performance of a transparent repeater with a regenerative repeater.

References

Bargellini, P.L. (1984). 'A reassessment of satellite communications in the 20 and 30 GHz bands', *International Journal of Satellite Communications*, Vol. 2, pp 101–106.
Campanella, J.S., Pontano, B.A. and Chitre, D.M. (1990). 'A user's perspective of the ACTS hopping beam TDMA system', *AIAA Conference Record*, Paper AIAA-90-0833-CP, pp 484–489.
Casas, J.M. and Fromm, H.H. (1988). 'The role of satellites in the ISDN era', *Networks '88*, June, London.
Graebner, J.C. and Cashman, F. (1990). 'ACTS multibeam communication package: technology for the 1990s', *AIAA Conference Record*, Paper AIAA-90-0835-CP, pp 497–507.
Holmes, W. and Beck, G. (1984). 'The ACTS flight segment: cost effective advanced communications technology', *AIAA 10th Communications Satellite System Conference Record*, Orlando, Florida, March 19–22.
Ida, T., Shimada, M., Iwasaki, K., Ohkami, Y., Azuma, H., Kibe, S. and Kai, T. (1984). 'Japan's large experimental communications satellite (ACTS-E): its mission, model and technology', *Proc ICC*, pp 1098–1101.
Ida, T., Shimoseko, K.I. and Shimada, M. (1985). 'Satellite communica-

tions in the next decade', *Space Communication and Broadcasting*, Vol. 3, pp 27–38.

IEE (1989). *IEE Colloquium on Highly Elliptical Orbit Satellite Systems*, Digest no. 1989/86, 24 May.

International Journal of Satellite Communication (1988). Special issue on 'Inter Satellite Links', Vol. 6.

Lundberg, O. (Director General, INMARSAT) (1991). 'Project 21: a vision for the 21st century', *News Briefing*, September 12.

Mahle, C.E., Hyde, G. and Inukai, T. (1987). 'Satellite scenarios and technology', *IEEE Journal on Selected Areas of Communications*, Vol. SAC-5, No. 4, May, pp 556–570.

Maral, G., Ridder, J.D., Evans, B.G. and Richharia, M. (1991). 'Low earth orbit satellite systems for communications', *International Journal of Satellite Communications*, Vol. 9, pp 209–225.

Pelton, J.N. and Wu, W.W. (1987). 'The challenge of 21st century satellite communications: INTELSAT enters the second millennium', *IEEE Journal on Selected Areas in Communications*, Vol. SAC-5, No. 4, May, pp 571–591.

Rao, U.R., Pant, N., Kale, P.P., Narayan, K., Ramchandran, P. and Singh, J.P. (1987). 'The Indian national satellite system – INSAT', *Space Communication and Broadcasting*, Vol. 5, pp 339–358.

Richharia, M. and Evans, B.G. (1988). 'Synergy between land mobile satellite and terrestrial systems – possibilities in the European region', *Fourth International Conference on Satellite Systems for Mobile Communications and Navigation*, London, October.

Richharia, M. and Evans, B.G. (1990). 'Synergy between satellite land mobile and terrestrial cellular systems – towards a personal communication system', *International Satellite Conference*, AIAA, California, March.

Richharia, M., Hansel, P., Bousquet, P.W. and O'Donnel, M. (1989). 'A feasibility study of a mobile communication network using a constellation of low earth orbit satellites', *IEEE Global Telecommunications Conference, GLOBECOM '89*, Dallas, 27–30 November.

Appendix A: Useful Data

Earth gravitational parameter (GM) =	$398\ 600.5\ \text{km}^3/\text{s}^2$
Earth mass (M) =	$5.9733 \times 10^{24}\ \text{kg}$
Earth gravitational constant =	$6.673 \times 10^{-20}\ \text{km}^3/\text{kg s}^2$
Earth equatorial radius =	$6378.14\ \text{km}$
Velocity of light =	$299\ 792.458\ \text{km/s}$
Average radius of geostationary orbit =	$42\ 164.57\ \text{km}$
Velocity of geostationary satellite =	$3.074\ 689\ \text{km/s}$
Angular velocity of geostationary satellites =	$72.921\ 15 \times 10^{-6}\ \text{rad/s}$
Geostationary satellite orbital period =	$86\ 164.09$ s (23 hours, 56 minutes, 4.09 seconds)
Boltzmann constant =	1.3803×10^{-23} W/K Hz or -228.6 dB W/K
Maximum range of geostationary satellite (0° elevation) =	$41\ 680\ \text{km}$
Minimum range of geostationary satellite (90° elevation) =	$35\ 786\ \text{km}$
Half-angle subtended at the satellite by Earth =	$8.69°$
Coverage limit on Earth (0°. elevation) =	$81.3°$

Appendix B: Useful Orbit-related Formulas

Doppler effect

The equation set included here is general enough to provide Doppler shifts in non-geostationary orbits. Only some of the equations of the set are adequate for satellites in geostationary orbits.

The Doppler shift Δf_d at a frequency f_t is given by

$$\Delta f_d = \pm \frac{v_r}{c} f_t \tag{B.1}$$

where v_r = relative radial velocity between the observer and the transmitter
c = velocity of light
f_t = transmission frequency.

The sign of the Doppler shift is positive when the satellite is approaching the observer.

The relative velocity can be approximated as

$$v_r \sim \frac{\rho_2(t_2) - \rho_1(t_1)}{(t_2 - t_1)} \tag{B.2}$$

where $\rho_1(t_1)$ and $\rho_2(t_2)$ are satellite ranges at times t_1 and t_2 respectively; $(t_1 - t_2)$ is arbitrarily small.

$\rho(t)$ at any instant t can be obtained from the orbital parameters by using the technique given in a following section ('Satellite position from orbital parameters'). Range rate can then be obtained by using equation (B.2), at two successive instants.

The following equation set may be used for approximate estimation of the range rate of a geostationary satellite. We note that range rate is a function of orbital eccentricity, inclination and satellite drift rate. The range rate for each of these components is given as (Morgan and Gordon, 1989):

(a) *Eccentricity*

$$\dot{\rho}_e = \frac{e \, a^2 \, \dot{\omega} \sin(\dot{\omega} t_p)}{\rho_m} \tag{B.3}$$

where $\dot{\rho}_e$ = range rate due to eccentricity

e = eccentricity
a = semi-major axis
$\dot{\omega}$ = angular velocity
ρ_m = mean range
t_p = time from perigee.

(b) *Inclination*

$$\dot{\rho}_i = -\frac{i \, a \, R \, \dot{\omega}}{\rho_m} \sin \theta \cos (\dot{\omega} \, t_i) \tag{B.4}$$

where $\dot{\rho}_i$ = range rate due to inclination
 i = inclination
 R = Earth radius
 θ = latitude of earth station
 t_i = time from ascending node.

(c) *Drift*

$$\dot{\rho}_d = \frac{DaR}{\rho_m} \cos \theta \sin \Delta\phi \tag{B.5}$$

where D = drift rate in radians/s
 $\dot{\rho}_d$ = range rate due to satellite drift
 $\Delta\phi$ = difference in longitude between satellite and earth station.

The total range rate at any given time is the sum of range rates due to each of the above components.

CCIR Report 214 gives the following approximate relationship for estimating the maximum Doppler shift:

$$\Delta f_{dm} \approx \pm 3.0(10)^{-6} f_t s \tag{B.6}$$

where f_t = operating frequency
 s = number of revolutions/24 hour of the satellite with respect to a fixed point on the Earth.

For a more precise treatment of the subject the reader is referred to the literature (e.g. Slabinski, 1974).

Near geostationary satellites

On various occasions, communication satellites are in near geostationary orbits. Examples are: (a) when orbit inclination is intentionally left uncorrected to conserve on-board fuel and thereby prolong the satellite's useful lifetime and (b) when a satellite is being relocated to another position or a newly launched satellite is being moved to the operational location (such a drifting satellite is sometimes used for communication provided the transmissions do not interfere with other systems).

When the satellite orbit is lower than the geostationary orbit altitude, the angular velocity of the satellite is greater than the angular velocity of the Earth. Consequently the satellite drifts in an eastward direction with respect to an earth station. When the satellite altitude is higher than the geostationary height, the satellite drifts westward (Morgan and Gordon, 1989).

The following relationships apply:

$$\frac{\Delta P}{P} = -\frac{\Delta \omega}{\omega} \tag{B.7}$$

where ΔP = change in orbital period
       P = orbital period
      $\Delta \omega$ = change in angular velocity
      ω = angular velocity

and

$$\frac{\Delta r}{r} = -\left(\frac{2}{3}\right)\frac{\Delta \omega}{\omega} \tag{B.8}$$

where r = orbital radius
      Δr = change in orbital radius.

For example, a change in radius of $+1$ km from the nominal causes a westward drift of 0.0128°/day.

The required change in satellite velocity Δv_c to correct the drift is given by

$$\Delta v_c = \frac{1}{3} v \frac{\Delta \omega}{\omega} \tag{B.9a}$$

or

$$\Delta v_c = \frac{1}{3} a \, \Delta \omega \tag{B.9b}$$

where a = satellite velocity.

Effect of inclination

The main effect of inclination on a geostationary satellite is to cause north–south oscillation of the sub-satellite point, with an amplitude of i and period of a day. When the inclination is small (the condition is, tan (i) ≈ i in radians), the motion can be approximated as a sinusoid in a right ascension–declination coordinate system. An associated relatively minor effect is an east–west oscillation with a period of *half a day*. This is caused by the change in rate of variation of the right ascension relative to the average rate. The satellite appears to drift west for the first 3 hours and then east for the next half quarter. The satellite continues to move eastward during the next half quarter and then westward, completing the cycle in half a day. The maximum amplitude of such east–west oscillation for a circular orbit is given by

$$\Delta EW_i = \arcsin \left[\tan^2 \left(\frac{i}{2} \right) \right] \tag{B.10a}$$

$$\approx \frac{1}{229} i^2 \tag{B.10b}$$

where i is in degrees.
Usually the east–west oscillation is very small (e.g. for i = 2.5°, ΔEW_i = 0.027°).
The net effect of these two motions is the often-quoted figure-of-eight motion of the sub-satellite point.

Effect of eccentricity

The effect of eccentricity in a geostationary orbit is to cause east–west oscillation with a period of a day. The satellite is to the east of its nominal position between perigee and apogee and to the west between apogee and perigee. The amplitude of the oscillation is given by

$$\Delta EW_e = 2e \text{ radians} \tag{B.11}$$

For example, an eccentricity of 0.001 produces an oscillation of 0.1145°.

Coverage contours

It is often necessary to plot the coverage contours of geostationary satellites on the surface of the Earth. The satellite antenna boresight (the centre of coverage area) and a specified antenna power beamwidth (usually, half-power beam-

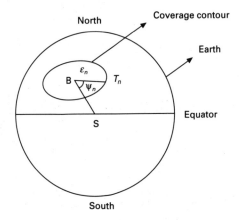

Figure B.1   Coverage contours geometry. S = sub-satellite point,
              B = boresight point on Earth, T_n = nth point on the
              coverage contour.

width) are known. In the case of an elliptical antenna beam shape, the sizes of
the major and minor axes together with the orientation of the major axis are
known. The coverage contour on the Earth is obtained by calculating the lati-
tude/longitude of n points on the periphery of the coverage (Siocos, 1973).
    Let us first define the following angles:

$\gamma_{\rm B}$, γ_n = tilt angles of antenna boresight and the nth point on the coverage
          contour, respectively
ε_n = angular antenna beamwidth of the specified power (e.g. half-power)
          in the direction of the nth point. For a circular beam, ε_n is a con-
          stant.

    To specify the nth coverage point we further define ψ_n as the angle of
rotation, the rotation being referenced to the plane containing the sub-satellite and boresight points (see figure B.1).
    The following steps are used to specify the nth coverage point T_n :

Obtain $\gamma_{\rm B}$ using the following equation set

$$\beta = \text{arccos} \ (\cos \theta_{\rm B} \cos \phi_{\rm SB}) \tag{B.12a}$$

$$\gamma_{\rm B} = \text{arctan} \ [\sin \beta/(6.6235 - \cos \beta)] \tag{B.12b}$$

where $\theta_{\rm B}$ = latitude of boresight
         $\phi_{\rm SB}$ = longitude of boresight with respect to sub-satellite point, taken
              positive when to the west of the sub-satellite point.

Then

$$\cos \gamma_n = \cos \gamma_B \cos \varepsilon_n + \sin \gamma_B \sin \varepsilon_n \cos \psi_n \qquad \text{(B.13a)}$$

$$\ctn \phi_n = (\sin \gamma_B \ctn \varepsilon_n - \cos \gamma_B \cos \psi_n)/\sin \psi_n \qquad \text{(B.13b)}$$

$$\xi_n = \arctan(\sin \phi_{SB}/\tan \theta_B) + \phi_n \qquad \text{(B.13c)}$$

$$\beta_n = \arcsin (6.6235 \sin \gamma_n) - \gamma_n \qquad \text{(B.13d)}$$

$$\theta_n = \arcsin (\sin \beta_n \cos \xi_n) \qquad \text{(B.13e)}$$

$$\phi_{Sn} = \arctan (\tan \beta_n \sin \xi_n) \qquad \text{(B.13f)}$$

where ϕ_{Sn} = longitude of nth point relative to sub-satellite point
θ_n = latitude of nth point.
When the beam is elliptical, ε_n depends on ψ_n as follows:

$$\ctn^2\varepsilon_n = \ctn^2\varepsilon_1 \cos^2(\alpha + \psi_n) + \cot^2\varepsilon_2 \sin^2(\alpha + \psi_n) \qquad \text{(B.14)}$$

where α = rotation of ε_1 away from the direction of the azimuth of the boresight
ε_1 and ε_2 are the semi-major and semi-minor axes.
ψ_n can be varied from 0° to 360° to obtain as many points on the coverage contour as desired.

Sun transit time

Around the equinox periods (March and September), the Sun is directly behind the geostationary orbit and therefore appears within earth stations' antenna beam. Sun transit through an earth station's antenna causes disruption to communication services because of a large increase in system noise temperature caused by the Sun. The transit time of the Sun through an antenna is predictable, giving the earth station operator the option to make alternative communication arrangements or at least not be taken by surprise when communication is disrupted.

The position of astronomical bodies such as the Sun is published in a readily available annual publication called the *Nautical Almanac* (US Government Printing Office). The position is given in the right ascension–declination coordinate system. Sun-caused outage occurs when the ascension and declination of the satellite and the Sun become equal at an earth station (or nearly equal so that the Sun appears in the beamwidth of the earth station antenna). The position of the satellite at an earth station is usually given in the celestial horizon system, as azimuth and elevation. Therefore it is only necessary to convert the satellite azimuth and the elevation to the ascension–declination coordinate system and determine from the *Nautical Almanac* the day and the time when the Sun has the same ascension and declination. The equations for this conversion are (Siocos, 1973):

Declination D is given by

$$\sin D = \sin \theta \sin \eta - \cos \theta \cos \eta \cos \xi \qquad (B.15)$$

where θ = latitude of earth station
  η = satellite elevation
  ξ = satellite azimuth (positive when the denomination is west)
  D is positive when denomination is north.

The ascension α of the earth station in hour angle relative to the satellite meridian is obtained from

$$\sin \alpha = \cos \eta \sin \xi / \cos D \qquad (B.16)$$

In the *Nautical Almanac*, the ascension of the Sun is given with respect to the Greenwich meridian. α is converted to HA_G from

$$HA_G = \phi_e - \alpha \qquad (B.17)$$

where HA_G = hour angle with respect to Greenwich
  ϕ_e = longitude of earth station
  ϕ_e is positive when the earth station is to the west of the satellite.

Note that the right ascensions of astronomical objects are expressed in hour angle, where 1 hour = 15°.

Solar eclipse caused by the Moon

The occurrence of solar eclipse on a geostationary satellite caused by the Moon is irregular. It may be recalled that Earth-induced eclipses are predictable, occurring within ±21 days of equinoxes. It is also necessary to predict the duration and the extent of occurrences of Moon-induced eclipses for spacecraft operations' planning. The technique given here (Siocos, 1981) makes use of Sun and Moon position data available from the *Nautical Almanac*.

An eclipse occurs when the azimuth/elevation coordinates of the Sun and the Moon from the satellite position are equal or close enough to cause the Moon disk to mask the Sun partially or completely.

The effective elevation H of the Sun or Moon from the satellite location can be obtained from the following equation set:

$$\cos \beta = \cos d \cos LHA \tag{B.18a}$$

$$\tan H = \left[\cos \beta - \left(\frac{R_0}{R_0 + R_s}\right)\right]\bigg|\sin \beta \tag{B.18b}$$

where d = declination of the stellar object (Sun or Moon)
LHA = local horizon angle
$LHA = HA_G + \theta$
HA_G = hour angle with respect to Greenwich, available from the *Nautical Almanac*
θ = longitude of the earth station (0° to 180°, positive when to the east of Greenwich)
R_0 = geostationary orbit height from geocentre ≈ 6.62 R (where R is earth radius)
$(R_0 + R_s)$ = distance of Sun or Moon from geocentre

and

$$\frac{R_0}{R_0 + R_s} = 6.62 \sin(HP) \tag{B.19}$$

where HP = horizontal parallex (the maximum difference in geocentric and satelli-centric altitude of the stellar object).
For the Sun:

$$HP \approx 8.85 \text{ seconds}$$

For the Moon, the hourly horizontal parallex can be obtained from the *Nautical Almanac*.

The azimuth of the Sun and the Moon observed from the satellite location is determined by the equation

$$\tan z = \sin LHA/\tan d \tag{B.20}$$

where z = 180° − Az
Az = azimuth of the Sun or the Moon
d = declination of the Sun or the Moon
z is easterly when $LHA > 180°$
z is westerly when $LHA < 180°$
and when d is negative, (B.20) gives the value $z + 180°$ rather than z.

An eclipse occurs whenever the centre-to-centre distance between the Sun disk and the Moon disk, as viewed from the geostationary orbit, is less than the sum of their radii (see figure B.2):

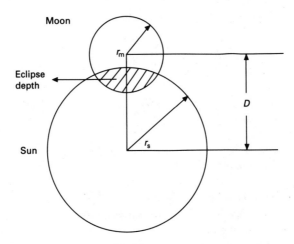

Figure B.2   Solar eclipse on geostationary satellite caused by the Moon
– view from geostationary orbit.

$$D < r_s + r_m \tag{B.21}$$

where r_s and r_m are the radii of the Sun and the Moon obtained from

$$r = \frac{1 - \sin (HP)}{[1 - 5.52 \sin (HP)]} \sin r_c \tag{B.22}$$

and

$$D = \text{arccos } (\cos \Delta H \cos \Delta Z) \tag{B.23}$$

where ΔH and ΔZ are the differences between the effective elevations and
azimuths, respectively.
r_c is the semi-diameter of the celestial object, as observed on the
surface of the Earth, available from the *Nautical Almanac*
HP is obtained from the *Nautical Almanac*.

Eclipse depth

The covered area of Sun's disk or the depth of eclipse (see the hatched
portion in figure B.2) can be obtained from the equation

$$\varepsilon_d = \left[\frac{2A}{360} - \frac{\sin(2A)}{2\pi} \right] + (r_m / r_s)^2 \left[\frac{2B}{360} - \frac{\sin(2B)}{2\pi} \right] \tag{B.24}$$

where

$$\cos A = \frac{\cos r_m - \cos r_s \cos D}{\sin r_s \sin D} \tag{B.25a}$$

$$\cos B = \frac{\cos r_s - \cos r_m \cos D}{\sin r_m \sin D} \tag{B.25b}$$

Satellite-referred coordinates to earth coordinates

Sometimes the antenna pattern of a satellite is referred to the satellite centred coordinate system. In such a coordinate system the satellite location is taken as the origin. The latitude and longitude are referred to an imaginary sphere around the satellite. The following equation set is used to transform the satellite-centred coordinate system to earth coordinates:

$$\gamma_e = \arccos[\cos \theta_s \cos \phi_s] \tag{B.26a}$$

$$\xi_e = \arctan[\sin \phi_s / \tan \theta_s] \tag{B.26b}$$

$$\beta_e = \arcsin(6.617 \sin \gamma_e) - \gamma_e \tag{B.26c}$$

$$\theta_e = \arcsin(\sin \beta_e \sin \xi_e) + \phi_0 \tag{B.26d}$$

$$\phi_e = \arctan(\tan \beta_e \sin \xi_e) + \phi_0 \tag{B.26e}$$

where ϕ_0 = longitude of sub-satellite point
θ_s, ϕ_s = satelli-centric latitude and longitude respectively
θ_e, ϕ_e = transformed latitude and longitude on Earth respectively.

Map projections

Earth coverage from a satellite is most commonly shown as satellite antenna pattern contours (referenced from the beam centre) on a suitable map. A coverage contour is obtained by plotting the latitude and longitude of the coverage periphery on a map. The coverage contours appear distorted in many types of map projections such as Albers and Maracots, whereas in several projections the shape of the coverage is undistorted. In general, the choice of map depends on the type of orbit and the users. For example, polar projections are popular with radio amateurs because of advantages such as simplicity in plotting ground tracks.

In satellite communications, rectangular projections are often used. One commonly used projection represents the X-axis as longitude and the Y-axis as latitude. However, in such projections the shape of the coverage contours appears distorted. For planning, it is simpler to use maps which retain the angle information of the contours. If a projection is made on a plane which is at right-angles to the satellite–Earth vector, the shape of the beams is retained (Chouinard, 1981; CCIR, 1982). Distances on such a projection are linearly related to the angles. The following set of equations transforms a point P_i on Earth to a satelli-centric sphere:

$$\gamma = \arctan \left[\sin \beta / (6.6235 - \cos \beta) \right] \tag{B.27}$$

where

$$\beta = \arccos \left[(\cos \theta_i \cos(\phi_i - \phi_0) \right] \tag{B.28a}$$

$$\xi = \arctan \left[\sin (\phi_i - \phi_0) / \tan \theta_i \right] \tag{B.28b}$$

Here θ_i and ϕ_i are the latitude and longitude of point P_i
   ϕ_0 is the longitude of the sub-satellite point.
   Finally, the transformed latitude θ_i' and ϕ_i' on a satelli-centric unit sphere are given by

$$\theta_i' = \arcsin (\sin \gamma \cos \xi) \tag{B.29}$$

$$\phi_i' = \arctan (\tan \gamma \sin \xi) \tag{B.30}$$

Because θ_i' and ϕ_i' are less than $8°41'$ ($\sim 1/2$ of the angular diameter of Earth from a geostationary orbit), mapping them in Cartesian coordinates is quite adequate. On such a map, if the two scales are equal, angles are almost preserved.

Off-axis angles

To facilitate interference calculations between satellite networks, it becomes necessary to develop expressions for off-axis angles. An off-axis angle is defined here as the angle between the wanted direction and the undesired direction which gives rise to interference. Figure B.3 shows two modes of interference encountered in practice. Figure B.3(a) shows the interference mode, where interference is either received at the satellite (the 'wanted' satellite) serving the desired network from an earth station of another network, or caused at an earth station of another network by the desired satellite. Figure B.3(b) shows the interference mode where interference is either received by an earth station (a 'wanted' earth station) in the desired network from a satellite serving another network (the 'external' satellite) or

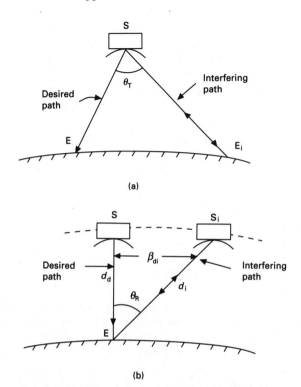

(a)

(b)

Figure B.3   (a) Interference received or caused by a satellite;
(b) interference received or caused by an earth station.
(S = wanted satellite, E = wanted earth station,
S_i = satellite causing or susceptible to interference,
E_i = earth station causing or susceptible to interference).

caused by a wanted earth station external to the satellite.

Referring to figure B.3(a), the interference received or caused by a satellite is given by the off-axis angle (Siocos, 1973)

$$\cos\theta_T = \frac{\rho_d^2 + \rho_i^2 - 2(1 - \cos\beta_{di})}{2\rho_d\rho_i} \tag{B.31}$$

where ρ_d = range between the satellite and its boresight on Earth
ρ_i = range between the satellite and the interfered point
β_{di} = great circle arc between boresight and interfered point
Range is given in terms of Earth radius (equation 2.20b).

$$\cos\beta_{di} = \sin\theta_d \sin\theta_i + \cos\theta_d \cos\theta_i \cos(\Delta\phi_i - \Delta\phi_d) \tag{B.32}$$

where θ_d, θ_i = latitude of points d and i respectively
  $\Delta\phi_i$ = longitude of point i with respect to the sub-satellite point
  $\Delta\phi_i$ and $\Delta\phi_d$ are positive when the point is to the west of the sub-
  satellite point.
  $\Delta\phi_d$ = longitude of point d with respect to the sub-satellite point
The off-axis angle θ_R, figure B.3(b), is given by (Morgan and Gordon, 1989)

$$\cos \theta_R = \frac{\rho_d^2 + \rho_i^2 - \left[84\,332 \sin \left(\frac{\Delta\phi_{SI}}{2}\right)^2\right]}{2\rho_d\rho_i} \tag{B.33}$$

where $\Delta\phi_{SI}$ = orbital separation of interfering satellite from wanted satellite (degrees of longitude).
Here ranges ρ_d and ρ_i are in *km* (equation 2.20a).

Satellite position from orbital parameters

To estimate the orbital parameter of a satellite, the satellite control centre
measures satellite positions regularly. There are a number of techniques for
estimating orbital parameters from such measurements (e.g. see Morgan and
Gordon, 1989). Orbital parameters are made available to earth station operators and used to estimate useful system parameters such as look angles
and Doppler shifts. The method for estimating satellite position, velocity
and look angle from any specified location presented here is suited for
computer solution (Morgan and Gordon, 1989).

There are three broad steps involved in the process. In the first step,
satellite position is estimated in the orbital plane; the second step involves
transforming the satellite coordinates to the three-dimensional earth-centred
coordinate system; finally, the earth-centred coordinates of the satellite are
transformed to an earth-station-centred coordinate system for obtaining the
look angle of the satellite from the earth station.

The following orbital parameters are assumed known: eccentricity, ascending node, inclination, mean anomaly at a reference time called epoch
(mean anomaly = 0 if epoch is taken at perigee pass), and argument of
perigee.

Some useful relationships involving eccentric anomaly E, true anomaly
v and mean anomaly M are:

$$\cos E = \frac{\cos v + e}{1 + e \cos v} \tag{B.34}$$

$$\cos v = \frac{\cos E - e}{1 - e \cos E} \tag{B.35}$$

where e is the orbit eccentricity.
The mean anomaly M at time t is given by

$$M = M_0 + \dot{\omega}(t - t_0) \tag{B.36}$$

where M_0 is the mean anomaly at a reference time t_0 (epoch) and $\dot{\omega}$ is the angular velocity of the satellite.

Step I

(a) The mean anomaly at the specified time is determined from equation (B.36).
(b) The eccentric anomaly is determined by solving Kepler's equation

$$M = E - e \sin E \tag{B.37}$$

For eccentricity < 0.001 the mean anomaly can be approximated as

$$E \sim M + e \sin M + \tfrac{1}{2} e^2 \sin(2M) \tag{B.38}$$

For larger values, equation (B.37) must be solved. The equation, being non-linear, requires a numerical solution technique. The Newton–Raphson method provides a quick and accurate estimate. The following steps are involved:

- Obtain an initial estimate of E using equation (B.38)
- Obtain the mean anomaly M^* using equation (B.37)
- The difference $M - M^*$ must be made ~ 0 by trial and error.

The increment ΔE^* is obtained from

$$\Delta E^* = \frac{M - M^*}{1 - e \cos E^*} \tag{B.39}$$

where $(1 - e \cos E^*)$ is the slope of the curve $M^* = E^* - e \sin E$. The process is repeated until the difference $M - M^*$ is as small as desirable. Note that M and E in the above equations are in radians.
When the true anomaly and eccentricity are known, the eccentric anomaly can be determined by using equation (B.34). Steps (a) and (b) are then not necessary.

(c) The position of the satellite in the orbital plane is given by

$$x_0 = a (\cos E - e) \tag{B.40a}$$

$$y_0 = a \left(1 - e^2\right)^{\frac{1}{2}} \sin E \tag{B.40b}$$

Step 2

The inclination of the satellite, the right ascension of the ascending node and the argument of perigee are used to transform the perifocal coordinate system to the geocentric equatorial coordinate system. The following equation set can be used for this transformation:

$$P_x = \cos \omega \cos \Omega - \sin \omega \sin \Omega \cos i \tag{B.41a}$$

$$P_y = \cos \omega \sin \Omega + \sin \omega \cos \Omega \cos i \tag{B.41b}$$

$$P_z = \sin \omega \sin i \tag{B.41c}$$

$$Q_x = -\sin \omega \cos \Omega - \cos \omega \sin \Omega \cos i \tag{B.41d}$$

$$Q_y = -\sin \omega \sin \Omega + \cos \omega \cos \Omega \cos i \tag{B.41e}$$

$$Q_z = \cos \omega \sin i \tag{B.41f}$$

Satellite position in the geocentric coordinate system is given by

$$x = P_x x_0 + Q_x y_0 \tag{B.42a}$$

$$y = P_y x_0 + Q_y y_0 \tag{B.42b}$$

$$z = P_z x_0 + Q_z y_0 \tag{B.42c}$$

Step 3

Finally, the following set of equations can be used to obtain satellite azimuth and elevation from a specified earth station:

Right ascension, $\alpha = \arctan (y/x)$ \hfill (B.43)

Declination, $\delta = \arctan \left(\dfrac{z}{\sqrt{x^2 + y^2}} \right)$ \hfill (B.44)

Elevation, $\eta = \arctan \left(\dfrac{\sin \eta_s - \dfrac{R}{r}}{\cos \eta_s} \right)$ \hfill (B.45) where

$$\eta_s = \arcsin \left[\sin \delta \sin \theta_e + \cos \delta \cos\theta_e \cos \phi_{es} \right] \qquad \text{(B.46)}$$

and R = Earth radius
r = satellite distance from Earth centre
θ_e = earth station latitude
$\phi_{se} = \phi_s - \phi_e$
ϕ_s = satellite longitude
ϕ_e = earth station longitude

$$\text{Azimuth, } A = \arctan \left[\frac{\sin \phi_{se}}{\cos \theta_e \tan \delta - \sin \theta_e \cos \phi_{se}} \right] \qquad \text{(B.47)}$$

Use the convention given in chapter 2, section 2.6 to obtain the azimuth quadrant.

The equations given above assume no perturbation in satellite orbit. The accuracy in these equations can be improved by including the effects of perturbations. Equations (2.13) and (2.14) can be used as a first approximation.

As a corollary, the range rate at a given location can be obtained from (B.2) and the Doppler shift from (B.1). The time increment $(t_2 - t_1)$ can be made as small as necessary.

Range

The distance ρ of a satellite from a given point on the Earth is given as

$$\rho = \sqrt{r^2 - R^2 \cos^2\eta} - R \sin \eta \qquad \text{(B.48)}$$

Look angle from earth station

Because of the combined effects of inclination and eccentricity, a near geostationary satellite appears to traverse an ellipse in the sky when viewed from the ground. From basic electronics it is well known that this type of shape (Lissajous' figure) consists of two sinusoidal components orthogonal to each other.

As mentioned, in addition to the effect of inclination and eccentricity, the non-uniform gravitational force caused by the oblate shape of the Earth causes a geosynchronous satellite to drift towards one of the two stable locations on the geostationary arc – 79°E and 252.4°E. The acceleration caused by this force depends on the longitude of the satellite, the maximum value being $\sim 0.0018°/day^2$. To an earth station antenna, the drift appears as a linear displacement in the satellite position.

The most accurate estimate of satellite look angles from an earth station is obtained by using the orbital parameters. For most practical applications the azimuth and elevation components of the satellite motion viewed from the ground may be approximated as (Richharia, 1984):

$$\theta_a(t) = \theta_{ai} + A_m \cos\left[\frac{\pi}{12}(t - T_a)\right] + A_i t + \xi_1 \qquad (B.49)$$

$$\theta_e(t) = \theta_{ei} + E_m \cos\left[\frac{\pi}{12}(t - T_e)\right] + E_i t + \xi_2 \qquad (B.50)$$

where $\theta_a(t)$ = satellite azimuth from an earth station at time t (in hours)
        θ_{ai} = initial azimuth of the satellite
        $\theta_e(t)$ = satellite elevation from the earth station at time t
        θ_{ei} = initial elevation of the satellite.
        A_i and E_i are the linear components of the azimuth and elevation angles respectively
        A_m and E_m are the maximum excursions in the azimuth and the elevation respectively
        ξ_1 and ξ_2 are the uncertainties in the position estimates of the satellite for the two axes respectively
The period of the sinusoid is 24 hours.

The cosine terms in equations (B.49) and (B.50) can be expanded in a series form to facilitate development of the model from real-time position data obtained from a tracking system (Richharia, 1984).

Optimal satellite constellation

To date, satellites in non-geostationary orbit have not been considered favourably for communication applications because of the complexity of the network architecture. (The exception is the Russian *Molniya* satellite system). However, currently there is considerable interest in the use of low and medium earth orbit (LEO and MEO) satellite constellations for mobile satellite communications.

One of the important issues in network architectures using LEO or MEO is the design of an optimal constellation. Network optimization involves a number of factors such as the effects of radiation, launch cost, traffic density/distribution in the coverage area, total number of satellites in the constellation, terrestrial connectivity, constellation maintenance, operating cost and complexity. The complete treatment of the subject is outside the scope of the book. However here one aspect in the optimization process – that of

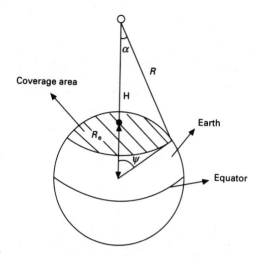

Figure B.4   Geometry for optimizing a satellite constellation.

minimizing the number of satellites at a given orbit altitude – is considered. The results of Beste (Beste, 1978) are given here as an illustration. Beste formulates the minimal number of polar orbit satellites to provide coverage below a specified latitude. There are other studies treating the problem using different optimization criteria and techniques. The interested reader should refer to the literature for a more thorough investigation (e.g. Luders, 1961; Walker, 1973; Ballard, 1980; Dondl, 1984).

Beste's result provides the optimal number of satellites in polar orbit for continuous coverage globally or above a specified latitude, in terms of the Earth-centred half-cone angles of each satellite (see figure B.4). The optimization is based on the use of the relationship at the equator:

$$(n - 1) \, \psi + (n + 1) \, \Delta = \pi \tag{B.51}$$

where n is the number of orbital planes
ψ is the angular radius of coverage

$$\Delta = \arccos \left[\cos \, \psi / \cos \, (\pi/m) \right] \tag{B.52}$$

The n orbital planes are separated by angle ϕ given by

$$\phi = \psi + \Delta \tag{B.53}$$

where m is the number of satellites/orbit.
For a simple formulation, the optimization gives the result

$N = nm$

$$\approx \frac{4}{(1 - \cos \psi)}, \quad 1.3n < m < 2.2n \qquad \text{(B.54)}$$

When the coverage is only required above the latitude θ, the optimization equation is given by

$$(n \overset{+}{-} 1)\, \psi + (n + 1)\, \Delta = \pi \cos \theta \qquad \text{(B.55)}$$

References

Ballard, A.H. (1980). 'Rosette constellations of earth satellites', *IEEE Trans. Aerosp. and Elect. Systems*, Vol. AES-16, No. 5, September, pp 656–673.

Beste, D.C. (1978). 'Design of satellite constellation for optimal continuous coverage', *IEEE Trans. Aerosp. and Elect. Systems*, Vol. AES-14, No. 3, May, pp 466–473.

CCIR (1982). *Report of Interim Working Party*, PLEN/3, CCIR, XVth Plenary Assembly, Geneva.

Chouinard, G. (1981). 'Satellite beam optimization for the broadcasting satellite service', *IEEE Trans. Broadcasting*, Vol. BC-27, No. 1, pp 7–20.

Dondl, P. (1984). 'Loopus opens a new dimension in satellite communications', *International Journal of Satellite Communications*, Vol. 2, pp 241–250.

Luders, R.D. (1961). 'Satellite networks for continuous zonal coverage', *ARSJ*, Vol. 31, February, pp 179–184.

Morgan, W.L. and Gordon, G.D. (1989). *Communications Satellite Handbook, Nautical Almanac* (yearly). Superintendent of Documents, US Government Printing Office, Washington DC, 20402. Wiley, New York.

Richharia, M. (1984). 'An optimal strategy for tracking geosynchronous satellites', *JIETE (India)*, Vol. 30, No. 5, pp 103–108.

Siocos, C.A. (1973). 'Broadcasting satellite coverage – geometric considerations', *IEEE Trans. Broadcasting*, Vol. BC-19, No. 4, December, pp 84–87.

Siocos, C.A. (1981). 'Broadcasting satellite power blackouts from solar eclipses due to moon', *IEEE Trans. Broadcasting*, Vol. BC-27, No. 2, June, pp 25–28.

Slabinski, V.J. (1974). 'Variations in range, range-rate, propagation time delay and Doppler shift in a nearly geostationary satellite' *Prog. Astronaut. Aeronaut.*, Vol. 33, No. 3.

Walker, J.G. (1973). 'Continuous whole earth coverage by circular orbit satellites', presented at the *IEE Satellite Systems for Mobile Communications Conf.*, March 13–15, Paper 95.

Index